Signaling and Communication in Plants

Series Editor

František Baluška, IZMB, Department of Plant Cell Biology, University of Bonn,
Bonn, Nordrhein-Westfalen, Germany

More information about this series at http://www.springer.com/series/8094

Sanjib Kumar Panda · Yoshiharu Y. Yamamoto
Editors

Redox Homeostasis in Plants

From Signalling to Stress Tolerance

Springer

Editors
Sanjib Kumar Panda
Department of Life Science and Bioinformatics
Assam University
Silchar, India

Yoshiharu Y. Yamamoto
Plant Molecular Physiology Lab
Gifu University
Gifu, Japan

ISSN 1867-9048 ISSN 1867-9056 (electronic)
Signaling and Communication in Plants
ISBN 978-3-319-95314-4 ISBN 978-3-319-95315-1 (eBook)
https://doi.org/10.1007/978-3-319-95315-1

Library of Congress Control Number: 2018965459

This Springer imprint is published by the registered company Springer Nature Switzerland AG
The registered company address is: Gewerbestrasse 11, 6330 Cham, Switzerland

Contents

About the Editors

Prof. Sanjib K. Panda obtained his master's, Ph.D., and D.Sc. degrees from Utkal University, Vani Vihar, Bhubaneswar, Odisha, India. He has been working as a Full Professor in the Department of Life Science & Bioinformatics, Assam University (a Central University), Silchar, India, since 2011. He has been in various fellowships like JSPS, Japan, BOYSCAST, USA, and IUSSTF, USA, as Visiting Researcher and has been Visiting Professor to various Japanese, European, and Russian universities. His research areas are molecular biology and functional genomics of plant abiotic stress tolerance. He has numerous international research publications and is an editorial member of various international and national research journals. He is Fellow of Royal Society of Biologist (FRSB), London.

Prof. Yoshi Yamamoto obtained his bachelor's degree in Science in 1989 from Kyoto University and obtained his Ph.D. degree from Graduate School of Science at Kyoto University in 1994. After working as Postdoc/Research Associate in Hokkaido University, Yale University, Riken, and also Nagoya University, he became a PI at Faculty of Applied Biological Sciences at Gifu University since 2009. His research areas are photosynthesis-related stress physiology and genome-wide promoter studies of plants including its basic aspects.

Thiol-based Redox Regulation in Plant Chloroplasts

Keisuke Yoshida and Toru Hisabori

Abstract To cope with fluctuating environmental cues, plants must regulate their own biological systems in a flexible manner. Thiol-based redox regulation is an important strategy to control the activity of target proteins in response to changes in local redox environments. In chloroplasts, this regulatory system is linked to the excitation of photosynthetic electron transport, allowing light-responsive control of chloroplast functions. A simple redox cascade mediated by the thioredoxin (Trx) has been accepted as the molecular basis of chloroplast redox regulation. However, it is becoming increasingly apparent that chloroplasts have a complicated redox network with divergent composition of redox-mediator proteins and their target proteins. Next, major challenges should be directed to comprehensively clarify how the overall system is organized in chloroplasts and works toward environmental fluctuations. This chapter gives an overview of the recent advances in understanding the biochemical basis and physiological significance of redox-based regulatory network in chloroplasts.

1 Introduction

Thiol-based redox regulation is a post-translational mechanism to control enzymatic activity by modifying the Cys residue on the target protein (e.g., formation/cleavage of disulfide bond). This regulatory system transmits environmental signals as a reducing power to targets, ensuring rapid adjustment of cellular functions. A key player for the redox regulation is the thioredoxin (Trx), which is a small protein first discovered in *Escherichia coli* as a hydrogen donor to ribonucleotide reductase (Laurent et al. 1964). Trx possesses a redox-sensitive Cys pair in the active site sequence of WCGPC. By using this Cys pair, Trx catalyzes the dithiol-disulfide exchange reac-

K. Yoshida (✉) · T. Hisabori (✉)
Laboratory for Chemistry and Life Science, Tokyo Institute of Technology, Nagatsuta 4259-R1-8, Midori-Ku, Yokohama 226-8503, Japan
e-mail: yoshida.k.ao@m.titech.ac.jp

T. Hisabori
e-mail: thisabor@res.titech.ac.jp

S. K. Panda and Y. Y. Yamamoto (eds.), *Redox Homeostasis in Plants*,
Signaling and Communication in Plants, https://doi.org/10.1007/978-3-319-95315-1_1

tion and thereby acts as the redox mediator to its target proteins. Redox regulation system driven by Trx is ubiquitously found in all kingdoms of life.

In plant chloroplasts, redox regulation has a unique property that is associated with the photosynthetic electron transport chain. The canonical regulatory pathway is shown in Fig. 1a. A part of reducing power generated by photochemical reactions is transferred from a mobile electron carrier protein ferredoxin (Fd) to Trx via Fd-Trx reductase (FTR). A reduced form of Trx then transfers reducing power to specific target proteins. Four enzymes in the Calvin-Benson cycle (glyceraldehyde-3-phosphate dehydrogenase (GAPDH), fructose-1,6-bisphosphatase (FBPase), sedoheptulose-1,7-bisphosphatase (SBPase), and phosphoribulokinase (PRK)) are classically well known as representative Trx targets. Because these enzymes are activated upon reduction, a series of reducing power transfer via FTR/Trx redox cascade makes it possible to turn on the carbon fixation metabolism in concert with the excitation of electron transport. This is a classic model for chloroplast redox regulation, established by Buchanan et al. in 1970s (Buchanan et al. 1979; Buchanan 1980; Buchanan and Balmer 2005). Chloroplast redox regulation dynamically works in vivo in response to changes in light conditions (Konno et al. 2012; Yoshida et al. 2014).

Now, we are at a turning point for the field of redox study. Owing to growing availability of omics data, a large number of proteins have been identified as the possible constituent of redox regulation system in chloroplasts (Fig. 1b). This emerging protein catalog indicates that chloroplasts have evolved a complex redox-based regulatory network, controlling a diverse array of functions in a flexible and sophisticated manner. In this regard, it is of great importance to understand how the redox regulation system is organized in chloroplasts and has impacts on plant viability. In this chapter, we review the current knowledge of the biochemical basis and physiological significance of chloroplast redox regulation. Recent other reviews are also available to gain insights into these issues (Michelet et al. 2013; Richter and Grimm 2013; Serrato et al. 2013; Balsera et al. 2014; Nikkanen and Rintamaki 2014; Geigenberger et al. 2017; Nikkanen et al. 2017).

2 Overview of Protein Multiplicity in Chloroplast Redox Regulation

The completion of the *Arabidopsis thaliana* genome sequence (Arabidopsis Genome Initiative 2000) has led to the identification of multiple genes encoding redox-mediator proteins, including Trxs and other related proteins. In *Arabidopsis*, seven subtypes of Trx (*f*-, *m*-, *x*-, *y*-, *z*-, *h*-, and *o*-type) encoded by totally twenty genes have been found in various subcellular compartments. Among them, five (*f*-, *m*-, *x*-, *y*-, and *z*-type) are localized in chloroplasts (Fig. 1b) (Lemaire et al. 2007; Serrato et al. 2013). They have different molecular characteristics, such as the protein surface charge and midpoint redox potential (Collin et al. 2003; Michelet et al. 2005; Toivola et al. 2013; Yoshida et al. 2015; Yoshida and Hisabori 2016), which possibly confers

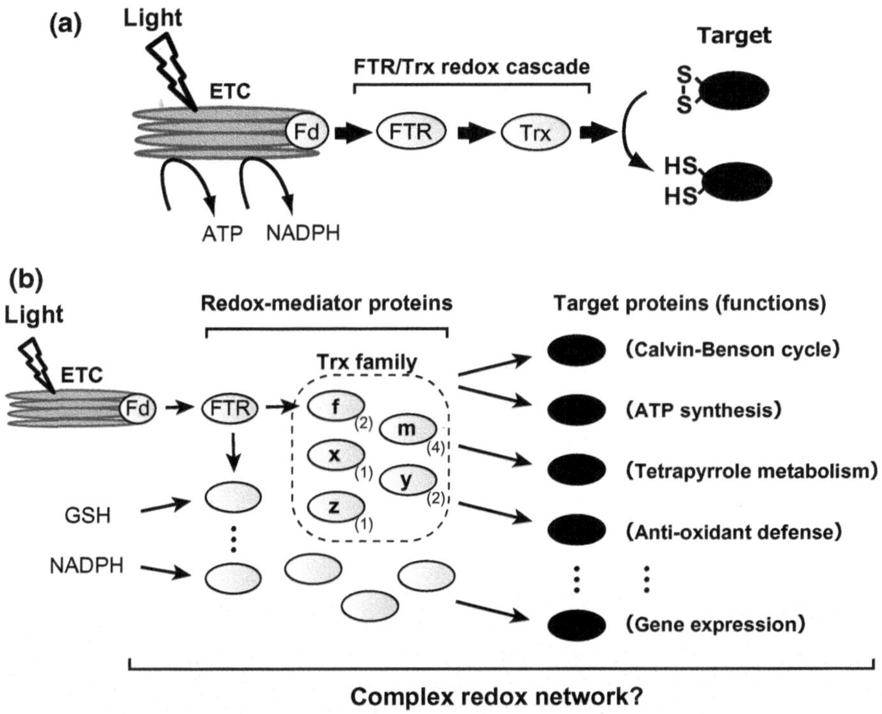

Fig. 1 Overview of thiol-based redox regulation in plant chloroplasts. **a** Classically known regulatory pathway. Reducing power is transferred from ferredoxin (Fd) in the electron transport chain (ETC) to specific target proteins *via* ferredoxin-thioredoxin reductase (FTR)/thioredoxin (Trx) redox cascade, enabling light-responsive control of chloroplast functions. **b** Newly emerging regulatory network, that is organized by multiple redox-mediator proteins and target proteins. This is possibly helpful for flexibly controlling a range of chloroplast functions; however, its whole picture remains undescribed. The number of gene encoding each Trx subtype in *Arabidopsis* is shown in the parenthesis. Abbreviations: GSH, glutathione

the functional versatility on chloroplast Trx family (see below). Besides, some novel redox-mediator proteins are also located in chloroplasts. The best-studied example is the NADPH-Trx reductase C (NTRC), a hybrid protein composed of an NADPH-Trx reductase (NTR) domain and a Trx domain (Serrato et al. 2004). Other proteins containing the Trx-like motif have been also predicted to participate in chloroplast redox regulation, although their functions remain to be characterized in detail.

Before the beginning of this century, a limited number of chloroplast proteins have been reported as the target of redox regulation. They included four Calvin-Benson cycle enzymes (GAPDH, FBPase, SBPase, and PRK), NADP-malate dehydrogenase (NADP-MDH, involved in the malate valve), glucose-6-phosphate dehydrogenase (G6PDH, involved in the oxidative pentose phosphate pathway), acetyl-CoA carboxylase (ACCase, involved in the fatty acid metabolism), Rubisco activase (RCA), and ATP synthase CF_1-γ subunit (Buchanan et al. 1979; Buchanan 1980; Mills et al.

1980; Scheibe and Anderson 1981; Sasaki et al. 1997; Zhang and Portis 1999). The major breakthrough was brought by the development of strategies for systematically screening Trx target candidates (Motohashi et al. 2001; Yano et al. 2001). For example, the Trx affinity chromatography using monocysteinic Trx as a bait allowed identification of previously unrecognized Trx targets in chloroplast stroma as well as classically known ones (Motohashi et al. 2001; Balmer et al. 2003). The newly discovered targets contain numerous proteins involved in nitrogen metabolism, starch synthesis, tetrapyrrole metabolism, antioxidant defense system, and other metabolic pathways in chloroplasts. Furthermore, target screening from the thylakoid membrane or chloroplast inner envelope has been challenged (Balmer et al. 2006; Motohashi and Hisabori 2006; Bartsch et al. 2008; Hall et al. 2010). These studies have indicated that Trx associates with the protein import machineries, proteases, and photosynthetic electron transport complexes. Thus, novel methods of so-called redox proteomics have identified a number of chloroplast proteins as the Trx target candidate (Fig. 1b). The integral lists for these proteins are provided by other reviews (Lindahl and Kieselbach 2009; Montrichard et al. 2009).

3 Functional Diversity of Chloroplast Trx Family: Insights from Biochemical Studies

Protein multiplicity in chloroplast redox regulation has raised a question: How do a variety of redox-mediator proteins and target proteins communicate in chloroplasts? Several biochemical studies have addressed functional specificity and redundancy of chloroplast Trx family. Before the genomic era, only Trx-*f* and Trx-*m* were known to exist in chloroplasts. As named according to their properties, Trx-*f* and Trx-*m* were originally defined as the efficient reductive activator for FBPase and NADP-MDH, respectively (Wolosiuk et al. 1979; Schurmann et al. 1981). High specificity of Trx-*f* for regulating FBPase has been well established by following biochemical studies (Geck et al. 1996; Collin et al. 2003; Yoshida et al. 2015). By contrast, the initial view for NADP-MDH redox regulation should be revised; NADP-MDH was shown to undergo Trx-*f*-dependent reductive activation with the efficiency comparable with or even higher than that of Trx-*m* (Hodges et al. 1994; Geck et al. 1996; Collin et al. 2003). NADP-MDH contains two redox-active Cys pairs at N- and C-terminal extensions (Miginiac-Maslow and Lancelin 2002). Our recent study revealed that Trx-*m* can reduce only C-terminal disulfide bond, whereas Trx-*f* can reduce both N- and C-terminal ones, allowing full activation of NADP-MDH (Yoshida et al. 2015). This highlights a unique aspect of NADP-MDH redox regulation; two Cys pairs in a single polypeptide have different Trx selectivity.

It has been shown that, compared to Trx-*m*, Trx-*f* is more effective in activating other Calvin-Benson cycle enzymes (GAPDH, SBPase, and PRK), RCA, and ATP synthase (Schwarz et al. 1997; Zhang and Portis 1999; Marri et al. 2009; Yoshida et al. 2015). Furthermore, other metabolic enzymes involved in the oxidative pentose phosphate pathway (G6PDH), starch synthesis or degradation (ADP-glucose pyrophosphorylase (AGPase), starch synthase, α- or β-amylase, and starch phosphorylase), and lipid synthesis (ACCase and monogalactosyldiacylglycerol synthase) are redox-regulated by Trx-*f* with higher efficiency than that of Trx-*m* (Sasaki et al. 1997; Ballicora et al. 2000; Mikkelsen et al. 2005; Sparla et al. 2006; Yamaryo et al. 2006; Nee et al. 2009; Seung et al. 2013; Silver et al. 2013; Thormahlen et al. 2013; Skryhan et al. 2015). Taken together, Trx-*f* can be regarded as the major redox regulator for a range of photosynthetic reactions and other metabolisms in chloroplasts (Fig. 2a). By contrast, the specific role of Trx-*m* remains elusive. As a Trx-*m*-specific target, HCF164 was identified; HCF164 is reduced by Trx-*m*, but not Trx-*f* (Motohashi and Hisabori 2006). HCF164 is a Trx-like protein anchored to the thylakoid membrane, and its redox-active site faces the lumenal side. Together with the thylakoid protein CcdA (Motohashi and Hisabori 2010), HCF164 is supposed to act as a reducing power transmitter from stroma to lumenal proteins. Some lumenal proteins involved in the photosynthetic electron transport, xanthophylls cycle, and immunophilins were suggested to be under redox regulation (Gopalan et al. 2004; Motohashi and Hisabori 2006; Hall et al. 2010; Simionato et al. 2015). It is therefore possible that Trx-*m* indirectly controls key reactions in the thylakoid membrane. This may be related to growth inhibition in Trx-*m*-deficient plants (see below).

In 2000s, genomic and phylogenic studies have identified additional three types of chloroplast Trx (Trx-*x*, -*y*, -*z*) (Fig. 2a). Trx-*x* and Trx-*y* are efficient electron donor to 2-Cys peroxiredoxin (Prx) and Prx Q, respectively, both of which are involved in the antioxidant defense system (Collin et al. 2003, 2004; Yoshida et al. 2015). In addition, we recently showed that Trx-*y* can reduce Mg-chelatase I subunit (CHLI, involved in the tetrapyrrole metabolism) with high efficiency (Yoshida and Hisabori 2016). Trx-*z* was also reported to serve to transfer reducing power to several antioxidant enzymes, such as Prx Q and Met sulfoxide reductase (MSR) (Chibani et al. 2011; Yoshida et al. 2015). More importantly, Trx-*z* may play a critical role in regulating plastidial transcription, although its regulatory mechanism is somewhat unclear (Arsova et al. 2010; Wimmelbacher and Bornke 2014). Further biochemical characterization is required for better understanding of Trx-*z*-mediated redox regulation.

Given that each Trx subtype differentially controls diverse target proteins, the electron partitioning from FTR to Trx family is thought to be the critical step for determining the consequence of chloroplast redox regulation. We recently addressed the electron transfer from FTR to ten Trx isoforms in *Arabidopsis* (Yoshida and Hisabori 2017). The results showed that all Trxs can be reduced by FTR, but their reduction kinetics are largely variable (Fig. 2b). In addition to the above-described target selectivity, these data also highlight an aspect of highly organized circuits of chloroplast Trx family.

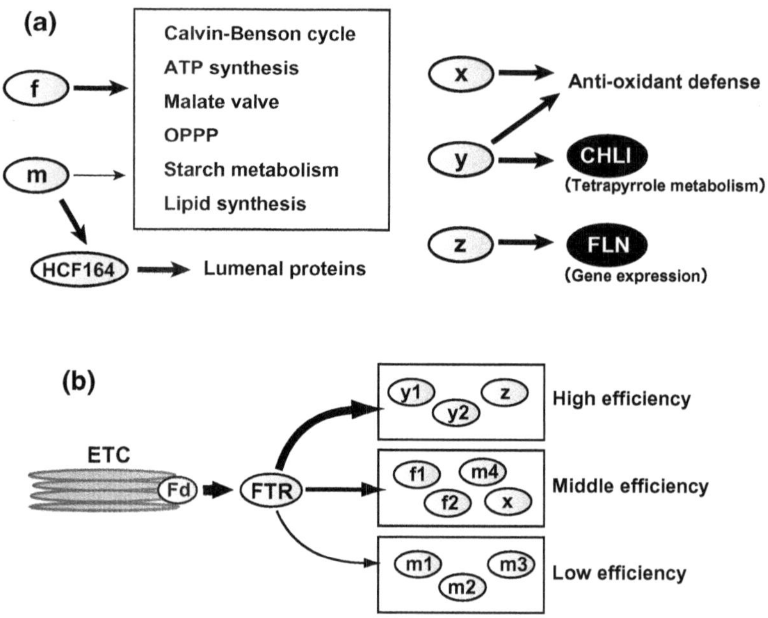

Fig. 2 Functional diversity of chloroplast thioredoxin (Trx) family. **a** Target selectivity of five Trx subtypes (Trx-*f*, -*m*, -*x*, -*y*, and -*z*). Pathways for transferring reducing power from each Trx to target proteins or functions are described. An arrow with a thin line indicates low efficiency in reducing power transfer. See text for details. **b** Distinct reducing power transfer from ferredoxin-thioredoxin reductase (FTR) to ten Trx isoforms in *Arabidopsis*. Ten Trx isoforms can be clustered into three classes based on the efficiencies in FTR-dependent reduction. Modified from Yoshida and Hisabori (2017). Abbreviations: CHLI, Mg-chelatase I subunit; ETC, electron transport chain; Fd, ferredoxin; FLN, fructokinase-like protein; OPPP, oxidative pentose phosphate pathway

4 Physiological Impact of Trx: Insights from Reverse-Genetic Studies

Biochemical data have provided valuable insights into molecular mechanisms of redox regulation, which is, however, still weak to discuss their in vivo relevance. Instead, reverse-genetic studies using *Arabidopsis* mutant plants have been adopted to understand physiological significances of each Trx subtype. Their results are summarized in Table 1.

In the Trx-*f*-deficient mutant (*trxf1* single or *trxf1/trxf2* double mutant), light-dependent reduction of FBPase, RCA, and AGPase was partially impaired, suggesting that Trx-*f* donates reducing power to these target proteins in vivo (Thormahlen et al. 2013, 2015; Yoshida et al. 2015; Naranjo et al. 2016a). Accompanied by the impairment of AGPase reduction, starch accumulation level was lowered in *trxf1* mutant (Thormahlen et al. 2013, 2015). By contrast, Trx-*f*-overexpressed tobacco plants showed an elevated level of starch accumulation (Sanz-Barrio et al. 2013). The *trxf1/trxf2* double mutant showed impaired photosynthesis and growth at high

Table 1 Phenotypes of *Arabidopsis* mutants deficient in each Trx subtype

Arabidopsis mutant	Notable phenotype	References
Trx-f-deficient mutant		
trxf1	Impairment of light-dependent redox regulation of AGPase	Thormahlen et al. (2013)
trxf1/trxf2	Impairment of light-dependent redox regulation of FBPase	Yoshida et al. (2015)
trxf1/trxf2	Growth impairment under high-light and short-day conditions	Naranjo et al. (2016a)
	Impairment of light-dependent redox regulation of FBPase and RCA	
	Decrease in photosynthetic electron transport efficiency	
Trx-m-deficient mutant		
trxm4	Enhancement of NDH-dependent cyclic electron transport	Courteille et al. (2013)
trxm1/trxm2/trxm4	Growth impairment with pale-green leaves	Wang et al. (2013)
	Impaired biogenesis of photosystem II	
trxm1/trxm2/trxm4	Growth impairment with pale-green leaves	Okegawa and Motohashi (2015)
trxm1/trxm2	Impairment of light-dependent activation of NADP-MDH	Thormahlen et al. (2017)
trxm1/trxm2/trxm4	Impaired chlorophyll synthesis	Da et al. (2017)
trxm1/trxm2/trxm4	Elevated NPQ concomitant with disturbance of xanthophyll cycle	Da et al. (2018)
Trx-x-deficient mutant		
trxx	Not observed	Pulido et al. (2010)
trxx	Not observed	Ojeda et al. (2017)
Trx-y-deficient mutant		
trxy1/trxy2	Lowered capacity of MSR	Laugier et al. (2013)
Trx-z-deficient mutant		
trxz	Albino phenotype	Arsova et al. (2010)
	Modified expression of chloroplast-encoded genes	

Abbreviations *AGPase* ADP-glucose pyrophosphorylase; *FBPase* fructose-1,6-bisphosphatase; *MSR* Met sulfoxide reductase; *NADP-MDH* NADP-malate dehydrogenase; *NDH* NADH dehydrogenase-like complex; *NPQ* non-photochemical quenching; *RCA* Rubisco activase; *Trx* thioredoxin light intensity (Naranjo et al. 2016a). These all data suggest that Trx-*f* has a significant impact on the adjustment of photosynthetic metabolism in illuminated leaves. On the other hand, the Trx-*x*-deficient mutant did not show any notable phenotype, possibly due to functional redundancy with other redox systems (Pulido et al. 2010; Ojeda et al. 2017). As well, the deficiency of Trx-*y* did not cause drastic alteration of growth, but MSR capacity was slightly lowered (Laugier et al. 2013).

In *Arabidopsis*, suppressed expression of three Trx-*m* isoforms (Trx-*m*1, Trx-*m*2, and Trx-*m*4) resulted in growth impairment even under normal growth conditions (Wang et al. 2013; Okegawa and Motohashi 2015), which is apparently distinct from phenotypes of other Trx-deficient mutants. Such growth phenotype was observed in Trx-*m* knockdown mutants in rice (Chi et al. 2008). Therefore, Trx-*m* is thought to have some specific roles in plant development, but its detailed functions are still undetermined. The *Arabidopsis trxm1/trxm2/trxm4* triple mutants failed to assemble photosystem II complex properly (Wang et al. 2013). Besides, other studies using *trxm* mutants suggest that Trx-*m* is involved in the regulation of cyclic electron transport around photosystem I and non-photochemical quenching (Courteille et al. 2013; Da et al. 2018). Considering these reports, the main function of Trx-*m* may be related to the biogenesis and maintenance of photosynthetic electron transport machineries, rather than the metabolic control in the stroma. This possibility is partly supported by the proteomic study indicating that Trx-*m* is localized adjacent to the thylakoid membrane (Peltier et al. 2002), but further biochemical evidence is needed to conclude this. It should be mentioned that another isoform of Trx-*m* (Trx-*m*3) is expressed in non-green plastids and is essential for meristem maintenance (Benitez-Alfonso et al. 2009).

In *Arabidopsis* and tobacco, reduced expression of Trx-*z* resulted in a severe albino phenotype (Arsova et al. 2010). Based on the yeast two-hybrid assay, Trx-*z* was shown to bind to fructokinase-like protein (FLN). The association of Trx-*z* with FLN may be essential for plastid-encoded RNA polymerase (PEP)-dependent gene expression. This hypothesis is supported by proteomic studies showing that both Trx-*z* and FLN are the component of PEP complex (Schroter et al. 2010; Steiner et al. 2011). However, the importance of redox regulation during this crosstalk is still a matter of debate. Because redox activity of Trx-*z* and FLN may be dispensable for plant growth, it cannot be excluded that these proteins are only structural components of PEP complex (Wimmelbacher and Bornke 2014).

5 NTRC: A Notable Redox-Mediator Protein

NTR is a redox-mediator protein that contains a flavin adenine dinucleotide (FAD) cofactor and a redox-active disulfide bond (Jacquot et al. 2009). In *Arabidopsis*, two isoforms of NTR (NTRA and NTRB) are known to be present in cytosol and mitochondria; however, NTRC was exceptionally identified in chloroplasts as a unique bimodular NTR which harbors Trx domain at a C-terminus (Serrato et al. 2004). As using NADPH as a source of reducing power, NTRC can work independently from

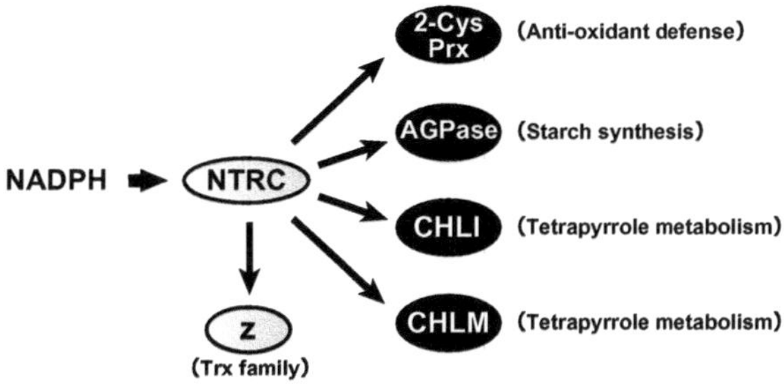

Fig. 3 Target proteins of NADPH-thioredoxin reductase C (NTRC) in chloroplasts. Pathways for NTRC-mediated reducing power transfer are described. Abbreviations: AGPase, ADP-glucose pyrophosphorylase; CHLI, Mg-chelatase I subunit; CHLM, Mg protoporphyrin IX methyltransferase; 2-Cys Prx, 2-Cys peroxiredoxin

light-driven FTR/Trx redox cascade. NTRC-dependent redox regulation is thought to have some indispensable roles for plants, because an *Arabidopsis* NTRC-deficient mutant (*ntrc* mutant) shows an impaired growth phenotype with pale-green leaves. To date, a number of studies have addressed biochemical and physiological aspects of this novel protein (Cejudo et al. 2012).

NTRC-targeted chloroplast proteins, which have been suggested based on the biochemical studies, are described in Fig. 3. Impacts of the absence of NTRC on *Arabidopsis* plant phenotype are summarized in Table 2. The first-reported and the most well-known target of NTRC is 2-Cys Prx (Moon et al. 2006; Perez-Ruiz et al. 2006). The biochemical reaction mode of NTRC with 2-Cys Prx has been clarified in detail (Perez-Ruiz and Cejudo 2009; Bernal-Bayard et al. 2012). Prior to these findings, 2-Cys Prx was shown to receive reducing power from Trxs (Konig et al. 2002; Collin et al. 2003) and CDSP32, a protein containing two Trx modules (Broin et al. 2002). However, NTRC emerges higher efficiency in reducing 2-Cys Prx than these redox-mediator proteins (Moon et al. 2006; Perez-Ruiz et al. 2006; Bernal-Bayard et al. 2014; Yoshida and Hisabori 2016). Given the hypersensitivity to several abiotic stresses in *ntrc* mutants (Table 2), NTRC/2-Cys Prx redox pathway is likely to constitute the major antioxidant system in chloroplasts. As present in some cyanobacteria (Sueoka et al. 2009; Pascual et al. 2011; Sanchez-Riego et al. 2016; Mihara et al. 2017), this is considered to be an evolutionally conserved system.

The phenotype of *ntrc* mutant is more severe than that of 2-Cys Prx-deficient plant, indicating additional functions of NTRC (Pulido et al. 2010). It has been demonstrated by in vitro studies that AGPase, CHLI, and Mg-protoporphyrin IX methyltransferase (CHLM, involved in the tetrapyrrole metabolism) are subjected to NTRC-dependent redox regulation (Fig. 3) (Michalska et al. 2009; Richter et al. 2013; Perez-Ruiz et al. 2014; Yoshida and Hisabori 2016). AGPase and CHLI are also redox-regulated by Trx (Ballicora et al. 2000; Ikegami et al. 2007; Luo et al. 2012; Thormahlen et al. 2013), but at least for CHLI reduction, NTRC seems to have higher efficiency than Trx (Perez-Ruiz et al. 2014; Yoshida and Hisabori 2016). In accordance with these biochemical results, *ntrc* mutants show the phenotypes of

Table 2 Phenotypes of *Arabidopsis* NTRC-deficient mutants and multiple mutants deficient in NTRC and FTR/Trx pathways

Arabidopsis mutant	Notable phenotype	References[a]
ntrc	Growth impairment with pale-green leaves	Serrato et al. (2004), Perez-Ruiz et al. (2006)
	Hypersensitivity to abiotic stresses	Serrato et al. (2004), Perez-Ruiz et al. (2006)
	Photoperiod-dependent growth inhibition	Perez-Ruiz et al. (2006), Lepisto et al. (2009)
	Accumulation of oxidized 2-Cys Prx	Perez-Ruiz et al. (2006), Kirchsteiger et al. (2009)
	Impaired redox regulation of AGPase and lowered starch content	Michalska et al. (2009)
	Impaired chlorophyll synthesis	Stenbaek et al. (2008), Richter et al. (2013)
	Elevated NPQ	Carrillo et al. (2016), Naranjo et al. (2016b)
ntrc/trxf1	Severe growth impairment	Thormahlen et al. (2015)
	Changes in metabolite profiles	
ftrb/ntrc	Lethal phenotype under autotrophic conditions	Yoshida and Hisabori (2016)
	Disruption of photosynthetic system	
ntrc/trxf1/trxf2	Severe growth impairment	Ojeda et al. (2017)
	Abnormal structure of chloroplasts	
ntrc/trxx	Severe growth impairment	Ojeda et al. (2017)
	Abnormal structure of chloroplasts	
ntrc/trxm1/trxm2/trxm4	Severe growth impairment	Da et al. (2017)
	Impairment of tetrapyrrole metabolism	

[a]Only representative references are described

Abbreviations *AGPase* ADP-glucose pyrophosphorylase; *2-Cys Prx* 2-Cys peroxiredoxin; *NPQ* non-photochemical quenching

lowered starch content and impaired chlorophyll synthesis (Table 2). Furthermore, other proteins including glutamyl-transfer RNA reductase (GluTR, involved in the tetrapyrrole metabolism), FBPase, PRK, and ATP synthase CF_1-γ subunit have been also suggested to interact with NTRC, based on the bimolecular fluorescence complementation or co-immunoprecipitation experiments (Richter et al. 2013; Nikkanen et al. 2016). However, it is still unconcluded whether they are actually capable of receiving reducing power from NTRC. It should be mentioned that some studies showed the inability of NTRC to reduce FBPase (Yoshida and Hisabori 2016; Ojeda et al. 2017).

In recent years, the exchange of reducing power between NTRC and Trxs has been suggested (Toivola et al. 2013; Nikkanen et al. 2016). Using a chromatography-based method, we revealed that NTRC associates with Trx-*z* with high affinity. The following biochemical assay indicated that NTRC can reduce Trx-*z* but not other Trx subtypes (Yoshida and Hisabori 2016). We therefore concluded that Trx-*z* is a unique Trx targeted by NTRC (Fig. 3); however, it should be investigated in more detail how this interplay works *in planta*.

6 Significance of Cooperative Redox Regulation by FTR/Trx and NTRC Pathways

As documented above, accumulating evidence suggests that NTRC has redox-regulatory functions distinct from those of FTR/Trx system. Recent studies have addressed the functional coordination of FTR/Trx and NTRC pathways, by constructing multiple mutants in *Arabidopsis* (Table 2).

Combined deficiency of NTRC and one Trx subtype (Trx-*f*, Trx-*m*, or Trx-*x*) results in more remarkable growth phenotype than that of *ntrc* single mutants. This severe growth impairment is accompanied by pleiotropic phenotypes, including changes in metabolite profiles and abnormal ultrastructure of chloroplasts (Thormahlen et al. 2015; Da et al. 2017; Ojeda et al. 2017). Most strikingly, double mutants impaired in FTR and NTRC expression display lethal phenotype under autotrophic growth conditions (Yoshida and Hisabori 2016). These mutants can survive in the sucrose-containing medium, but suffer from a drastic loss of photosynthetic performance and the resulting growth retardation. These all data indicate that FTR/Trx and NTRC pathways cooperatively control a range of chloroplast functions, which is critical for plant development and viability.

Apart from FTR/Trx redox cascade, NTRC can function even under non-photosynthetic conditions by using NADPH produced by the oxidative pentose phosphate pathway. This property is possibly beneficial at the early stage of plastid differentiation when the electron transport system is still immature. It therefore seems to be attractive to study how the engagement of each FTR/Trx or NTRC pathway changes depending on the developmental phase. The functional dynamics of two redox systems should be comprehensively dissected by future researches.

7 Concluding Remarks

It has been increasingly uncovered that chloroplast redox network is highly organized, dynamically responds to environments, and plays a critical role in plant biomass production. This regulatory system confers the elegant strategy on plants for surviving under ever-fluctuating natural conditions. However, the whole picture of chloroplast redox network is not yet understood. For example, the biochemical characterization of putative redox-mediator proteins (Trx-like proteins) remains to be fully attained, although some studies have addressed this issue (Dangoor et al. 2009, 2012; Chibani et al. 2012; Eliyahu et al. 2015). Furthermore, it is of an intriguing issue to study how a complex set of these proteins responds to environmental stresses and serves to control a wide variety of chloroplast functions. More comprehensive and in-depth studies will reveal an expanded map of chloroplast redox network and its biological functions.

References

Arabidopsis Genome Initiative (2000) Analysis of the genome sequence of the flowering plant *Arabidopsis thaliana*. Nature 408:796–815

Arsova B, Hoja U, Wimmelbacher M, Greiner E, Ustun S, Melzer M, Petersen K, Lein W, Bornke F (2010) Plastidial thioredoxin *z* interacts with two fructokinase-like proteins in a thiol-dependent manner: evidence for an essential role in chloroplast development in *Arabidopsis* and *Nicotiana benthamiana*. Plant Cell 22:1498–1515

Ballicora MA, Frueauf JB, Fu Y, Schurmann P, Preiss J (2000) Activation of the potato tuber ADP-glucose pyrophosphorylase by thioredoxin. J Biol Chem 275:1315–1320

Balmer Y, Koller A, del Val G, Manieri W, Schurmann P, Buchanan BB (2003) Proteomics gives insight into the regulatory function of chloroplast thioredoxins. Proc Natl Acad Sci USA 100:370–375

Balmer Y, Vensel WH, Hurkman WJ, Buchanan BB (2006) Thioredoxin target proteins in chloroplast thylakoid membranes. Antioxid Redox Signal 8:1829–1834

Balsera M, Uberegui E, Schurmann P, Buchanan BB (2014) Evolutionary development of redox regulation in chloroplasts. Antioxid Redox Signal 21:1327–1355

Bartsch S, Monnet J, Selbach K, Quigley F, Gray J, von Wettstein D, Reinbothe S, Reinbothe C (2008) Three thioredoxin targets in the inner envelope membrane of chloroplasts function in protein import and chlorophyll metabolism. Proc Natl Acad Sci USA 105:4933–4938

Benitez-Alfonso Y, Cilia M, San Roman A, Thomas C, Maule A, Hearn S, Jackson D (2009) Control of *Arabidopsis* meristem development by thioredoxin-dependent regulation of intercellular transport. Proc Natl Acad Sci USA 106:3615–3620

Bernal-Bayard P, Hervas M, Cejudo FJ, Navarro JA (2012) Electron transfer pathways and dynamics of chloroplast NADPH-dependent thioredoxin reductase C (NTRC). J Biol Chem 287:33865–33872

Bernal-Bayard P, Ojeda V, Hervas M, Cejudo FJ, Navarro JA, Velazquez-Campoy A, Perez-Ruiz JM (2014) Molecular recognition in the interaction of chloroplast 2-Cys peroxiredoxin with NADPH-thioredoxin reductase C (NTRC) and thioredoxin *x*. FEBS Lett 588:4342–4347

Broin M, Cuine S, Eymery F, Rey P (2002) The plastidic 2-cysteine peroxiredoxin is a target for a thioredoxin involved in the protection of the photosynthetic apparatus against oxidative damage. Plant Cell 14:1417–1432

Buchanan BB (1980) Role of light in the regulation of chloroplast enzymes. Annu Rev Plant Physiol 31:341–374

Buchanan BB, Balmer Y (2005) Redox regulation: a broadening horizon. Annu Rev Plant Biol 56:187–220

Buchanan BB, Wolosiuk RA, Schurmann P (1979) Thioredoxin and enzyme regulation. Trends Biochem Sci 4:93–96

Carrillo LR, Froehlich JE, Cruz JA, Savage LJ, Kramer DM (2016) Multi-level regulation of the chloroplast ATP synthase: the chloroplast NADPH thioredoxin reductase C (NTRC) is required for redox modulation specifically under low irradiance. Plant J 87:654–663
Cejudo FJ, Ferrandez J, Cano B, Puerto-Galan L, Guinea M (2012) The function of the NADPH thioredoxin reductase C-2-Cys peroxiredoxin system in plastid redox regulation and signalling. FEBS Lett 586:2974–2980
Chi YH, Moon JC, Park JH, Kim HS, Zulfugarov IS, Fanata WI, Jang HH, Lee JR, Lee YM, Kim ST, Chung YY, Lim CO, Kim JY, Yun DJ, Lee CH, Lee KO, Lee SY (2008) Abnormal chloroplast development and growth inhibition in rice thioredoxin m knock-down plants. Plant Physiol 148:808–817
Chibani K, Tarrago L, Gualberto JM, Wingsle G, Rey P, Jacquot JP, Rouhier N (2012) Atypical thioredoxins in poplar: the glutathione-dependent thioredoxin-like 2.1 supports the activity of target enzymes possessing a single redox active cysteine. Plant Physiol 159:592–605
Chibani K, Tarrago L, Schurmann P, Jacquot JP, Rouhier N (2011) Biochemical properties of poplar thioredoxin *z*. FEBS Lett 585:1077–1081
Collin V, Issakidis-Bourguet E, Marchand C, Hirasawa M, Lancelin JM, Knaff DB, Miginiac-Maslow M (2003) The *Arabidopsis* plastidial thioredoxins: new functions and new insights into specificity. J Biol Chem 278:23747–23752
Collin V, Lamkemeyer P, Miginiac-Maslow M, Hirasawa M, Knaff DB, Dietz KJ, Issakidis-Bourguet E (2004) Characterization of plastidial thioredoxins from *Arabidopsis* belonging to the new *y*-type. Plant Physiol 136:4088–4095
Courteille A, Vesa S, Sanz-Barrio R, Cazale AC, Becuwe-Linka N, Farran I, Havaux M, Rey P, Rumeau D (2013) Thioredoxin m4 controls photosynthetic alternative electron pathways in Arabidopsis. Plant Physiol 161:508–520
Da Q, Sun T, Wang M, Jin H, Li M, Feng D, Wang J, Wang HB, Liu B (2018) M-type thioredoxins are involved in the xanthophyll cycle and proton motive force to alter NPQ under low-light conditions in Arabidopsis. Plant Cell Rep 37:279–291
Da Q, Wang P, Wang M, Sun T, Jin H, Liu B, Wang J, Grimm B, Wang HB (2017) Thioredoxin and NADPH-dependent thioredoxin reductase C regulation of tetrapyrrole biosynthesis. Plant Physiol 175:652–666
Dangoor I, Peled-Zehavi H, Levitan A, Pasand O, Danon A (2009) A small family of chloroplast atypical thioredoxins. Plant Physiol 149:1240–1250
Dangoor I, Peled-Zehavi H, Wittenberg G, Danon A (2012) A chloroplast light-regulated oxidative sensor for moderate light intensity in *Arabidopsis*. Plant Cell 24:1894–1906
Eliyahu E, Rog I, Inbal D, Danon A (2015) ACHT4-driven oxidation of APS1 attenuates starch synthesis under low light intensity in *Arabidopsis* plants. Proc Natl Acad Sci USA 112:12876–12881
Geck MK, Larimer FW, Hartman FC (1996) Identification of residues of spinach thioredoxin *f* that influence interactions with target enzymes. J Biol Chem 271:24736–24740
Geigenberger P, Thormahlen I, Daloso DM, Fernie AR (2017) The unprecedented versatility of the plant thioredoxin system. Trends Plant Sci 22:249–262
Gopalan G, He Z, Balmer Y, Romano P, Gupta R, Heroux A, Buchanan BB, Swaminathan K, Luan S (2004) Structural analysis uncovers a role for redox in regulating FKBP13, an immunophilin of the chloroplast thylakoid lumen. Proc Natl Acad Sci USA 101:13945–13950
Hall M, Mata-Cabana A, Akerlund HE, Florencio FJ, Schroder WP, Lindahl M, Kieselbach T (2010) Thioredoxin targets of the plant chloroplast lumen and their implications for plastid function. Proteomics 10:987–1001
Hodges M, Miginiac-Maslow M, Decottignies P, Jacquot JP, Stein M, Lepiniec L, Cretin C, Gadal P (1994) Purification and characterization of pea thioredoxin f expressed in *Escherichia coli*. Plant Mol Biol 26:225–234
Ikegami A, Yoshimura N, Motohashi K, Takahashi S, Romano PG, Hisabori T, Takamiya K, Masuda T (2007) The CHLI1 subunit of *Arabidopsis thaliana* magnesium chelatase is a target protein of the chloroplast thioredoxin. J Biol Chem 282:19282–19291

Jacquot JP, Eklund H, Rouhier N, Schurmann P (2009) Structural and evolutionary aspects of thioredoxin reductases in photosynthetic organisms. Trends Plant Sci 14:336–343

Kirchsteiger K, Pulido P, Gonzalez M, Cejudo FJ (2009) NADPH thioredoxin reductase C controls the redox status of chloroplast 2-Cys peroxiredoxins in *Arabidopsis thaliana*. Mol Plant 2:298–307

Konig J, Baier M, Horling F, Kahmann U, Harris G, Schurmann P, Dietz KJ (2002) The plant-specific function of 2-Cys peroxiredoxin-mediated detoxification of peroxides in the redox-hierarchy of photosynthetic electron flux. Proc Natl Acad Sci USA 99:5738–5743

Konno H, Nakane T, Yoshida M, Ueoka-Nakanishi H, Hara S, Hisabori T (2012) Thiol modulation of the chloroplast ATP synthase is dependent on the energization of thylakoid membranes. Plant Cell Physiol 53:626–634

Laugier E, Tarrago L, Courteille A, Innocenti G, Eymery F, Rumeau D, Issakidis-Bourguet E, Rey P (2013) Involvement of thioredoxin y2 in the preservation of leaf methionine sulfoxide reductase capacity and growth under high light. Plant, Cell Environ 36:670–682

Laurent TC, Moore EC, Reichard P (1964) Enzymatic synthesis of deoxyribonucleotides. IV. Isolation and characterization of thioredoxin, the hydrogen donor from *Escherichia coli B*. J Biol Chem 239:3436–3444

Lemaire SD, Michelet L, Zaffagnini M, Massot V, Issakidis-Bourguet E (2007) Thioredoxins in chloroplasts. Curr Genet 51:343–365

Lepisto A, Kangasjarvi S, Luomala E-M, Brader G, Sipari N, Keranen M, Keinanen M, Rintamaki E (2009) Chloroplast NADPH-thioredoxin reductase interacts with photoperiodic development in *Arabidopsis*. Plant Physiol 149:1261–1276

Lindahl M, Kieselbach T (2009) Disulphide proteomes and interactions with thioredoxin on the track towards understanding redox regulation in chloroplasts and cyanobacteria. J Proteomics 72:416–438

Luo T, Fan T, Liu Y, Rothbart M, Yu J, Zhou S, Grimm B, Luo M (2012) Thioredoxin redox regulates ATPase activity of magnesium chelatase CHLI subunit and modulates redox-mediated signaling in tetrapyrrole biosynthesis and homeostasis of reactive oxygen species in pea plants. Plant Physiol 159:118–130

Marri L, Zaffagnini M, Collin V, Issakidis-Bourguet E, Lemaire SD, Pupillo P, Sparla F, Miginiac-Maslow M, Trost P (2009) Prompt and easy activation by specific thioredoxins of Calvin cycle enzymes of *Arabidopsis thaliana* associated in the GAPDH/CP12/PRK supramolecular complex. Mol Plant 2:259–269

Michalska J, Zauber H, Buchanan BB, Cejudo FJ, Geigenberger P (2009) NTRC links built-in thioredoxin to light and sucrose in regulating starch synthesis in chloroplasts and amyloplasts. Proc Natl Acad Sci USA 106:9908–9913

Michelet L, Zaffagnini M, Marchand C, Collin V, Decottignies P, Tsan P, Lancelin JM, Trost P, Miginiac-Maslow M, Noctor G, Lemaire SD (2005) Glutathionylation of chloroplast thioredoxin f is a redox signaling mechanism in plants. Proc Natl Acad Sci USA 102:16478–16483

Michelet L, Zaffagnini M, Morisse S, Sparla F, Perez-Perez ME, Francia F, Danon A, Marchand CH, Fermani S, Trost P, Lemaire SD (2013) Redox regulation of the Calvin-Benson cycle: something old, something new. Front Plant Sci 4:470

Miginiac-Maslow M, Lancelin JM (2002) Intrasteric inhibition in redox signalling: light activation of NADP-malate dehydrogenase. Photosynth Res 72:1–12

Mihara S, Yoshida K, Higo A, Hisabori T (2017) Functional significance of NADPH-thioredoxin reductase C in the antioxidant defense system of cyanobacterium *Anabaena* sp. PCC 7120. Plant Cell Physiol 58:86–94

Mikkelsen R, Mutenda KE, Mant A, Schurmann P, Blennow A (2005) α-Glucan, water dikinase (GWD): a plastidic enzyme with redox-regulated and coordinated catalytic activity and binding affinity. Proc Natl Acad Sci USA 102:1785–1790

Mills JD, Mitchell P, Schurmann P (1980) Modulation of coupling factor ATPase activity in intact chloroplasts: the role of the thioredoxin system. FEBS Lett 112:173–177

Montrichard F, Alkhalfioui F, Yano H, Vensel WH, Hurkman WJ, Buchanan BB (2009) Thioredoxin targets in plants: the first 30 years. J Proteomics 72:452–474

Moon JC, Jang HH, Chae HB, Lee JR, Lee SY, Jung YJ, Shin MR, Lim HS, Chung WS, Yun DJ, Lee KO (2006) The C-type *Arabidopsis* thioredoxin reductase ANTR-C acts as an electron donor to 2-Cys peroxiredoxins in chloroplasts. Biochem Biophys Res Commun 348:478–484

Motohashi K, Hisabori T (2006) HCF164 receives reducing equivalents from stromal thioredoxin across the thylakoid membrane and mediates reduction of target proteins in the thylakoid lumen. J Biol Chem 281:35039–35047

Motohashi K, Hisabori T (2010) CcdA is a thylakoid membrane protein required for the transfer of reducing equivalents from stroma to thylakoid lumen in the higher plant chloroplast. Antioxid Redox Signal 13:1169–1176

Motohashi K, Kondoh A, Stumpp MT, Hisabori T (2001) Comprehensive survey of proteins targeted by chloroplast thioredoxin. Proc Natl Acad Sci USA 98:11224–11229

Naranjo B, Diaz-Espejo A, Lindahl M, Cejudo FJ (2016a) Type-*f* thioredoxins have a role in the short-term activation of carbon metabolism and their loss affects growth under short-day conditions in *Arabidopsis thaliana*. J Exp Bot 67:1951–1964

Naranjo B, Mignee C, Krieger-Liszkay A, Hornero-Mendez D, Gallardo-Guerrero L, Cejudo FJ, Lindahl M (2016b) The chloroplast NADPH thioredoxin reductase C, NTRC, controls non-photochemical quenching of light energy and photosynthetic electron transport in *Arabidopsis*. Plant Cell Environ 39:804–822.

Nee G, Zaffagnini M, Trost P, Issakidis-Bourguet E (2009) Redox regulation of chloroplastic glucose-6-phosphate dehydrogenase: a new role for f-type thioredoxin. FEBS Lett 583:2827–2832

Nikkanen L, Rintamaki E (2014) Thioredoxin-dependent regulatory networks in chloroplasts under fluctuating light conditions. Philos Trans R Soc Lond B Biol Sci 369:20130224

Nikkanen L, Toivola J, Diaz MG, Rintamaki E (2017) Chloroplast thioredoxin systems: prospects for improving photosynthesis. Philos Trans R Soc Lond B Biol Sci 372:20160474

Nikkanen L, Toivola J, Rintamaki E (2016) Crosstalk between chloroplast thioredoxin systems in regulation of photosynthesis. Plant, Cell Environ 39:1691–1705

Ojeda V, Perez-Ruiz JM, Gonzalez M, Najera VA, Sahrawy M, Serrato AJ, Geigenberger P, Cejudo FJ (2017) NADPH thioredoxin reductase C and thioredoxins act concertedly in seedling development. Plant Physiol 174:1436–1448

Okegawa Y, Motohashi K (2015) Chloroplastic thioredoxin *m* functions as a major regulator of Calvin cycle enzymes during photosynthesis in vivo. Plant J 84:900–913

Pascual MB, Mata-Cabana A, Florencio FJ, Lindahl M, Cejudo FJ (2011) A comparative analysis of the NADPH thioredoxin reductase C-2-Cys peroxiredoxin system from plants and cyanobacteria. Plant Physiol 155:1806–1816

Peltier JB, Emanuelsson O, Kalume DE, Ytterberg J, Friso G, Rudella A, Liberles DA, Soderberg L, Roepstorff P, von Heijne G, van Wijk KJ (2002) Central functions of the lumenal and peripheral thylakoid proteome of *Arabidopsis* determined by experimentation and genome-wide prediction. Plant Cell 14:211–236

Perez-Ruiz JM, Cejudo FJ (2009) A proposed reaction mechanism for rice NADPH thioredoxin reductase C, an enzyme with protein disulfide reductase activity. FEBS Lett 583:1399–1402

Perez-Ruiz JM, Guinea M, Puerto-Galan L, Cejudo FJ (2014) NADPH thioredoxin reductase C is involved in redox regulation of the Mg-chelatase I subunit in *Arabidopsis thaliana* chloroplasts. Mol Plant 7:1252–1255

Perez-Ruiz JM, Spinola MC, Kirchsteiger K, Moreno J, Sahrawy M, Cejudo FJ (2006) Rice NTRC is a high-efficiency redox system for chloroplast protection against oxidative damage. Plant Cell 18:2356–2368

Pulido P, Spinola MC, Kirchsteiger K, Guinea M, Pascual MB, Sahrawy M, Sandalio LM, Dietz KJ, Gonzalez M, Cejudo FJ (2010) Functional analysis of the pathways for 2-Cys peroxiredoxin reduction in *Arabidopsis thaliana* chloroplasts. J Exp Bot 61:4043–4054

Richter AS, Grimm B (2013) Thiol-based redox control of enzymes involved in the tetrapyrrole biosynthesis pathway in plants. Front Plant Sci 4:371

Richter AS, Peter E, Rothbart M, Schlicke H, Toivola J, Rintamaki E, Grimm B (2013) Post-translational influence of NADPH-dependent thioredoxin reductase C on enzymes in tetrapyrrole synthesis. Plant Physiol 162:63–73

Sánchez-Riego AM., Mata-Cabana A, Galmozzi CV, Florencio FJ (2016) NADPH-thioredoxin reductase C mediates the response to oxidative stress and thermotolerance in the cyanobacterium *Anabaena* sp. PCC7120. Front Microbiol 7:1283

Sanz-Barrio R, Corral-Martinez P, Ancin M, Segui-Simarro JM, Farran I (2013) Overexpression of plastidial thioredoxin f leads to enhanced starch accumulation in tobacco leaves. Plant Biotechnol J 11:618–627

Sasaki Y, Kozaki A, Hatano M (1997) Link between light and fatty acid synthesis: thioredoxin-linked reductive activation of plastidic acetyl-CoA carboxylase. Proc Natl Acad Sci USA 94:11096–11101

Scheibe R, Anderson LE (1981) Dark modulation of NADP-dependent malate dehydrogenase and glucose-6-phosphate dehydrogenase in the chloroplast. Biochim Biophys Acta 636:58–64

Schroter Y, Steiner S, Matthai K, Pfannschmidt T (2010) Analysis of oligomeric protein complexes in the chloroplast sub-proteome of nucleic acid-binding proteins from mustard reveals potential redox regulators of plastid gene expression. Proteomics 10:2191–2204

Schurmann P, Maeda K, Tsugita A (1981) Isomers in thioredoxins of spinach chloroplasts. Eur J Biochem 116:37–45

Schwarz O, Schurmann P, Strotmann H (1997) Kinetics and thioredoxin specificity of thiol modulation of the chloroplast H^+-ATPase. J Biol Chem 272:16924–16927

Serrato AJ, Fernandez-Trijueque J, Barajas-Lopez JD, Chueca A, Sahrawy M (2013) Plastid thioredoxins: a "one-for-all" redox-signaling system in plants. Front Plant Sci 4:463

Serrato AJ, Perez-Ruiz JM, Spinola MC, Cejudo FJ (2004) A novel NADPH thioredoxin reductase, localized in the chloroplast, which deficiency causes hypersensitivity to abiotic stress in *Arabidopsis thaliana*. J Biol Chem 279:43821–43827

Seung D, Thalmann M, Sparla F, Abou Hachem M, Lee SK, Issakidis-Bourguet E, Svensson B, Zeeman SC, Santelia D (2013) *Arabidopsis thaliana* AMY3 is a unique redox-regulated chloroplastic α-amylase. J Biol Chem 288:33620–33633

Silver DM, Silva LP, Issakidis-Bourguet E, Glaring MA, Schriemer DC, Moorhead GB (2013) Insight into the redox regulation of the phosphoglucan phosphatase SEX4 involved in starch degradation. FEBS J 280:538–548

Simionato D, Basso S, Zaffagnini M, Lana T, Marzotto F, Trost P, Morosinotto T (2015) Protein redox regulation in the thylakoid lumen: the importance of disulfide bonds for violaxanthin de-epoxidase. FEBS Lett 589:919–923

Skryhan K, Cuesta-Seijo JA, Nielsen MM, Marri L, Mellor SB, Glaring MA, Jensen PE, Palcic MM, Blennow A (2015) The role of cysteine residues in redox regulation and protein stability of *Arabidopsis thaliana* starch synthase 1. PLoS ONE 10:e0136997

Sparla F, Costa A, Lo Schiavo F, Pupillo P, Trost P (2006) Redox regulation of a novel plastid-targeted β-amylase of Arabidopsis. Plant Physiol 141:840–850

Steiner S, Schroter Y, Pfalz J, Pfannschmidt T (2011) Identification of essential subunits in the plastid-encoded RNA polymerase complex reveals building blocks for proper plastid development. Plant Physiol 157:1043–1055

Stenbaek A, Hansson A, Wulff RP, Hansson M, Dietz K-J, Jensen PE (2008) NADPH-dependent thioredoxin reductase and 2-Cys peroxiredoxins are needed for the protection of Mg-protoporphyrin monomethyl ester cyclase. FEBS Lett 582:2773–2778

Sueoka K, Yamazaki T, Hiyama T, Nakamoto H (2009) The NADPH thioredoxin reductase C functions as an electron donor to 2-Cys peroxiredoxin in a thermophilic cyanobacterium *Thermosynechococcus elongatus* BP-1. Biochem Biophys Res Commun 380:520–524

Thormahlen I, Meitzel T, Groysman J, Ochsner AB, von Roepenack-Lahaye E, Naranjo B, Cejudo FJ, Geigenberger P (2015) Thioredoxin *f*1 and NADPH-dependent thioredoxin reductase C have overlapping functions in regulating photosynthetic metabolism and plant growth in response to varying light conditions. Plant Physiol 169:1766–1786

Thormahlen I, Ruber J, von Roepenack-Lahaye E, Ehrlich SM, Massot V, Hummer C, Tezycka J, Issakidis-Bourguet E, Geigenberger P (2013) Inactivation of thioredoxin *f*1 leads to decreased light activation of ADP-glucose pyrophosphorylase and altered diurnal starch turnover in leaves of *Arabidopsis* plants. Plant, Cell Environ 36:16–29

Thormahlen I, Zupok A, Rescher J, Leger J, Weissenberger S, Groysman J, Orwat A, Chatel-Innocenti G, Issakidis-Bourguet E, Armbruster U, Geigenberger P (2017) Thioredoxin play a crucial role in dynamic acclimation of photosynthesis in fluctuating light. Mol Plant 10:168–182

Toivola J, Nikkanen L, Dahlstrom KM, Salminen TA, Lepisto A, Vignols HF, Rintamaki E (2013) Overexpression of chloroplast NADPH-dependent thioredoxin reductase in *Arabidopsis* enhances leaf growth and elucidates *in vivo* function of reductase and thioredoxin domains. Front Plant Sci 4:389

Wang P, Liu J, Liu B, Feng D, Da Q, Wang P, Shu S, Su J, Zhang Y, Wang J, Wang HB (2013) Evidence for a role of chloroplastic *m*-type thioredoxins in the biogenesis of photosystem II in *Arabidopsis*. Plant Physiol 163:1710–1728

Wimmelbacher M, Bornke F (2014) Redox activity of thioredoxin *z* and fructokinase-like protein 1 is dispensable for autotrophic growth of *Arabidopsis thaliana*. J Exp Bot 65:2405–2413

Wolosiuk RA, Crawford NA, Yee BC, Buchanan BB (1979) Isolation of three thioredoxins from spinach leaves. J Biol Chem 254:1627–1632

Yamaryo Y, Motohashi K, Takamiya K, Hisabori T, Ohta H (2006) In vitro reconstitution of monogalactosyldiacylglycerol (MGDG) synthase regulation by thioredoxin. FEBS Lett 580:4086–4090

Yano H, Wong JH, Lee YM, Cho MJ, Buchanan BB (2001) A strategy for the identification of proteins targeted by thioredoxin. Proc Natl Acad Sci USA 98:4794–4799

Yoshida K, Hara S, Hisabori T (2015) Thioredoxin selectivity for thiol-based redox regulation of target proteins in chloroplasts. J Biol Chem 290:14278–14288

Yoshida K, Hisabori T (2016) Two distinct redox cascades cooperatively regulate chloroplast functions and sustain plant viability. Proc Natl Acad Sci USA 113:E3967–E3976

Yoshida K, Hisabori T (2017) Distinct electron transfer from ferredoxin-thioredoxin reductase to multiple thioredoxin isoforms in chloroplasts. Biochem J 474:1347–1360

Yoshida K, Matsuoka Y, Hara S, Konno H, Hisabori T (2014) Distinct redox behaviors of chloroplast thiol enzymes and their relationships with photosynthetic electron transport in *Arabidopsis thaliana*. Plant Cell Physiol 55:1415–1425

Zhang N, Portis AR Jr (1999) Mechanism of light regulation of Rubisco: a specific role for the larger Rubisco activase isoform involving reductive activation by thioredoxin-f. Proc Natl Acad Sci USA 96:9438–9443

H_2O_2-Mediated Biotic and Abiotic Stress Responses in Plants

Ayaka Hieno, Naznin Hushuna Ara and Yoshiharu Y. Yamamoto

Abstract Hydrogen peroxide (H_2O_2) is an important signaling molecule for various physiological processes that take place in higher plants during environmental adaptation. This article addresses the role of H_2O_2 in signal transduction and gene regulation in the context of a variety of biotic and abiotic stresses. Based on a comparative analysis using public and our own transcriptome data, we consider H_2O_2 as an independent mediator that acts in response to any of the plant stresses and also as having a role in the synergetic response to combinations of stress factors. Here, we propose H_2O_2 as a signal molecule that is able to trigger gene activation of a conserved set of genes that make a universal transcriptional subnetwork comprising tens of transcription factors.

1 Introduction

Plants have developed sophisticated mechanisms for acclimation to a variety of environmental conditions that are often not optimal or suitable for plants to grow in. Hydrogen peroxide (H_2O_2) is one of the signal molecules involved in environmental adaptation and acts as a universal player in all the known stress responses in plants.

H_2O_2 in plants is produced under stressful conditions. Its major source is the electron transport system (ETS) in chloroplasts and mitochondria, and peroxisomal glycolate oxidase in the photorespiration pathway. H_2O_2 production by these sources is promoted by many plant stressors, including high and low temperature, high-light, starvation, and stomatal closure caused by drought, high salinity, high osmolarity,

A. Hieno
River Basin Research Center, Gifu University, 1-1 Yanagido, Gifu City, Gifu 501-1193, Japan

N. Hushuna Ara · Y. Y. Yamamoto (✉)
Faculty of Applied Biological Sciences, Gifu University, 1-1 Yanagido, Gifu City, Gifu 501-1193, Japan
e-mail: yyy@gifu-u.ac.jp

Y. Y. Yamamoto
RIKEN CSRS, 1-7-22 Suehiro-cho, Tsurumi-ku, Yokohama, Kanagawa 230-0045, Japan

S. K. Panda and Y. Y. Yamamoto (eds.), *Redox Homeostasis in Plants*,
Signaling and Communication in Plants, https://doi.org/10.1007/978-3-319-95315-1_2

and pathogen infection. In addition to these, plants have systems to activate H_2O_2 production by internal stress signals.

Accumulated H_2O_2 has the potential to damage plant tissue, but at the same time it functions to trigger a stress signal that leads to transcriptional activation of hundreds of anti-stress factors, including enzymes for ROS detoxification and regeneration of their cofactors, defense proteins and secondary metabolites against diseases, pests and xenobiotics, cell and protein protection factors, and also signaling and transcription factors. One of the stress responses includes activation of H_2O_2 production by the hydrogen peroxide signal itself, which is generated by multiple systems. Raising and lowering of H_2O_2 levels triggered by the H_2O_2 signal lead to oxidative bursts (Zurbriggen et al. 2010), ROS homeostasis (Miller et al. 2010), and also H_2O_2 waving for long-distance signaling in a plant body (Mittler et al. 2011).

Another complex aspect of hydrogen peroxide is cellular redox. H_2O_2 itself is an oxidizer of proteins because of its chemical nature. However, the H_2O_2 signal activates enzymatic protein reduction systems in plant cells. This bidirectional promotion challenges the idea of the so-called cellular redox state.

In this chapter, we would like to present an overview of the roles of H_2O_2 in stress damage and response to highlight how ubiquitous the H_2O_2 signal is in plant stress responses. By reviewing various biotic and abiotic stresses of plants, we present a concise understanding of the role of H_2O_2 in plant stress.

2 H_2O_2 in Biotic and Abiotic Stress Responses in Plants

2.1 Pathogens

Infection by pathogens induces pathogen resistance in plants. One of the resistance pathways involves programmed cell death at the infected tissue to suppress the spread of the pathogen, which is known as the hypersensitive response (HR). H_2O_2 initiates the HR after infection (Levine et al. 1994). Another pathway is systemic acquired resistance (SAR), which elevates resistance level at uninfected tissues of the infected plants. This systemic signaling is mediated by salicylic acid (SA) (Gaffney et al. 1993; Delaney et al. 1994) and H_2O_2 (Alvarez et al. 1998).

After pathogen-associated molecular patterns (PAMPs) or microbe-associated molecular patterns (MAMPs) are recognized by the corresponding plant receptors, H_2O_2 accumulates as a signal molecule for the plant's responses to the pathogen. The accumulation is enzymatically achieved by the respiratory burst oxidase homolog (RBOH). Promoter analysis of two of the ten Arabidopsis RBOH genes, *AtRbohD* and *AtRbohF*, demonstrated that these two promoters are activated by the hemibiotrophic bacterial pathogen *Pseudomonas syringae* pv. *tomato* DC3000 and the necrotrophic fungal pathogen *Plectosphaerella cucumerina*, respectively. These results reveal that the activation of H_2O_2 biosynthesis by pathogen infection is achieved, in part, by transcriptional stimulation of the RBOH genes (Morales et al. 2016). Spatial analysis of promoter activation by Morales et al. also suggested that both genes are involved in the HR and SAR. These results imply crosstalk between the HR and SAR through RBOH, which complicates understanding of the pathogen response in plants and increases the physiological significance of the spatial accumulation profiles of H_2O_2.

In addition to RBOH, peroxisomal glycolate oxidase of the photorespiration pathway (Rojas et al. 2012) and periplasmic peroxidases [Arabidopsis PRX33 and PRX34, (O'Brien et al. 2012)] is also reported to have roles in H_2O_2 production in the HR. Photorespiration is likely promoted by stomatal closure as a response to pathogen infection, which will be explained later. Furthermore, chloroplasts are additional sources of H_2O_2 (Liu et al. 2007; Zurbriggen et al. 2009).

As mentioned above, stomatal closure, which blocks invasion by pathogens, is another protective response against infection. In this process, abscisic acid (ABA) signaling is known to stimulate H_2O_2 accumulation at guard cells to promote stomatal closure through regulation of several channels and transporters (Kwak et al. 2003; Khokon et al. 2011). For stomatal regulation, H_2O_2 evolution by ABA is achieved through transcriptional activation of *AtRbohD* and *AtRbohF* (Kwak et al. 2003), and also by elevation of activity at the protein level (Sirichandra et al. 2009; Zhang et al. 2009; Baxter et al. 2014).

Studies on transcriptional regulation by biotic stress have a long history. The biggest breakthrough was the identification of the master switch of SAR, *NONEXPRESSOR OF PR GENE 1* (*NPR1*) (Cao et al. 1997). In uninfected plants, NPR1 localizes in the cytoplasm as multimers connected by disulfide bonds, and the SA signal activated by biotic stress causes a reduction of NPR1, causing release of its monomers (Mou et al. 2003; Tada et al. 2008). These then go into the nucleus and regulate DNA-binding activity of the TGA class of basic domain/leucine zipper (bZIP) transcription factors to activate downstream defense genes (Fan and Dong 2002; Despres et al. 2003; Choi et al. 2010). Almost all of the SA-responsive genes are controlled by NPR1 [99% of BTH-responsive genes (Wang et al. 2006)], demonstrating that NPR1 is the master switch of the SA signal.

H_2O_2 produced by SAR can directly cause protein oxidation, but NPR1 is thought to be reduced by an H_2O_2-stimulated protein reduction system (Mou et al. 2003; Tada et al. 2008).

2.2 *High-Light*

Under high-light stress, electrons overflow forms the ETS at the thylakoid membrane, and the superoxide radicals (O_2^-) evolved are metabolized into H_2O_2 by superoxide dismutase as a first step of ROS detoxification. H_2O_2 is then removed by catalases and peroxidases (Allahverdiyeva and Aro 2012). Major damage to photosynthetic tissues under high-light stress is caused by these ROS species. High-light also enhances photorespiration, which includes H_2O_2 biosynthesis at the peroxisomes (Waszczak et al. 2016), providing an additional source of ROS production.

Response to high-light stress in Arabidopsis includes gene activation of catalase and peroxidase genes (Karpinski et al. 1997). Of these, *ASCORBATE PEROXIDASE 2* (*APX2*) is remarkably activated by high-light and this activation is mediated by H_2O_2 (Karpinski et al. 1999). High-light stress causes activation of many small and large heat shock protein (HSP) genes (Rossel et al. 2002; Kimura et al. 2003) that are also activated by H_2O_2, suggesting this triggering under high-light stress is mediated by H_2O_2 (Yamamoto et al. 2004). This pathway activates genes for small and large HSPs (HSP17-20 and HSP70), cell protection (LEA and lipid transfer proteins), defense factors, and ROS detoxification including catalases, peroxidases, glutathione S-transferase (GST), glutathione reductase (GR), and thioredoxin (TRX) (Yamamoto et al. 2004) (Yamamoto et al., unpublished results).

In addition to the H_2O_2-dependent signaling pathway, there is also an H_2O_2-independent pathway for the high-light stress response regulating a photoprotection related gene, *ELIP2* (Kimura et al. 2001, 2003). This pathway regulates genes for photoprotection including the protection and recovery of photosystem II (ELIP and FtsH), flavonoids, sinapate ester and carotenoid biosynthesis, and a component of plastoglobules (fibrillin).

Comparative microarray analysis between high-light and H_2O_2 responses revealed three high-light-activated gene groups. These groups are H_2O_2-up-regulated, H_2O_2-down-regulated, and H_2O_2-independent, and they all have different characteristics of induction kinetics and dose response to high-light stress (Yamamoto et al. 2004). For example, the H_2O_2-down-regulated group shows slower induction and less response under the higher intensity of high-light (350 W/m^2) than the H_2O_2-independent group. This suggests that H_2O_2 functions not only as a direct signal mediator but also as a modulator of the stress response.

A zinc finger-type transcription factor *RHL41*, which is also known as *ZAT12*, was identified as one of the high-light-responsive genes, and its overexpression caused a partial high-light stress resistance (Iida et al. 2000). This report demonstrated that *RHL41/ZAT12* mediates a part of the high-light stress response. It is also induced by heat and wounding stresses, and by H_2O_2, showing a multi-stress response *via* H_2O_2 (Rizhsky et al. 2004).

Some of the transcriptional response to high-light stress in Arabidopsis are mediated by a blue light receptor, HY4/CRY1, *via* the transcription factor HY5 (Kleine et al. 2007; Hayami et al. 2015). As this high-light signaling pathway controls Arabidopsis *ELIP2*, it is thought to represent a high-light-specific pathway, different from the H_2O_2-dependent one. Interestingly, HY5 is also known to transduce the UV-B signal from an UV-B photoreceptor, UVR8 (Ulm et al. 2004; Brown et al. 2005; Hayami et al. 2015).

2.3 Cold

Cold stress induces H_2O_2 accumulation in plants, and the mechanism is well understood (Danon 2012). Under low temperatures, the Calvin cycle in the chloroplast slows down and the consumption of NADPH and ATP is reduced (Falk et al. 1996).

This fall pushes the redox state of the photosynthetic ETS toward the reduced side and causes an overflow of electrons from the ETS, resulting in the generation of a large amount of toxic ROS species ($^3O_2$ and O_2^-) in the chloroplast. Low temperature also causes crystallization of the thylakoid membrane and disturbs the electron transport, also promoting ROS generation. The superoxide radical (O_2^-) is metabolized into H_2O_2 as a stable intermediate of the detoxification process, as mentioned in the section on high-light.

Studies on the transcriptional response to cold stress have revealed that some known H_2O_2-inducible genes, including ZAT12, are found amongst the known cold-inducible genes (Fowler and Thomashow 2002; Doherty et al. 2009). These results strongly suggest that H_2O_2 mediates some of the cold stress responses.

There are two signal transduction pathways reported for the cold stress response: the *CBF/DREB1*-dependent pathway for activation of *COR15A* (Kurepin et al. 2013), and the H_2O_2-dependent pathway where *ZAT12* activation is involved. The former pathway is mediated by transcriptional activation of *DREB1* genes, and they target the drought-responsive element (DRE/C-Repeat (CRT), which is also the target of the drought stress response *via DREB2* (Nakashima et al. 2014) and shared with promoters of genes for cell protection against frost, drought, and high osmolarity.

H_2O_2 regulates not only the H_2O_2-dependent cold response pathway but also *DREB1*-dependent pathway, demonstrated by expression analysis of suppressors of a thylakoid membrane-bound ascorbate peroxidase (tAPX) gene (Maruta et al. 2012). However, H_2O_2's role in the regulation of the *DREB1* genes appears to be modulating the response rather than direct mediation of the cold signal (Maruta et al. 2012). This idea of no direct involvement of H_2O_2 in the *DREB1* pathway also fits with the observations that cold activation of DRE/CRT occurs in a light-dependent way, and that DCMU (3-(3,4-dichlorophenyl)-1,1-dimethyl urea), an inhibitor of photosynthetic ETS between photosystem II and plastoquinone, does not stop the cold activation (Kim et al. 2002).

2.4 UV-B

Plant exposure to UV-B causes damage to DNA, photosystem II, membranes, and IAA, a phytohormone, through direct absorption of damaged molecules (Jansen et al. 1998). The most conspicuous damage is to DNA that leads to irreversible growth arrest in the exposed tissue. UV-B irradiation of plant tissue also causes the generation of ROS from photosystem II (Jansen et al. 1998) and UV-B-absorbing molecules including pigments, phenolic compounds, and aromatic amino acids (Brosché and Strid 2003). The ROS species generated also cause damage to plant tissue.

Transcriptional response to UV-B stress in Arabidopsis is in part mediated by the UV-B photoreceptor, UVR8 (Rizzini et al. 2011). UVR8 activates many UV-B-responsive genes for photoprotection including those involved in flavonoid biosynthesis and the recovery of photosystem II, DNA repair, and ROS scavenging (Brown et al. 2005). The majority of the UVR8-regulated genes are controlled through HY5 (Brown and Jenkins 2008), which is also the central regulator of the high-light stress response by CRY1.

A UVR8-independent transcriptional response to UV-B, which is more dominant under higher fluence, has also been reported (Brown and Jenkins 2008). UV-B is directly absorbed by DNA and damages it, which is another trigger for the UV-B response. This idea is supported by expression analysis of *UDPgtfp*, which is activated by UV-B in a UVR8-independent manner that revealed application of an inhibitor of DNA repair, hydroxyurea and enhanced the UV-B response of the gene (Biever et al. 2014).

In addition to these UVR8- and DNA damage-dependent signal transduction pathways, another pathway where H_2O_2 plays a role is also present (Czegeny et al. 2014).

Under UV-B stress, ROS are generated by photosystem II and UV-absorbing molecules as mentioned. In addition to these, the involvement of enzymatic oxidative bursts is also reported. Studies on Arabidopsis mutants have shown that the activation of enzymatic biosynthesis by *AtRbohD* and *AtRbohF* is a major source of H_2O_2 accumulation under UV-B, demonstrating the involvement of RBOH (Kalbina and Strid 2006). This dominance, however, may depend on low fluence.

2.5 Heat

The major response to heat in plants is the so-called heat shock response, which is conserved in organisms ranging from bacteria to mammals. In Arabidopsis, high temperature (37 °C in experiments) cancels HSP90/70 repression of HSFA1s, and the activated HSFA1s bind to the heat shock element (HSE) in the promoter region of heat shock protein genes (HSPs) resulting in transcriptional activation (Ohama et al. 2017). The HSP superfamily is composed of several subfamilies: small HSP (sHSP, ~20 iDa), HSP60/GroEL/chaperonin, HSP70/DnaK, HSP90, and HSP100/Clp protease. Their major functions include the prevention of protein aggregation, assisting in protein refolding, protein translocation, and protein degradation (Wang et al. 2004).

For activation of HSPs, H_2O_2 is also involved in heat stress signaling, demonstrated by the suppression of the response after application of antioxidants (Volkov et al. 2006). Treatment of plants with salicylic acid (SA) is reported to enhance heat stress tolerance of spinach seedlings, suggesting that the SA-stimulated H_2O_2 response causes a partial heat stress response (Dat et al. 1998). We also found that as many as 34 sHSP genes are activated by H_2O_2 treatment in Arabidopsis seedlings (Yamamoto et al. 2004). Dickinson et al. have recently reported induction of binding of HSFA1a to the promoter region of heat-inducible Arabidopsis *HSP70* after the application of H_2O_2, and the suppressive and inducible effects on the binding by DCMU and 2,5-dibromo-3-methyl-6-isopropyl-p-benzoquinone (DBMIB), respectively. These results suggest that H_2O_2 amplifies the early heat signal evolved by the heat-dependent release of HSFA1s from HSP90/70. According to this idea, heat and H_2O_2 independently stimulate this early part of the heat shock signaling, which is protein activation of HSFA1s, and application of both gives an additive effect on the response. The apparent conflict with the observed dependence of H_2O_2 on the heat shock response, as mentioned above, can be resolved if the observation is re-interpreted as an additive effect of heat shock and heat-generated H_2O_2.

There are several sources of H_2O_2 generation under heat stress as reviewed by Mohanty et al. (2012). The activity of the Calvin cycle is reduced by heat due to the high-heat sensitivity of the enzymes involved, leading to overflow of electrons from the photosynthetic ETS. Photorespiration is increased by heat due to the nature of RuBisCO to produce H_2O_2 in the peroxisomes. Heat sensitivity of photosystem II leads to ROS production under high temperatures. Balancing the energy distribution between the two photosystems is modified under high temperature, so that ROS are generated from them. Energy transfer from the antenna to the core antenna complexes is also arrested, leading to additional ROS production. The evolved H_2O_2 attacks macromolecules and further spreads the stress damage. At the same time, it promotes the heat shock signaling as mentioned above and also ROS signaling.

The H_2O_2 generated under heat stress is subjected to the general ROS homeostasis, which includes H_2O_2 quenching via H_2O_2-dependent transcriptional activation of peroxidase and catalase genes (Grene 2002). In addition, direct activation of an Arabidopsis ascorbate peroxidase gene *APX1*, by the heat shock signaling through HSE in its promoter is also reported (Storozhenko et al. 1998).

2.6 *Drought and Salt*

Major drought stress signaling pathways are either ABA-dependent or DREB2-dependent (Shinozaki and Yamaguchi-Shinozaki 2007). The salt stress response has similarities with that of drought stress, suggesting a large overlap in their signaling pathways (Seki et al. 2002). In addition, the involvement of ROS in the drought stress signaling is also known (Noctor et al. 2014).

Biochemical studies decades ago observed the production of ROS, including H_2O_2 under drought stress, and enzymes and their cofactors for ROS quenching (Smirnoff 1993). Application of H_2O_2 to soybean plants or detached maize leaves elevated drought stress resistance through partial activation of the stress response of metabolites, demonstrating H_2O_2 as a signal molecule of this response (Ishibashi et al. 2011; Terzi et al. 2014).

The mechanism of H_2O_2 generation under drought stress is not well understood (Noctor et al. 2014). One possible major source is photorespiration (Smirnoff 1993; Noctor et al. 2014). Stomata are closed by ABA under drought conditions, and the closure results in limited CO_2 that increases photorespiration. As a result, a large amount of H_2O_2 is produced at the peroxisomes by glycolate oxidase in the photorespiration cycle. CO_2 limitation by stomatal closure also reduces the speed of the Calvin cycle and thus consumption of NADPH. This situation promotes the generation of a large amount of ROS from the photosynthetic ETS.

The ABA signal for stomatal closure in Arabidopsis is known to promoter H_2O_2 production through activation of RBOH (AtRbohD and AtRbohF) at the gene expression level (Kwak et al. 2003; Bright et al. 2006) and also protein activity (Sirichandra et al. 2009; Zhang et al. 2009). Therefore, there is the possibility that ABA activates H_2O_2 production by RBOH in other tissues as well. Once H_2O_2 is produced, the general systemic response through vascular tissue is activated by a positive feedback loop of H_2O_2 production (Baxter et al. 2014).

2.7 Wounding

Plants react to mechanical damage, often caused by pests, by activating a set of genes through wound signaling pathways, both locally in the damaged area and systemically in non-damaged areas (Leon et al. 2001). Some examples of wound-inducible genes are those that encode tomato proteinase inhibitor I and II (Pin1 and Pin2) for defense against pests (Graham et al. 1986), phenylpropanoid biosynthesis enzymes of French bean [phenylalanine ammonia-lyase (PAL) and chalcone synthase (CHS)] for wound sealing and formation of phenolic defense compounds (phytoalexin/phaseollin) (Dixon and Lamb 1979; Lawton et al. 1983).

Several signaling molecules have been identified for the wound stress response: ethylene, jasmonic acid, the peptide hormone systemin for Solanaceae plants, and oligosaccharides for Arabidopsis (Leon et al. 2001). Electronic stimuli are also reported to evoke the wound signal (Wildon et al. 1992). In addition to these, the induction of H_2O_2 accumulation is reported in many plant species (Orozco-Cardenas and Ryan 1999). After wounding, plants accumulate H_2O_2 in damaged and also undamaged tissue, suggesting the involvement of H_2O_2 in both local and systemic responses (Orozco-Cardenas and Ryan 1999).

This induction of H_2O_2 is achieved by RBOH, as revealed by studies using its inhibitor, diphenylene iodonium chloride (DPI) (Orozco-Cardenas and Ryan 1999), and also gene expression and mutant analyses of Arabidopsis *RbohD* (Takahashi et al. 2011). Wounding-dependent accumulation of H_2O_2 by RBOH in tomato is mediated by systemin and jasmonic acid (Orozco-Cardenas et al. 2001). In Arabidopsis, RbohD-dependent H_2O_2 production by wounding stress is regulated by the MAP kinase cascade (Takahashi et al. 2011).

2.8 Stress Combination

In nature, multiple abiotic stressors may act at the same time. Drought stress in savanna and desert areas is often accompanied by salt and/or heat stresses. Sunlight on sunny days challenges plants with high-light, UV-B, and heat. In high mountain areas, it is not uncommon to have cold, high light, UV-B, and drought stresses at the same time. Therefore, understanding the effects combinations of multiple stresses have is important in ecophysiology and also agriculture.

The effects of combinations of cold, heat, high-light, salt, and flagellin treatments on transcriptional responses were investigated, and various effects were identified, and categorized into "combinatorial," "canceled," "prioritized," "independent," and "similar" types (Rasmussen et al. 2013). As H_2O_2 is involved in all the stress responses, there is the possibility that it is also involved in the synergetic effects ("combinatorial") induced by combinations of stresses, along with other potentially common signaling elements like nitric oxide (NO), stress-related phytohormones, the MABK cascade, and some multi-stress-responsive transcription factors (Fujita et al. 2006; Mittler et al. 2011). The idea of the involvement of H_2O_2 accumulation in this combinatorial effect is supported by the finding that the synergetic effect of a combination of heat and drought stresses was enhanced by the mutation of Arabidopsis *ASCORBATE PEROXIDASE 1* (*APX1*) encoding a cytosolic peroxidase (Koussevitzky et al. 2008). In addition to H_2O_2, a recent report showed that ABA also plays a role in the synergetic effect of salt and heat stresses (Suzuki et al. 2016). Studies on the molecular mechanisms of the synergetic effects of stress combination started years ago, and further studies are expected to help the understanding of this complex but important phenomenon.

3 H_2O_2 in Systemic Response

A systemic response is where tissue challenged by a stressor stimulates a stress response in the whole body of the individual including unchallenged tissue. This was first reported at the molecular level for SAR in the pathogen response (Ward et al. 1991), and the involvement of SA and H_2O_2 in this intercellular signaling was revealed (Ward et al. 1991; Alvarez et al. 1998). Subsequently, systemic signaling in abiotic stress responses was also reported for wounding- (Orozco-Cardenas and Ryan 1999) and high-light (Karpinski et al. 1999) stress responses. In both cases, H_2O_2 was reported to be the molecule responsible for the systemic signaling. In systemic and long-distance signaling, a boost to the signal is theoretically required to avoid its decline before the extremities of the plant are reached. Actually, it has been found that Arabidopsis *RbohD* is involved in this process, both biotic and abiotic stress signaling, including heat, cold, and salt stresses (Miller et al. 2009).

One of the common features of these systemic signaling pathways appears to be mediation through vascular tissue, as evidenced by preferential accumulation of H_2O_2 in vascular tissue [wounding: (Orozco-Cardenas et al. 2001), high-light: (Fryer et al. 2003; Galvez-Valdivieso et al. 2009), and cold: (Dong et al. 2009)]. Vascular tissue-enriched accumulation of H_2O_2 is correlated with expression profiles of the RBOH genes (Morales et al. 2016). Therefore, this accumulation profile is a result of RBOH activity. These examples illustrate that such systemic responses are mediated by a common mechanism in the vascular tissue.

In addition to the autonomous self-amplification system of H_2O_2, phytohormone-assisted amplification systems are also possible. As an example, the ROS signal activated by high-light treatments stimulates the ABA response through ABA biosynthesis, as has been revealed by analysis of the hormone's biosynthesis mutants (Galvez-Valdivieso et al. 2009). We also observed that the activation of ABA-responsive genes by H_2O_2 treatment in Arabidopsis shoots was preceded by ABA accumulation, and the activation was canceled by mutation of an ABA biosynthesis gene (Hieno et al.). These observations demonstrate that H_2O_2 is able to activate the ABA signal *via* accumulation of ABA. This activation implies that some of the systemic response can be mediated *via* ABA accumulation and transport.

In contrast, the ABA signal produces H_2O_2 through activation of RbohF by a protein kinase, OST1, in ABA-dependent stomatal closure (Joshi-Saha et al. 2011). Stomatal regulation through the interaction of ABA and H_2O_2 has been reviewed recently (Mittler and Blumwald 2015). In addition, stomatal closure itself, which is under the control of ABA, leads to H_2O_2 production by the promotion of photorespiration and the limitation of the Calvin cycle, as mentioned above in the drought stress response section. This tells us that ABA is also able to promote H_2O_2 accumulation. The activation of both of ABA accumulation by H_2O_2, and of H_2O_2 production by ABA suggests that the ABA and H_2O_2 signals can activate each other to make a positive feedback loop, and this has the potential to support H_2O_2-stimulated systemic signaling.

SA is also known to have mutual activation with H_2O_2. There are several early reports in the 2000s showing sequential accumulation of H_2O_2 to SA in response to pathogen infection, ozone, and UV-B stresses, suggesting H_2O_2-triggered SA accumulation (Herrera-Vasquez et al. 2015). Later studies have revealed that mutation of a peroxisomal catalase gene, *CAT2*, causes accumulation of SA via accumulation of H_2O_2 (Chaouch et al. 2010), and silencing of a chloroplastic peroxidase gene, *tAPX*, causes up-regulation of a gene for a SA biosynthesis enzyme, *ICS2* (Noshi et al. 2012). These resorts strongly suggest that H_2O_2 promotes SA accumulation *via* activation of SA biosynthesis. There are also conflicting results where crosstalk from H_2O_2 to SA signaling occurred without elevation of the SA level, suggesting the crosstalk point is after SA accumulation before the action of NPR1 (Hieno et al.).

The reverse situation, H_2O_2 accumulation by SA, is also reported. Low SA accumulation mutants (*sid2-2*) and transgenic plants with low SA levels (NahG) of Arabidopsis showed less H_2O_2 accumulation, and high SA accumulation mutants (*cpr1-1*, *cpr5-1*, *cpr6-1*, and *dnd1-1*) had more (Mateo et al. 2006). In another report, SA was found to close stomata (Melotto et al. 2006), giving rise to enhanced H_2O_2 production. There is also direct evidence to show H_2O_2 accumulates in Arabidopsis leaves after SA application (Khokon et al. 2011). This, taken together with H_2O_2-dependent SA accumulation, suggests that hydrogen peroxide and salicylic acid have the potential to make a positive feedback loop, in a similar manner to hydrogen peroxide and ABA. Therefore, SA also has the potential to play a role in the systemic signaling of H_2O_2.

4 ROS Homeostasis and Positive Feedback Loop

H_2O_2 accumulates in response to pathogen infection, but the accumulation profile is not stable; it is transient with two distinct peaks at 1 and 6 h after infection (Lamb and Dixon 1997). Therefore, accumulated H_2O_2 is at times being quenched. In our experience, treatment of Arabidopsis seedlings with high-light causing strong H_2O_2 accumulation stimulated transient induction of H_2O_2-activated genes with a peak time of 1 h after start of the treatment with reduced expression at 3 h, and the same gene set had no reduction at 3 h in suppressors of CATALASE 2 (CAT2) (Vandenabeele et al. 2004; Yamamoto et al. 2004), demonstrating again tight homeostasis of the H_2O_2 level in wild type. We also observed that the treatment of Arabidopsis shoots with H_2O_2 did not cause detectable elevation of H_2O_2 accumulation in the treated tissue within a few hours after the treatment, and moderate elevation was detected only at 6 and 12 h after treatment. However, very drastic changes in gene expression profiles were observed from 1 to 24 h after treatment (Hieno et al.). These observations and others indicate that while temporal elevation of the H_2O_2 level does trigger long and sequential anti-stress responses, the H_2O_2 level itself is subjected to a strong ROS homeostasis so that initial accumulation by plant stressors soon disappears.

Part of the mechanism for ROS homeostasis is H_2O_2-dependent induction of H_2O_2-quenching enzymes and cofactors, including peroxidases, catalases, ascorbate-regenerating enzyme (MDAR), SOD activator (LSU: response to low sulfur), and glutathione-regeneration enzyme (GR) (Mittler et al. 2004; Miller et al. 2010; Noctor et al. 2014; Garcia-Molina et al. 2017). These responses explain the reduction of H_2O_2 levels elevated by stimulation of a stressor, giving transient accumulation profiles. In addition, because protein reduction systems including glutathione are also activated by the H_2O_2 signal, these responses bring a complex situation where an increase of H_2O_2 which is an oxidizer of proteins, also activates protein reduction.

On the other hand, the systemic response *via* H_2O_2 theoretically requires H_2O_2-dependent hydrogen peroxide production to avoid decay of this long-distance signaling, and this is achieved in part by H_2O_2-dependent activation of gene expression and of enzymatic activity of RBOHs (Miller et al. 2009; Morales et al. 2016) [reviewed in (Baxter et al. 2014)]. Involvement of other H_2O_2-producing enzymes, like glycolate oxidase and periplasmic peroxidase, is also reported as we have seen in the section on pathogens. In addition, stomatal closure, which is caused by H_2O_2, ABA, and SA, leads to H_2O_2 production, giving rise to the possibility of another positive feedback loop involving stomatal regulation. This potential system is suggested to function in the HR (as reviewed in the "Pathogens" section).

Recently, overexpression of a stress-responsive transcription factor, *RRTF1*, is reported to cause H_2O_2 production (Matsuo et al. 2015). Gene expression of *RRTF1* is strongly activated by H_2O_2 (Matsuo et al. 2015), and this provides a positive feedback system for H_2O_2 accumulation. As RBOH genes are not reported to be activated by *RRTF1*, this is an additional system to the RBOH-dependent positive feedback system. An interesting aspect of this new feedback is the involvement of a transcription factor. *RRTF1* is activated by many plant stressors, including salt, drought, cold, UV-B, heat, osmotic stress, high-light, and also PAMPs (Matsuo et al. 2015), supporting the idea of the presence of a multi-stress-activated H_2O_2 response and a positive feedback system.

Coexistence of H_2O_2 homeostasis and the positive feedback of H_2O_2 accumulation give rise to a complex situation. One simple idea is that H_2O_2 homeostasis is for transient accumulation, and the positive feedback extends the signal achieving systemic signal transduction. Together, ROS waving in a plant body would be built up (Mittler et al. 2011).

Other known potential positive feedback systems for hydrogen peroxide accumulation require phytohormones (ABA and SA) and are explained in the previous section (III).

5 H_2O_2-Triggered Transcriptional Response as Common Subnetwork in Plant Stress Response

As we have seen, the biotic and abiotic stresses overviewed in this chapter all produce H_2O_2 in plants. As H_2O_2 is a signal molecule that is able to cause a transcriptional response without assistance of other plant signals, it is reasonable to assume the common transcriptional response to all these stressors is mediated by H_2O_2. In this section, we would like to evaluate this possibility with the aid of the public and our own transcriptome data.

Here, we look at how many of the H_2O_2-responsive genes are also responsive to biotic and abiotic stresses. We analyzed the transcriptional response to H_2O_2 application in Arabidopsis seedlings (Hieno et al.). We then examined the cross-response of 246 H_2O_2-responsive genes to plant stressors, including high-light, cold, UV-B, salt, drought, wounding, and pathogens (*Pst* DC3000). The results have shown that 51–130 genes showed a dual response to H_2O_2 and a stressor (Fig. 1a). One feature of the results is the asymmetry of the crosstalk. For example, the 130 genes showing crosstalk with a UV-B response make up a small portion of the 3805 UV-B-responsive genes (3.4%). However, crosstalking genes account for the majority of the total 246 H_2O_2-responsive genes (53%). This asymmetry is conserved among all the analyzed stressors as shown in Fig. 1a, b. These results mean that each of the plant stressors causing an H_2O_2 response is a small portion of the whole response to every single stressor, and the triggered H_2O_2 response is common to all the plant stressors.

The crosstalk with stress-related phytohormones is also analyzed by the same approach (Fig. 2). The results show that the ratio of genes with crosstalk is small between each phytohormone-responsive genes (0.68–5.9%) and also among H_2O_2-responsive genes (6.9–17%). In comparison, genes with crosstalk with ABA, SA, and JA each show higher percentages among the H_2O_2-responsive genes (13–17%).

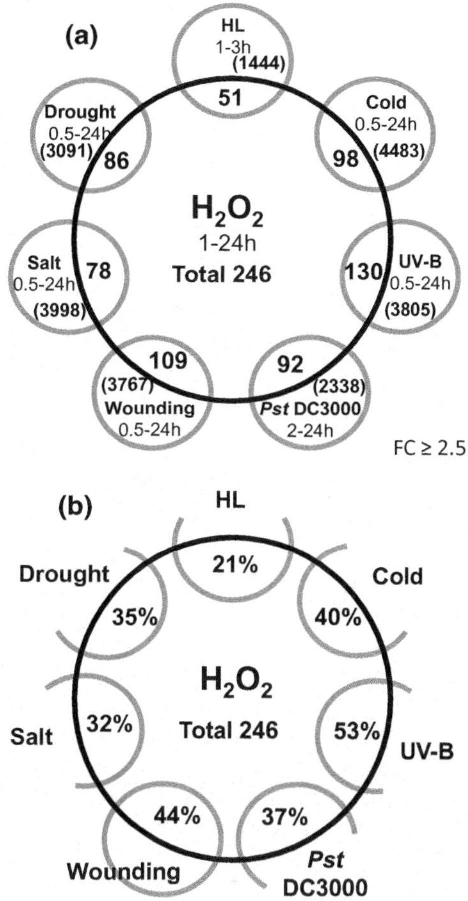

Fig. 1 Crosstalk of the H_2O_2 response with biotic and abiotic stress responses. **a** The number of genes showing crosstalk with each stress response is shown. The number in parentheses is the total number of responsive genes for each stress including ones showing crosstalk with the H_2O_2 response. The sources of transcripome data are as follows: H_2O_2: Hieno et al., HL: Yamamoto et al. 2004, cold: TAIR-ME00325, UV-B: TAIR-ME00329, *Pst* DC3000: TAIR-ME00331, wounding: TAIR-ME00330, salt: TAIR-ME00328, drought: TAIR-ME00338. **b** Percentage of genes showing crosstalk among the 246 H_2O_2-responsive genes is shown

These data indicate that crosstalk with any phytohormone is minor in the H_2O_2 response.

Figure 3 shows the number of crosstalks between the seven stressors and each H_2O_2-responsive gene. As shown in the figure, 60.5% of the genes show crosstalk with three or more stressors. Some of the genes show crosstalk with all the stressors (7 in the figure, 4.9%). These results also support the presence of a universal stress response unit that is mediated by H_2O_2.

Considering the results we have seen, we have classified the plant stress response into three conceptual groups according to the degree of involvement of H_2O_2 (Fig. 4). The first group represents the universal stress response activated by H_2O_2 alone, which we have discussed. Essentially, this group does not show any specificity to stressors. The second group represents the stressor-specific response that requires H_2O_2 accumulation. This group consists of a variety of subgroups, each of which responds to a specific stressor. The mechanism of these stressor-specific H_2O_2 sig-

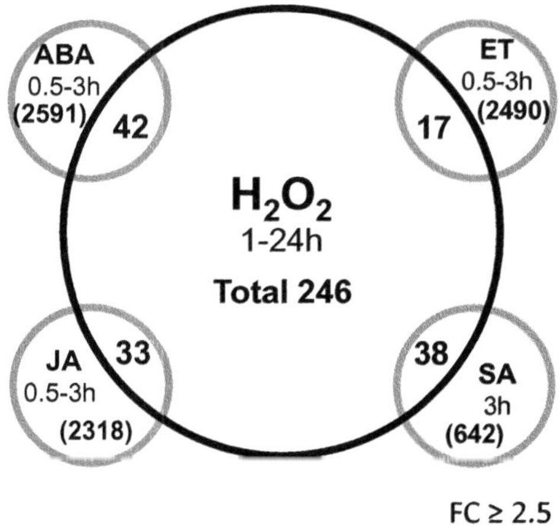

Fig. 2 Crosstalk of the H_2O_2 response with stress-related phytohormone responses. The number of genes showing crosstalk with each stress response is shown. The number in parentheses is the total number of responsive genes for each stress including ones showing crosstalk with the H_2O_2 response. The source of transcriptome data are as follows: ET (ethylene, ACC application): TAIR-ME00334, SA (salicylic acid)* TAIR-ME00364, JA (methyl jasmonate): TAIR-ME00337, ABA: TAIR-ME00333

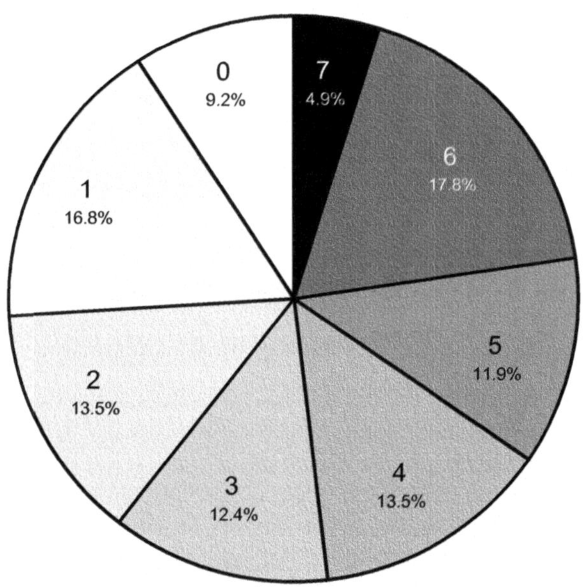

Fig. 3 Number of crosstalks of H_2O_2-responsive genes with biotic and abiotic stress responses. Data of H_2O_2-responsive 246 genes (Fig. 1) is summarized. The number of crosstalk with high-light, cold, UV-B, *Pst* DC3000, wounding, salt, and drought stress responses for each gene is shown

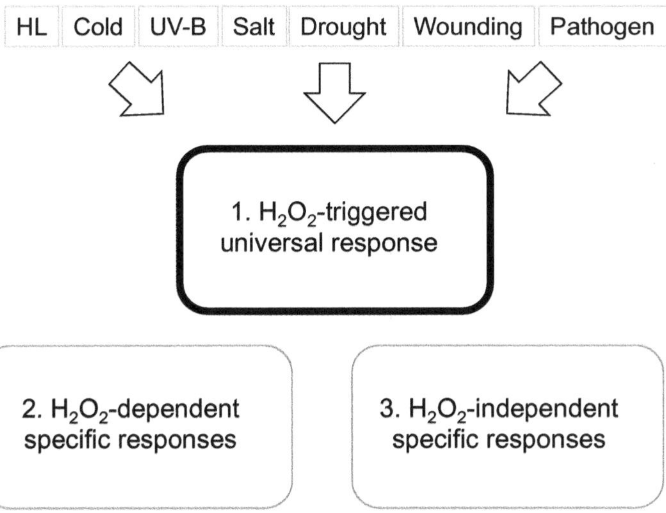

Fig. 4 Three conceptual groups of stress-responsive genes

naling pathways has been discussed (Mittler et al. 2011; Choudhury et al. 2017). This group is not activated by H_2O_2 treatment alone, and therefore, transcriptome analysis after H_2O_2 application does not detect this group. Manipulation of H_2O_2 accumulation by antioxidant feeding or mutant analysis for modulation of the ROS homeostasis would identify the corresponding genes. The third group is also made of a variety of subgroups that are specifically activated by each stressor, and H_2O_2 is not involved in the signaling. This group is expected to be very heterogeneous depending on the corresponding stress signaling, and even within signaling by a single stressor, multiple signaling pathways are expected including stress-sensing mechanisms. Stress-related phytohormones have the potential to have a role in all the three groups.

The H_2O_2-triggered gene set of Arabidopsis that we propose as the universal stress-responsive genes is illustrated in Table 1. The genes are categorized as ROS homeostasis and oxidative burst, HSP, defense-related, glycosylation/sugar metabolism/cell wall, extracellular protein, signal transduction, and transcription factor. The involvement of CBF2/DREB1C and DREB2A is a surprise, because they have been considered as specific regulators for cold and drought stress responses, respectively. The table shows several genes to help our understanding of the physiological aspects of the response.

Table 1 H_2O_2-activated universal stress-responsive genes

Gene categories and representative genes total ~250 genes
ROS homeostasis and oxidative burst ~20 genes GST, TRX, GPX, GRX, LSU, RBOH
HSP ~15 genes HSP17-20, HSP70, DNAJ, HSP100/Clp protease
Defense-related[a] ~35 genes Protease inhibitor[b], PDF family, LRR protein kinase, chitinase, anti-fungal lipid transfer protein EARLI1, flavonoid biosynthesis, P450 detoxification enzyme of xenobiotics
Glycosylation, sugar metabolism, cell wall ~15 genes UDP-glucosyltransferase, UDP-glycosyltransferase
Extracellular protein ~30 genes P450 family, aspartyl protease family, GPX, chitinase, protease inhibitor, PDF family, lipid transfer protein
Signal transduction[c] ~60 genes EF-hand family, CaM-like, protein kinase, receptor protein kinase, phytohormone metabolism
Transcription factor ~50 genes ZAT12/RHL41, RRTF1, RAP2.6, WRKY33, WRKY40, ERF1, HSFA2, CBF2/DREB1C, DREB2A

Notes The numbers of genes for each category are not mutually exclusive, but double counting is included

GST glutathione S-transferase, *TRX* thioredoxin, *GPX* glutathione peroxidase, *GRX* glutaredoxin, *LSU* response to low sulphur (plastidic superoxide dismutase activator), *RBOH* respiratory burst oxidase homolog, *HSP* heat shock protein, *LEA* late embryogenesis abundant family protein, *PDF* plant defensin, *CaM* calmodulin

[a]Genes for signal transducers and transcription factors, and also genes with unknown function are excluded

[b]Trypsin inhibitor: Kunitz family trypsin and protease inhibitor, cysteine protease inhibitor, and serine-type endopeptidase inhibitor/potato inhibitor I-type protein

[c]Transcription factors are excluded

6 Initial Events After H_2O_2 Accumulation and Triggered Transcriptional Subnetwork

The universal stress-responsive group consists of hundreds of genes that make a transcriptional subnetwork and tens of regulation sets (regulons) are included within it. Here, we would like to view current knowledge of the subnetwork and its triggering mechanism.

As we mentioned, we have identified ~250 H_2O_2-responsive genes in Arabidopsis shoots that are categorized into several distinct induction profiles (Fig. 1). They include ~50 transcription factor genes, such as *ZAT12/RHL41*, *ERF1*, *RRTF1*, *RAP2.6*, *WRKY33*, *WRKY40*, *CBF2/DREB1C*, *DREB2A*, and *HSFA2* (Table 1), which have been identified to have important roles in biotic and/or abiotic stress responses. These transcription factor genes make a transcriptional subnetwork by mutual regulation between them. Therefore, regulons (pairs of transcription factors

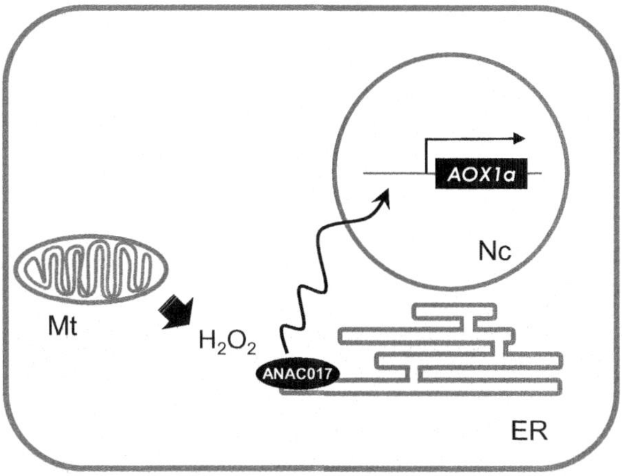

Fig. 5 Initial events triggering the H_2O_2 transcriptional network inhibition of mitochondrial ETS by antimycin A, myxothiazol, and suppression of the TCA cycle by monofluoroacetate activate one of the so-called retrograde signals. This signal, targeting *AOX1a* activation, requires ANAC017. Recent findings suggest that ANAC017 anchored to the ER by its C-terminal transmembrane domain is digested by a rhomboid protease and the released ANAC017 translocates into the nucleus to transcriptionally activate *AOX1a* through direct binding to the promoter (Ng et al. 2013)

and their target transcriptional regulatory sequences) within the subnetwork are not single but heterogeneous, or diverse. We have succeeded in partial identification of the subnetwork through the combination of in vitro promoter—transcription factor binding analysis and in vivo identification of regulatory hierarchy using transcriptome data of corresponding mutants and/or overexpressors. Results reveal a network circuit showing the presence of several hub transcription factors which should be crucial to the network architecture, and divergent and convergent regulatory flows give rise to the generation of heterogeneous induction profiles by a single H_2O_2 stimulation (Hieno et al.).

Initial events leading to activation of the subnetwork start from the perception of H_2O_2, but its mechanism is not yet well understood. As for the initial trigger of the transcriptional subnetwork, recent studies have revealed that the signal is mediated from the endoplasmic reticulum (ER) to the nucleus as explained below and shown in Fig. 5.

Estelle Giraud and her colleagues reported an interesting transcription factor, Arabidopsis ANAC017, which controls as much as >85% of H_2O_2-regulated genes (Ng et al. 2013). This protein is anchored at the ER by a C-terminal transmembrane domain, and an unidentified rhomboid protease removes the transmembrane domain. This promotes translocation of ANAC017 into the nucleus for transcriptional activation of its target genes, including a gene for mitochondrial alternative oxidase, *AOX1a*, and an ERF-type transcription factor *ATERF71*, both of which are activated by H_2O_2 (Ng et al. 2013; Yamamoto et al. 2017). These findings strongly suggest that the translocation of ANAC017 is one of the initial events to trigger the whole transcriptional subnetwork of the H_2O_2 response. What is not clearly known so far is the mechanism of digestion of ANAC017 at the ER, and also the input signals for the digestion.

The role of *ANAC017* was further investigated using its mutants, and several roles of the gene in addition to H_2O_2 signaling have been revealed: (1) retrograde signaling from mitochondria to the nucleus for activation of nuclear-encoded *AOX1a* by inhibitors of the mitochondrial ETS or TCA cycle, antimycin A, mixothiazol, and monofluoroacetate, (2) drought stress resistance (Ng et al. 2013), and (3) retrograde signaling activated by mitochondrial dysfunction due to mutation of a mitochondrial RNA polymerase gene, *RPOTMP*, or a gene for mitochondrial inner membrane protein, *ATPHB3* (Van Aken et al. 2016).

It is clear that ANAC017 has a role in H_2O_2 signaling, but attempts to detect its role in environmental stress responses have so far given negative results (Ng et al. 2013). Therefore, it is suggested that ANAC017 is not the central regulator for the whole H_2O_2 response, but a rather specific mediator of the retrograde signaling from mitochondria. This idea is also supported by the finding that the ANAC017-regulating genes do not represent the majority of our H_2O_2-responsive genes (Hieno et al.), though this is in apparent conflict with the report by Ng et al. These suggestions give rise to a model where the H_2O_2 signal is mediated by parallel paths including the ANAC017-transmitted one into the transcriptional subnetwork for the whole H_2O_2 response. One interpretation of this potential divergence of the signaling is that the H_2O_2-responsive subnetwork has expanded during evolution by merging physiologically different sub-subnetworks that are triggered by each mediator. This hypothetical signaling structure (a single H_2O_2 stimulus and multiple triggers) enables stimulation of a sub-subnetwork depending on tissue or physiological conditions by limiting the paths from H_2O_2 accumulation to nuclear gene regulation. In this sense, the "universal" stress response means a potential responsive set that does not necessarily fully activate in any tissue or situation, although the majority of the responsive genes are actually activated in response to many biotic and abiotic stressors.

7 Conclusion

All plant stressors stimulate H_2O_2 signaling, and how a stressor-specific response is achieved with a universal stress signal mediator has been a long-standing unanswered question. We propose that there is a universal stress response unit that is triggered by H_2O_2 alone, in addition to a variety of stressor-specific response unit that require H_2O_2. The universal stress unit in Arabidopsis consists of GSTs, small HSPs, protease inhibitors, UGTs, *ZAT12*, and *RRTF1*.

References

Allahverdiyeva Y, Aro EM (2012) Photosynthetic responses of plants to excess light: mechanisms and conditions for photoinhibition, excess energy dissipation and repair. In: Eaton-Rye J, Tripathy BC, Sharkey TD (eds) Photosynthesis: plastid biology, energy conversion and carbon assimilation. Springer, Dordrecht, pp 275–297

Alvarez ME, Pennell RI, Meijer PJ, Ishikawa A, Dixon RA, Lamb C (1998) Reactive oxygen intermediates mediate a systemic signal network in the establishment of plant immunity. Cell 92:773–784

Baxter A, Mittler R, Suzuki N (2014) ROS as key players in plant stress signalling. J Exp Bot 65:1229–1240

Biever JJ, Brinkman D, Gardner G (2014) UV-B inhibition of hypocotyl growth in etiolated *Arabidopsis thaliana* seedlings is a consequence of cell cycle arrest initiated by photodimer accumulation. J Exp Bot 65:2949–2961

Bright J, Desikan R, Hancock JT, Weir IS, Neill SJ (2006) ABA-induced NO generation and stomatal closure in Arabidopsis are dependent on H_2O_2 synthesis. Plant J 45:113–122

Brosché M, Strid A (2003) Molecular events following perception of ultraviolet-B radiation by plants. Physiol Plant 117:1–10

Brown BA, Jenkins GI (2008) UV-B signaling pathways with different fluence-rate response profiles are distinguished in mature Arabidopsis leaf tissue by requirement for UVR8, HY5, and HYH. Plant Physiol 146:576–588

Brown BA, Cloix C, Jiang GH, Kaiserli E, Herzyk P, Kliebenstein DJ, Jenkins GI (2005) A UV-B-specific signaling component orchestrates plant UV protection. Proc Natl Acad Sci U S A 102:18225–18230

Cao H, Glazebrook J, Clarke JD, Volko S, Dong X (1997) The Arabidopsis NPR1 gene that controls systemic acquired resistance encodes a novel protein containing ankyrin repeats. Cell 88:57–63

Chaouch S, Queval G, Vanderauwera S, Mhamdi A, Vandorpe M, Langlois-Meurinne M, Van Breusegem F, Saindrenan P, Noctor G (2010) Peroxisomal hydrogen peroxide is coupled to biotic defense responses by ISOCHORISMATE SYNTHASE1 in a daylength-related manner. Plant Physiol 153:1692–1705

Choi J, Huh SU, Kojima M, Sakakibara H, Paek KH, Hwang I (2010) The cytokinin-activated transcription factor ARR2 promotes plant immunity via TGA3/NPR1-dependent salicylic acid signaling in Arabidopsis. Dev Cell 19:284–295

Choudhury FK, Rivero RM, Blumwald E, Mittler R (2017) Reactive oxygen species, abiotic stress and stress combination. Plant J 90:856–867

Czegeny G, Wu M, Der A, Eriksson LA, Strid A, Hideg E (2014) Hydrogen peroxide contributes to the ultraviolet-B (280–315 nm) induced oxidative stress of plant leaves through multiple pathways. FEBS Lett 588:2255–2261

Danon A (2012) Environmentally-induced oxidative stress and its signaling. In: Eaton-RYe JJ, Tripathy BC, Sharkey TD (eds) Photosynthesis: plastid biology, energy conversion and carbon assimilation. Springer, Dordrecht, pp 319–330

Dat JF, Lopez-Delgado H, Foyer CH, Scott IM (1998) Parallel changes in H_2O_2 and catalase during thermotolerance induced by salicylic acid or heat acclimation in mustard seedlings. Plant Physiol 116:1351–1357

Delaney TP, Uknes S, Vernooij B, Friedrich L, Weymann K, Negrotto D, Gaffney T, Gut-Rella M, Kessmann H, Ward E, Ryals J (1994) A central role of salicylic Acid in plant disease resistance. Science 266:1247–1250

Despres C, Chubak C, Rochon A, Clark R, Bethune T, Desveaux D, Fobert PR (2003) The Arabidopsis NPR1 disease resistance protein is a novel cofactor that confers redox regulation of DNA binding activity to the basic domain/leucine zipper transcription factor TGA1. Plant Cell 15:2181–2191

Dixon RA, Lamb CJ (1979) Stimulation of de novo synthesis of L-phenylalanine ammonia-lyase in relation to phytoalexin accumulation in *Colletotrichum lindemuthianum* elicitor-treated cell suspension cultures of French bean (Phaseolus vulgaris). Biochim Biophys Acta 586:453–463

Doherty CJ, Van Buskirk HA, Myers SJ, Thomashow MF (2009) Roles for Arabidopsis CAMTA transcription factors in cold-regulated gene expression and freezing tolerance. Plant Cell 21:972–984

Dong CH, Zolman BK, Bartel B, Lee BH, Stevenson B, Agarwal M, Zhu JK (2009) Disruption of Arabidopsis CHY1 reveals an important role of metabolic status in plant cold stress signaling. Mol Plant 2:59–72

Falk S, Maxwell DP, Laudenbach DE, Huner NPA (1996) Photosynthetic adjustment to temperature. In: Baker NB (ed) Photosynthesis and the environment. Kluwer Academic Publishers, Dordrecht, pp 367–385

Fan W, Dong X (2002) In vivo interaction between NPR1 and transcription factor TGA2 leads to salicylic acid-mediated gene activation in Arabidopsis. Plant Cell 14:1377–1389

Fowler S, Thomashow MF (2002) Arabidopsis transcriptome profiling indicates that multiple regulatory pathways are activated during cold acclimation in addition to the CBF cold response pathway. Plant Cell 14:1675–1690

Fryer MJ, Ball L, Oxborough K, Karpinski S, Mullineaux PM, Baker NR (2003) Control of Ascorbate Peroxidase 2 expression by hydrogen peroxide and leaf water status during excess light stress reveals a functional organisation of Arabidopsis leaves. Plant J 33:691–705

Fujita M, Fujita Y, Noutoshi Y, Takahashi F, Narusaka Y, Yamaguchi-Shinozaki K, Shinozaki K (2006) Crosstalk between abiotic and biotic stress responses: a current view from the points of convergence in the stress signaling networks. Curr Opin Plant Biol 9:436–442

Gaffney T, Friedrich L, Vernooij B, Negrotto D, Nye G, Uknes S, Ward E, Kessmann H, Ryals J (1993) Requirement of salicylic Acid for the induction of systemic acquired resistance. Science 261:754–756

Galvez-Valdivieso G, Fryer MJ, Lawson T, Slattery K, Truman W, Smirnoff N, Asami T, Davies WJ, Jones AM, Baker NR, Mullineaux PM (2009) The high light response in Arabidopsis involves ABA signaling between vascular and bundle sheath cells. Plant Cell 21:2143–2162

Garcia-Molina A, Altmann M, Alkofer A, Epple PM, Dangl JL, Falter-Braun P (2017) LSU network hubs integrate abiotic and biotic stress responses via interaction with the superoxide dismutase FSD2. J Exp Bot 68:1185–1197

Graham JS, Hall G, Pearce G, Ryan CA (1986) Regulation of synthesis of proteinase inhibitors I and II mRNAs in leaves of wounded tomato plants. Planta 169:399–405

Grene R (2002) Oxidative stress and acclimation mechanisms in plants. Arabidopsis Book 1:e0036

Hayami N, Sakai Y, Kimura M, Saito T, Tokizawa M, Iuchi S, Kurihara Y, Matsui M, Nomoto M, Tada Y, Yamamoto YY (2015) The responses of arabidopsis ELIP2 to UV-B, high light, and cold stress are regulated by a transcriptional regulatory unit composed of two elements. Plant Physiol 169:840–855

Herrera-Vasquez A, Salinas P, Holuigue L (2015) Salicylic acid and reactive oxygen species interplay in the transcriptional control of defense genes expression. Front Plant Sci 6:171

Hieno A, Naznin HA, Inaba-Hasegawa K, Yokogawa T, Hayami N, Nomoto M, Tada Y, Yokogawa T, Higuchi-Takeuchi M, Hanada K, Matsui M, Hirayama T, Kusunoki K, Koyama H, Mitsuda N, Yamamoto YY Transcriptome analysis of the H_2O_2 response in Arabidopsis seedlings and partial identification of the corresponding transcription regulatory network (Submitted)

Iida A, Kazuoka T, Torikai S, Kikuchi H, Oeda K (2000) A zinc finger protein RHL41 mediates the light acclimatization response in *Arabidopsis*. Plant Journal 24:191–203

Ishibashi Y, Yamaguchi H, Yuasa T, Iwaya-Inoue M, Arima S, Zheng SH (2011) Hydrogen peroxide spraying alleviates drought stress in soybean plants. J Plant Physiol 168:1562–1567

Jansen MAK, Gaba V, Greenberg BM (1998) Higher plants and UV-B radiation: balancing damage, repair and acclimation. Trends Plant Sci 3:131–135

Joshi-Saha A, Valon C, Leung J (2011) A brand new START: abscisic acid perception and transduction in the guard cell. Sci Sig **4,** re4

Kalbina I, Strid A (2006) The role of NADPH oxidase and MAP kinase phosphatase in UV-B-dependent gene expression in Arabidopsis. Plant Cell Environ 29:1783–1793
Karpinski S, Escobar C, Karpinska B, Creissen G, Mullineaux PM (1997) Photosynthetic electron transport regulates the expression of cytosolic ascorbate peroxidase genes in Arabidopsis during excess light stress. Plant Cell 9:627–640
Karpinski S, Reynolds H, Karpinska B, Wingsle G, Creissen G, Mullineaux P (1999) Systemic signaling and acclimation in response to excess excitation energy in *Arabidopsis*. Science 284:654–657
Khokon AR, Okuma E, Hossain MA, Munemasa S, Uraji M, Nakamura Y, Mori IC, Murata Y (2011) Involvement of extracellular oxidative burst in salicylic acid-induced stomatal closure in Arabidopsis. Plant Cell Environ 34:434–443
Kim HJ, Kim YK, Park JY, Kim J (2002) Light signalling mediated by phytochrome plays an important role in cold-induced gene expression through the C-repeat/dehydration responsive element (C/DRE) in *Arabidopsis thaliana*. Plant J 29:693–704
Kimura M, Yoshizumi T, Manabe T, Yamamoto YY, Matsui M (2001) *Arabidopsis* transcriptional regulation by light stress *via* hydrogen peroxide-dependent and -independent pathways. Genes Cells 6:607–617
Kimura M, Yamamoto YY, Seki M, Sakurai T, Sato M, Abe T, Yoshida S, Manabe K, Shinozaki K, Matsui M (2003) Identification of *Arabidopsis* genes regulated by high light stress using cDNA microarray. Photochem Photobiol 77:226–233
Kleine T, Kindgren P, Benedict C, Hendrickson L, Strand A (2007) Genome-wide gene expression analysis reveals a critical role for CRYPTOCHROME1 in the response of Arabidopsis to high irradiance. Plant Physiol 144:1391–1406
Koussevitzky S, Suzuki N, Huntington S, Armijo L, Sha W, Cortes D, Shulaev V, Mittler R (2008) Ascorbate peroxidase 1 plays a key role in the response of *Arabidopsis thaliana* to stress combination. J Biol Chem 283:34197–34203
Kurepin LV, Dahal KP, Savitch LV, Singh J, Bode R, Ivanov AG, Hurry V, Huner NP (2013) Role of CBFs as integrators of chloroplast redox, phytochrome and plant hormone signaling during cold acclimation. Int J Mol Sci 14:12729–12763
Kwak JM, Mori IC, Pei ZM, Leonhardt N, Torres MA, Dangl JL, Bloom RE, Bodde S, Jones JD, Schroeder JI (2003) NADPH oxidase AtrbohD and AtrbohF genes function in ROS-dependent ABA signaling in Arabidopsis. EMBO J 22:2623–2633
Lamb C, Dixon RA (1997) The oxidative burst in plant disease resistance. Annu Rev Plant Physiol Plant Mol Biol 48:251–275
Lawton MA, Dixon RA, Hahlbrock K, Lamb C (1983) Rapid induction of the synthesis of phenylalanine ammonia-lyase and of chalcone synthase in elicitor-treated plant cells. Eur J Biochem 129:593–601
Leon J, Rojo E, Sanchez-Serrano JJ (2001) Wound signalling in plants. J Exp Bot 52:1–9
Levine A, Tenhaken R, Dixon R, Lamb C (1994) H_2O_2 from the oxidative burst orchestrates the plant hypersensitive disease resistance response. Cell 79:583–593
Liu Y, Ren D, Pike S, Pallardy S, Gassmann W, Zhang S (2007) Chloroplast-generated reactive oxygen species are involved in hypersensitive response-like cell death mediated by a mitogen-activated protein kinase cascade. Plant J 51:941–954
Maruta T, Noshi M, Tanouchi A, Tamoi M, Yabuta Y, Yoshimura K, Ishikawa T, Shigeoka S (2012) H_2O_2-triggered retrograde signaling from chloroplasts to nucleus plays specific role in response to stress. J Biol Chem 287:11717–11729
Mateo A, Funck D, Muhlenbock P, Kular B, Mullineaux PM, Karpinski S (2006) Controlled levels of salicylic acid are required for optimal photosynthesis and redox homeostasis. J Exp Bot 57:1795–1807
Matsuo M, Johnson JM, Hieno A, Tokizawa M, Nomoto M, Tada Y, Godfrey R, Obokata J, Sherameti I, Yamamoto YY, Bohmer FD, Oelmuller R (2015) High REDOX RESPONSIVE TRANSCRIPTION FACTOR1 levels result in accumulation of reactive oxygen species in *Arabidopsis thaliana* shoots and roots. Mol Plant 8:1253–1273

Melotto M, Underwood W, Koczan J, Nomura K, He SY (2006) Plant stomata function in innate immunity against bacterial invasion. Cell 126:969–980
Miller G, Suzuki N, Ciftci-Yilmaz S, Mittler R (2010) Reactive oxygen species homeostasis and signalling during drought and salinity stresses. Plant Cell Environ 33:453–467
Miller G, Schlauch K, Tam R, Cortes D, Torres MA, Shulaev V, Dangl JL, Mittler R (2009) The plant NADPH oxidase RBOHD mediates rapid systemic signaling in response to diverse stimuli. Sci Sig **2**, ra45
Mittler R, Blumwald E (2015) The roles of ROS and ABA in systemic acquired acclimation. Plant Cell 27:64–70
Mittler R, Vanderauwera S, Gollery M, Van Breusegem F (2004) Reactive oxygen gene network of plants. Trends Plant Sci 9:490–498
Mittler R, Vanderauwera S, Suzuki N, Miller G, Tognetti VB, Vandepoele K, Gollery M, Shulaev V, Van Breusegem F (2011) ROS signaling: the new wave? Trends Plant Sci 16:300–309
Mohanty P, Kreslavski VD, Klimov VV, Los DA, Mimuro M, Carpentier R, Allakhverdiev SI (2012) Heat stress: susceptibility, recovery, and regulation. In: Eaton-Rye J, Tripathy BC, Sharkey TD (eds) Photosynthesis: plastid biology, energy conversion and carbon assimilation. Springer, Dordrecht, pp 251–274
Morales J, Kadota Y, Zipfel C, Molina A, Torres MA (2016) The Arabidopsis NADPH oxidases RbohD and RbohF display differential expression patterns and contributions during plant immunity. J Exp Bot 67:1663–1676
Mou Z, Fan W, Dong X (2003) Inducers of plant systemic acquired resistance regulate NPR1 function through redox changes. Cell 113:935–944
Nakashima K, Yamaguchi-Shinozaki K, Shinozaki K (2014) The transcriptional regulatory network in the drought response and its crosstalk in abiotic stress responses including drought, cold, and heat. Front Plant Sci 5:170
Ng S, Ivanova A, Duncan O, Law SR, Van Aken O, De Clercq I, Wang Y, Carrie C, Xu L, Kmiec B, Walker H, Van Breusegem F, Whelan J, Giraud E (2013) A membrane-bound NAC transcription factor, ANAC017, mediates mitochondrial retrograde signaling in Arabidopsis. Plant Cell 25:3450–3471
Noctor G, Mhamdi A, Foyer CH (2014) The roles of reactive oxygen metabolism in drought: not so cut and dried. Plant Physiol 164:1636–1648
Noshi M, Maruta T, Shigeoka S (2012) Relationship between chloroplastic H 2O 2 and the salicylic acid response. Plant Sig Behav 7:944–946
O'Brien JA, Daudi A, Finch P, Butt VS, Whitelegge JP, Souda P, Ausubel FM, Bolwell GP (2012) A peroxidase-dependent apoplastic oxidative burst in cultured Arabidopsis cells functions in MAMP-elicited defense. Plant Physiol 158:2013–2027
Ohama N, Sato H, Shinozaki K, Yamaguchi-Shinozaki K (2017) Transcriptional regulatory network of plant heat stress response. Trends Plant Sci 22:53–65
Orozco-Cardenas M, Ryan CA (1999) Hydrogen peroxide is generated systemically in plant leaves by wounding and system in via the octadecanoid pathway. Proc Natl Acad Sci U S A 96:6553–6557
Orozco-Cardenas ML, Narvaez-Vasquez J, Ryan CA (2001) Hydrogen peroxide acts as a second messenger for the induction of defense genes in tomato plants in response to wounding, systemin, and methyl jasmonate. Plant Cell 13:179–191
Rasmussen S, Barah P, Suarez-Rodriguez MC, Bressendorff S, Friis P, Costantino P, Bones AM, Nielsen HB, Mundy J (2013) Transcriptome responses to combinations of stresses in Arabidopsis. Plant Physiol 161:1783–1794
Rizhsky L, Davletova S, Liang H, Mittler R (2004) The zinc finger protein Zat12 is required for cytosolic ascorbate peroxidase 1 expression during oxidative stress in Arabidopsis. J Biol Chem 279:11736–11743
Rizzini L, Favory JJ, Cloix C, Faggionato D, O'Hara A, Kaiserli E, Baumeister R, Schafer E, Nagy F, Jenkins GI, Ulm R (2011) Perception of UV-B by the Arabidopsis UVR8 protein. Science 332:103–106

Rojas CM, Senthil-Kumar M, Wang K, Ryu CM, Kaundal A, Mysore KS (2012) Glycolate oxidase modulates reactive oxygen species-mediated signal transduction during nonhost resistance in *Nicotiana benthamiana* and Arabidopsis. Plant Cell 24:336–352

Rossel JB, Wilson IW, Pogson BJ (2002) Global changes in expression in response to high light in Arabidopsis. Plant Physiol 130:1109–1120

Seki M, Narusaka M, Ishida J, Nanjo T, Fujita M, Oono Y, Kamiya A, Nakajima M, Enju A, Sakurai T, Satou M, Akiyama K, Taji T, Yamaguchi-Shinozaki K, Carninci P, Kawai J, Hayashizaki Y, Shinozaki K (2002) Monitoring the expression profiles of 7000 Arabidopsis genes under drought, cold and high-salinity stresses using a full-length cDNA microarray. Plant J 31:279–292

Shinozaki K, Yamaguchi-Shinozaki K (2007) Gene networks involved in drought stress response and tolerance. J Exp Bot 58:221–227

Sirichandra C, Gu D, Hu HC, Davanture M, Lee S, Djaoui M, Valot B, Zivy M, Leung J, Merlot S, Kwak JM (2009) Phosphorylation of the Arabidopsis AtrbohF NADPH oxidase by OST1 protein kinase. FEBS Lett 583:2982–2986

Smirnoff N (1993) The role of active oxygen in the response of plants to water deficit and desiccation. New Phytol 125:27–58

Storozhenko S, De Pauw P, Van Montagu M, Inze D, Kushnir S (1998) The heat-shock element is a functional component of the Arabidopsis APX1 gene promoter. Plant Physiol 118:1005–1014

Suzuki N, Bassil E, Hamilton JS, Inupakutika MA, Zandalinas SI, Tripathy D, Luo Y, Dion E, Fukui G, Kumazaki A, Nakano R, Rivero RM, Verbeck GF, Azad RK, Blumwald E, Mittler R (2016) ABA Is required for plant acclimation to a combination of salt and heat stress. PLoS ONE 11:e0147625

Tada Y, Spoel SH, Pajerowska-Mukhtar K, Mou Z, Song J, Wang C, Zuo J, Dong X (2008) Plant immunity requires conformational changes [corrected] of NPR1 via S-nitrosylation and thioredoxins. Science 321:952–956

Takahashi F, Mizoguchi T, Yoshida R, Ichimura K, Shinozaki K (2011) Calmodulin-dependent activation of MAP kinase for ROS homeostasis in Arabidopsis. Mol Cell 41:649–660

Terzi M, Kadioglu A, Kalaycioglu E, Saglam A (2014) Hydrogen peroxide pretreatment induces osmotic stress tolerance by influencing osmolyte and abscisic acid levels in maize leaves. J Plant Interact 9:559–565

Ulm R, Baumann A, Oravecz A, Mate Z, Adam E, Oakeley EJ, Schafer E, Nagy F (2004) Genome-wide analysis of gene expression reveals function of the bZIP transcription factor HY5 in the UV-B response of Arabidopsis. Proc Natl Acad Sci U S A 101:1397–1402

Van Aken O, Ford E, Lister R, Huang S, Millar AH (2016) Retrograde signalling caused by heritable mitochondrial dysfunction is partially mediated by ANAC017 and improves plant performance. Plant J 88:542–558

Vandenabeele S, Vanderauwera S, Vuylsteke M, Rombauts S, Langerbartels C, Seidlitz HK, Zabeau M, Van Montagu F, Inzé D, Van Breusegem F (2004) Catalase deficiency drastically affects gene expression induced by high light in *Arabidopsis thaliana*. Plant J 39:45–58

Volkov RA, Panchuk II, Mullineaux PM, Schoffl F (2006) Heat stress-induced H($_2$)O ($_2$) is required for effective expression of heat shock genes in Arabidopsis. Plant Mol Biol 61:733–746

Wang D, Amornsiripanitch N, Dong X (2006) A genomic approach to identify regulatory nodes in the transcriptional network of systemic acquired resistance in plants. PLoS Pathog 2:e123

Wang W, Vinocur B, Shoseyov O, Altman A (2004) Role of plant heat-shock proteins and molecular chaperones in the abiotic stress response. Trends Plant Sci 9:244–252

Ward ER, Uknes SJ, Williams SC, Dincher SS, Wiederhold DL, Alexander DC, Ahl-Goy P, Metraux JP, Ryals JA (1991) Coordinate gene activity in response to agents that induce systemic acquired resistance. Plant Cell 3:1085–1094

Waszczak C, Kerchev PI, Muhlenbock P, Hoeberichts FA, Van Der Kelen K, Mhamdi A, Willems P, Denecker J, Kumpf RP, Noctor G, Messens J, Van Breusegem F (2016) SHORT-ROOT deficiency alleviates the cell death phenotype of the arabidopsis catalase 2 mutant under photorespiration-promoting conditions. Plant Cell 28:1844–1859

Wildon DC, Thain JF, Minchin PEH, Gubb IR, Reilly AJ, Skipper YD, Doherty HM, O'Donnell PJ, Bowles DJ (1992) Electrical signal and systemic proteinase inhibitor induction in the wounded plant. Nature 360:62–65

Yamamoto YY, Shimada Y, Kimura M, Manabe K, Sekine Y, Matsui M, Ryuto H, Fukunishi N, Abe T, Yoshida S (2004) Global classification of transcriptional responses to light stress in *Arabidopsis thaliana*. Endocytobiosis Cell Res 15:438–452

Yamamoto YY, Ichida H, Hieno A, Obata D, Tokizawa M, Nomoto M, Tada Y, Kusunoki K, Koyama H, Hayami N (2017) Prediction of bipartite transcriptional regulatory elements using transcriptome data of Arabidopsis. DNA Res 24:271–278

Zhang Y, Zhu H, Zhang Q, Li M, Yan M, Wang R, Wang L, Welti R, Zhang W, Wang X (2009) Phospholipase dalpha 1 and phosphatidic acid regulate NADPH oxidase activity and production of reactive oxygen species in ABA-mediated stomatal closure in Arabidopsis. Plant Cell 21:2357–2377

Zurbriggen MD, Carrillo N, Hajirezaei MR (2010) ROS signaling in the hypersensitive response: when, where and what for? Plant Sig Behav 5:393–396

Zurbriggen MD, Carrillo N, Tognetti VB, Melzer M, Peisker M, Hause B, Hajirezaei MR (2009) Chloroplast-generated reactive oxygen species play a major role in localized cell death during the non-host interaction between tobacco and Xanthomonas campestris pv. vesicatoria. Plant J 60:962–973

The Interplay of ROS and Iron Signaling in Plants

Cham Thi Tuyet Le, Tzvetina Brumbarova and Petra Bauer

Abstract Reactive oxygen species (ROS) and iron (Fe) metabolic pathways are linked. The first part of this chapter summarizes how Fe contributes to ROS production and elimination in cells. We review shortly how plants acquire Fe. Comparative transcriptomic datasets reflecting high Fe, low Fe, and very low Fe responses are scanned for differentially expressed genes encoding enzymes for ROS generation and scavenging as well as ROS signature genes. These genes are assembled and their expression patterns are discussed in light of the importance of ROS production under high and very low Fe. In the second part of this chapter, we highlight different areas of research questions revealing the regulatory interconnections between ROS and Fe signaling, namely NADPH oxidase signaling and heme breakdown, glutathione-based signaling, retrograde chloroplast, and Fe–S cluster signaling, ferroptosis and ROS marker ZAT12-mediated signaling.

1 Introduction

The high oxygen level that occurred on the Earth with the evolution of photosynthesis necessarily caused a conflict with iron (Fe). Reactive oxygen species (ROS) arise from electron or energy transfer under atmospheric oxygen conditions. Iron (Fe) ranks among the four most prevalent elements on Earth and is biologically the most important transition metal catalyzing ROS generation. During evolution, ROS have turned into key signaling molecules for many developmental programs and stress responses in plants and elicit a package of so-called oxidative stress responses. This

C. Thi Tuyet Le
Vietnam National University of Agriculture, Hanoi, Vietnam
e-mail: lttcham@vnua.edu.vn

T. Brumbarova · P. Bauer (✉)
Heinrich Heine University, Düsseldorf, Germany
e-mail: petra.bauer@hhu.de

T. Brumbarova
e-mail: tzvetina.brumbarova@hhu.de

S. K. Panda and Y. Y. Yamamoto (eds.), *Redox Homeostasis in Plants*,
Signaling and Communication in Plants, https://doi.org/10.1007/978-3-319-95315-1_3

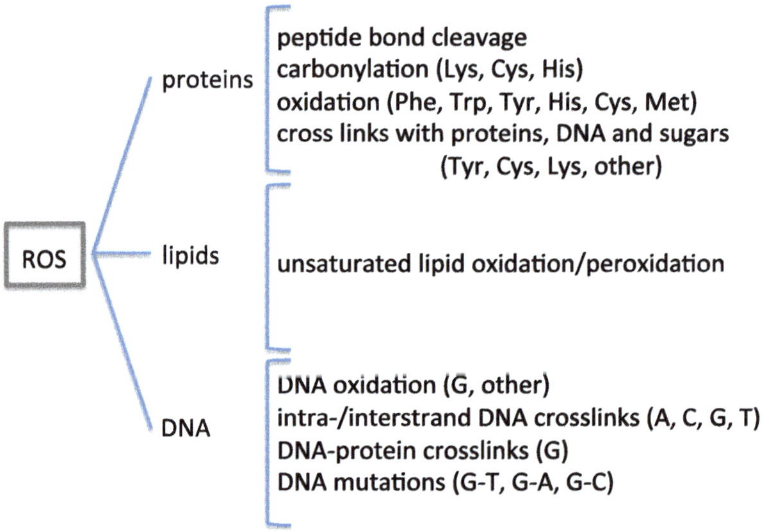

Fig. 1 Damaging effects of ROS on proteins, lipids, and DNA. The effects are explained in more detail for proteins by Stadtman and Levine (2003), for lipids by Niki (2009) and DNA by Jena (2012)

review will highlight the interconnections between ROS and iron (Fe) signaling in plants.

ROS are produced in various plant cell compartments, either via light excitation of photosensitizers or transfer of electrons in redox reactions, recently the subject of reviews (Czarnocka and Karpinski 2018). Four types of ROS are distinguished with relevance in cells, namely singlet oxygen ($^1O_2$), superoxide anion $\cdot O_2^-$, peroxides, especially hydrogen peroxide (H_2O_2) and lipid peroxides (LOOH), and radicals, especially in the form of the hydroxyl radical (·OH), lipid, lipid alkoxy and lipid peroxy radicals (L·, LO·, LOO·).

Radicals are the most reactive ROS and not readily eliminated by cells. Singlet oxygen ($^1O_2$) and superoxide anion ($\cdot O_2^-$) are generally more reactive than peroxides, whereby all three are targeted by cellular scavenging mechanisms (Czarnocka and Karpinski 2018). ROS oxidize various types of biomolecules. Particularly, protein, lipid, and DNA oxidation cause deleterious effects ranging from protein inactivation, protein decomposition, protein crosslinking to lipid decomposition, DNA mutation, and DNA crosslinking (Fig. 1; Stadtman and Levine 2003; Niki 2009; Jena 2012). ROS and ROS targets can be cellular signals, but become toxic in situations of unbalanced cellular ROS scavenging, e.g., under metal toxicity, high light, and temperature stress (Shahid et al. 2014; Pospisil 2016).

Plants acquire the micronutrient Fe through their root system and regulate Fe uptake in response to external and internal signals, which has been intensively reviewed (Brumbarova et al. 2015; Connorton et al. 2017). Fe has a relatively low

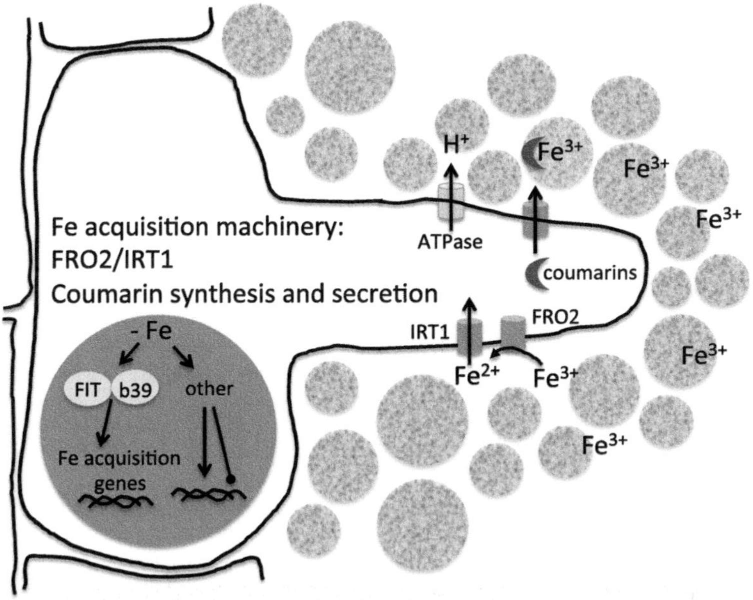

Fig. 2 Fe acquisition in the model system *Arabidopsis thaliana*. The scheme depicts Fe uptake and its nuclear regulation in an *Arabidopsis* root hair cell growing in the soil. The filled circles represent soil particles with ferric Fe. An ATPase contributes to soil acidification via proton extrusion. Coumarins are synthesized and secreted that may facilitate solubility and reduction of ferric Fe through chelation. FRO2 is a plasma membrane-bound NADPH-dependent Fe-chelate reductase. Ferrous Fe is taken up via the IRON-REGULATED TRANSPORTER1 (IRT1). This system is stimulated at transcriptional level upon low Fe by a cascade of transcription factors, ultimately leading to protein interaction and activation of bHLH039 (b39) and FIT transcription factors. Detailed explanations of this system are provided by Brumbarova et al. (2015). On the other side, Fe regulates a number of genes that are not dependent on FIT [details of regulated gene expression patterns are found in Mai et al. (2016) and Naranjo Arcos et al. (2017)]

solubility and bioavailability in soils, where it occurs mostly as precipitated and soil particle-bound ferric Fe (Fe^{3+}). In regular potting soils plants actively mobilize Fe via soil acidification, Fe chelation, and Fe reduction with specific adaptations in different plant lineages. Fe is taken up through root epidermal transporters either as reduced ferrous Fe (Fe^{2+}) or as chelated Fe^{3+} (Fig. 2; represented is the model system *Arabidopsis*). Fe bioavailability is several orders of magnitude higher in acidic soils compared with calcareous soils. For example, in acid paddy soils in Southeast Asia Fe toxicity is a frequent problem in rice agriculture, leading to leaf bronzing and cell death (Wu et al. 2017). On the other hand, calcareous soils limit agricultural production due to leaf chlorosis caused by low Fe for example in fruit trees in the Mediterranean region (Abadía et al. 2011).

ROS chemistry involves redox reactions, typically catalyzed by transition metal ions, explaining the interconnection of ROS and Fe metabolism. Cells must have mechanisms to measure Fe and ROS pools and coordinate their occurrence and reactivity in the cells. Surprisingly, only few studies directly addressed the molecular network connecting Fe and ROS signaling.

Here, we provide an overview on the various molecular links that exist between ROS and Fe and we highlight specific examples of the interconnections of the ROS and Fe signaling pathways in plants.

2 Involvement of Fe in the Generation and Elimination of ROS

Fe is involved in multiple ROS metabolic mechanisms, and therefore, Fe pools are decisive for the amount and activity of ROS. Here, we focus on the ROS-related aspects, which are in close context with Fe.

Under high Fe conditions, ROS are formed in the non-enzymatic Haber–Weiss reaction using Fe as a catalyzator (Fig. 3). In a first step, ferric Fe (Fe^{3+}) is reduced to ferrous Fe (Fe^{2+}), while $\cdot O_{2}-$ is oxidized to O_2. Then, Fe^{2+} catalyzes the so-called Fenton reaction, where it reacts with H_2O_2 to form ·OH and hydroxyl anion OH−. Instead of H_2O_2, other peroxides can trigger the formation of radicals. Lipid peroxides can yield in the Fenton reaction lipid alkoxy radicals (LO·). The ROS pool can be further affected by organic compounds that act as antioxidants or prooxidants depending on their concentrations, concentrations of redoxreactive metals and ROS in the cells and activities of enzymes that use these compounds as cofactors (Fig. 4). In addition to the non-enzymatic ROS reactions, enzymatic reactions are sources of $\cdot O_{2}-$, H_2O_2 and lipid peroxides, and also catalyze the elimination of ROS or the regeneration of redox-active cofactors, several of them dependent on Fe or heme-Fe (Fig. 5). Several enzymes act in concert in the ascorbate-glutathione and the glutathione/thioredoxin system (Czarnocka and Karpinski 2018).

Thus, Fe favors conditions suitable for ROS metabolism in cells, leading to either generation or elimination of ROS. Enzymes for ROS generation and elimination frequently contain Fe as cofactors. Later in this chapter, we will present some aspects of ROS metabolism in more detail and investigate gene expression in response to Fe.

3 Comparative Transcriptome Analysis of ROS-Related Genes in Response to High and Low Fe

From the role of Fe in ROS generation, we would predict that high Fe suppresses ROS-generating enzymes, but enhances ROS-scavenging mechanisms and ROS responses. We tested ROS-related gene expression in transcriptomic datasets reflecting different Fe nutrition states.

We had previously generated comparative transcriptome datasets of six-day-old *Arabidopsis thaliana* seedlings reflecting responses to high, low, and very low Fe

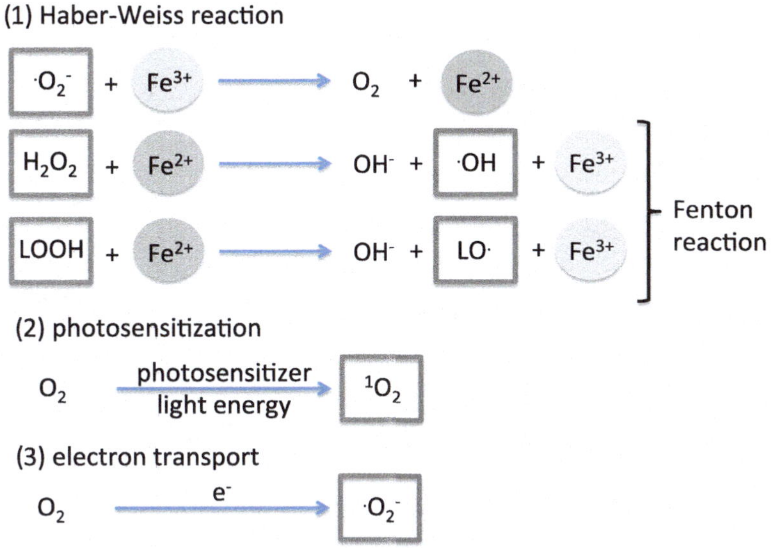

Fig. 3 Non-enzymatic ROS generation. Three examples for non-enzymatic ROS generation are shown; (1) the Haber–Weiss reaction, the second step of it being the Fenton reaction; (2) $\cdot O^{2-}$ generation via photosensitization, e.g., excited chlorophyll; (3) through electron deviation from electron transport

(Fig. 6). Wild-type plants grown under regular Fe conditions (50 μM Fe) have a low level of FIT activity. The bHLH protein FIT is an essential transcription factor up-regulating Fe acquisition genes. In the presence of Fe, low activity of FIT and low expression of Fe acquisition genes and proteins are sufficient to achieve nutrition with Fe, resulting in green plants (+Fe) (Jakoby et al. 2004). Under Fe deficiency (0 Fe) transcription factor bHLH039 is highly induced, interacts with FIT, resulting in a strong enhancement of Fe acquisition responses (Wang et al. 2007; Yuan et al. 2008). However, since Fe is not available, plants turn chlorotic and experience Fe deficiency (−Fe). The comparison of wild-type plants experiencing –versus +Fe represents the "low Fe" response [Fig. 6; dataset obtained from Mai et al. (2016)]. For the "high Fe" response, we compared transcriptomic changes between a constitutive bHLH039-overexpressing line (39Ox) and wild type at +Fe. 39Ox seedlings display high Fe uptake (+++Fe) and oxidative stress symptoms such as green-red leaves [Fig. 6; dataset obtained from Naranjo Arcos et al. (2017)]. For the "very low Fe" condition, we analyzed transcriptomes of loss of function *fit* mutants compared with wild type at −Fe. *fit* seedlings lack Fe, developing a severe leaf chlorosis under −Fe [−−−Fe, Fig. 6; dataset obtained from Mai et al. (2016)].

ROS metabolism genes were selected from recent literature (Berndt and Lillig 2017; Choudhury et al. 2017; Czarnocka and Karpinski 2018). ROS induction responses were followed using a list of 62 genes up-regulated by ROS that represent a ROS transcript signature (Vaahtera et al. 2014), see there the first 37 genes listed

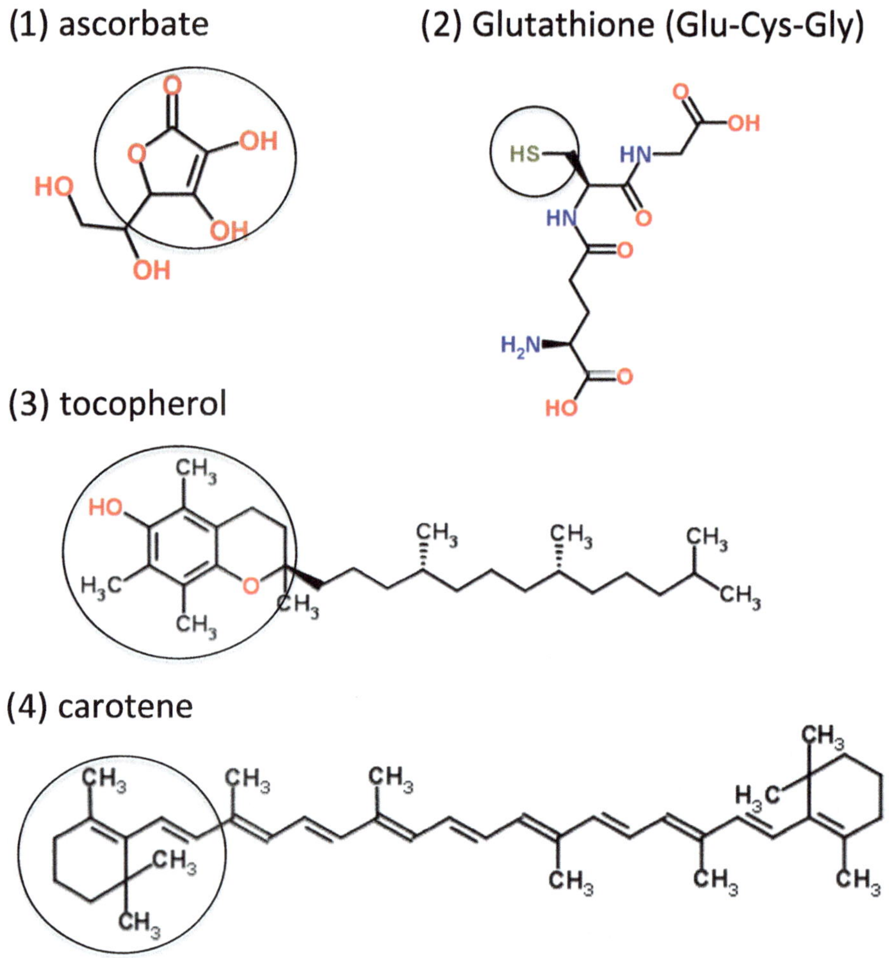

Fig. 4 Antioxidants. Four major antioxidants are depicted. The redoxreactive groups are circled. Chemical structures were obtained from ChemSpider (http://www.chemspider.com)

in Table 2 and all additional genes listed in Table 3, reflecting genes induced by high light, ozone, H_2O_2, $^1O_2$, $\cdot O_{2^-}$, and general ROS).

We scanned the "high Fe", "low Fe", and "very low Fe" comparative transcriptomics datasets for differentially expressed genes and identified 120 genes related to ROS metabolism and 35 genes described as ROS markers (Figs. 7, 8 and 9).

(a) Regulation of ROS metabolism genes

Among the 120 Fe-regulated ROS metabolism genes, 103 genes were regulated in the high Fe condition, nearly 3/5 being up-regulated, while only 27 and 42 genes were regulated in the low and very low Fe condition, nearly 1/2 down-regulated (Figs. 7 and 8).

$$1)\quad NADPH + 2\,O_2 \xrightarrow{RBOH} NADP^+ + H^+ + 2\,{\cdot}O_2^-$$

$$2)\quad LH + 2\,O_2 \xrightarrow{LOX} LOOH$$

$$3)\quad {\cdot}O_2^- + Me^{(n+1)+} + 2\,H^+ \xrightarrow{SOD} H_2O_2 + Me^{n+}$$

$$4)\quad 2\,H_2O_2 \xrightarrow{CAT} 2\,H_2O + O_2$$

$$5)\quad ROOR + 2\,e^- + 2\,H^+ \xrightarrow{PX} 2\,ROH$$

$$6)\quad 2\,GSH + H_2O_2 \xrightarrow{GPX} GSSG + H_2O$$

$$7)\quad AA + H_2O_2 \xrightarrow{APX} MDHA + 2\,H_2O$$

$$8)\quad LOOH + 2\,GSH \xrightarrow{GST} LOH + H_2O + GSSG$$

$$9)\quad TRX_{red} + H_2O_2 \xrightarrow{PRX} TRX_{ox} + 2\,H_2O$$

$$10)\quad Prot_{red} + 2\,GSH \xrightarrow{GRX} Prot_{ox} + GSSG$$

$$11)\quad Prot_{red} + \mathbf{Trx_{ox}} \longrightarrow Prot_{ox} + \mathbf{Trx_{red}}$$

$$12)\quad MDHA + NADPH \xrightarrow{MDHAR} AA + NADP^+$$

$$13)\quad DHA + 2\,GSH \xrightarrow{DHAR} AA + GSSG$$

$$14)\quad GSSG + NADPH + H^+ \xrightarrow{GR} 2\,GSH + NADP^+$$

$$15)\quad TRX_{ox} + 2\,NADPH + 2\,H^+ \xrightarrow{TRXR} TRX_{red} + 2\,NADP^+$$

Fig. 5 Enzymatic ROS metabolism. Examples of enzymatic ROS metabolism reactions are shown; ROS generation: (1) Respiratory burst oxidase homolog, NADPH-dependent (RBOH) as an example oxidase; (2) Lipoxygenase (LOX), Fe-dependent, (L = unsaturated fatty acid); ROS scavenging: (3) Superoxide dismutase (SOD), Cu/Zn, Mn or Fe, general Me (= metal)-dependent; (4) Catalase (CAT), cofactor heme-Fe; (5) Peroxidase (PX), cofactor often heme (general reaction), (R = organic remainder); (6) Glutathione peroxidase (GPX), generally Se-dependent, (GSH = glutathione reduced, GSSG = 2 glutathione molecules S-linked, reduced); (7) Ascorbate peroxidase (APX), heme-Fe cofactor, (AA = ascorbate, MDHA = monodehydroascorbate; (8) Glutathione-S-transferase (GST), glutathione-dependent GPX-like activity; (9) Peroxiredoxin (PRX), containing redox-active cysteines, uses TRX; redox-regulation and scavenger regeneration: (10) Glutaredoxin (GRX), GSH-dependent; redox-active cysteines, redox-active proteins (Prot) as substrates, Fe-S cluster binding; (11) Thiroredoxin (TRX), redox-active cysteines, redox-active proteins (Prot) as substrates. (12) Monodehydroascorbate reductase (MDHAR), NADPH-dependent; (13) Dehydroascorbate reductase (DHR), GSH-dependent; (14) Glutathione reductase (GR), NADPH and FAD-dependent. (15) Thioredoxin reductase (TRXR), FAD/NADPH-dependent, redox-active cysteines

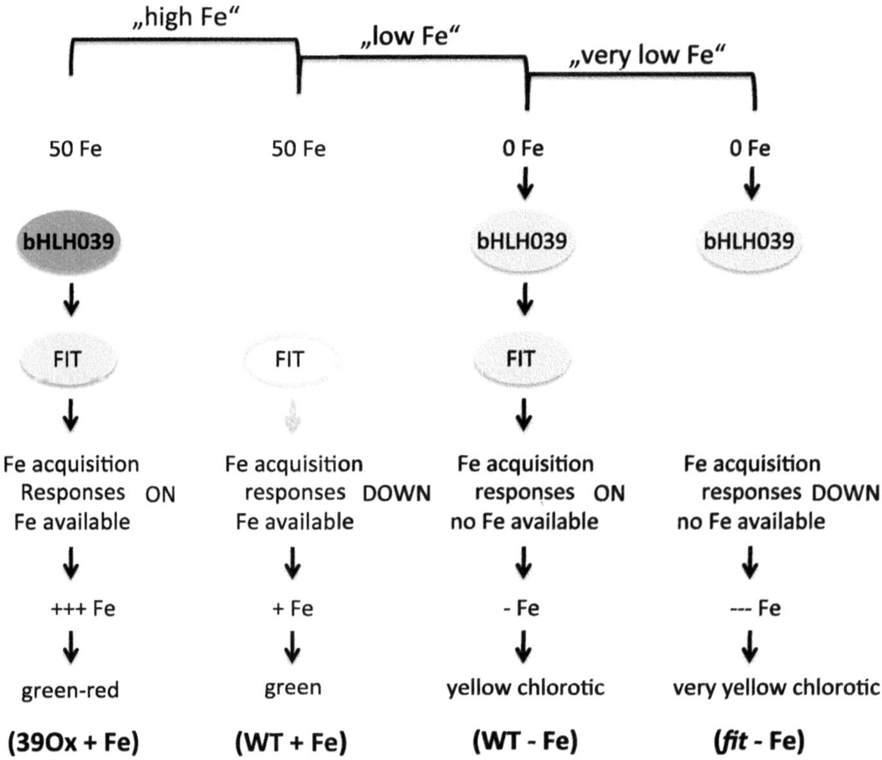

Fig. 6 Overview of comparative transcriptomic datasets used for analyzing ROS-related genes. The regulatory events leading to an onset of Fe deficiency responses via bHLH039/FIT are described in Fig. 2. Three transcriptomic datasets from 6-day-old seedlings of Arabidopsis were assembled. For the "high Fe" regulation, we used the differential transcriptomic profile of the constitutive bHLH039-overexpressing line (39Ox) versus wild type (WT), both grown at +Fe (50 Fe/+Fe = 50 μM Fe) (data obtained from Naranjo Arcos et al. 2017, Suppl. Table 1A). 39Ox plants displayed symptoms of Fe toxicity in the form of oxidative stress and elevated Fe content. WT plants were green and grew normally. Next, for the "low Fe" regulation we used genes differentially regulated in response to –Fe in wild type (WT 0 Fe/– Fe versus WT 50 Fe/+Fe). WT plants developed a leaf chlorosis at –Fe (data obtained from Mai et al. 2016). Finally, the "very low Fe" dataset is assembled from differential expression of genes, comparing the severely chlorotic *fit* loss of function mutant versus WT grown at 0 Fe/– Fe (data obtained from Mai et al. 2016)

The high number of PX genes up-regulated by high Fe seems at first very surprising, however, several of them encode known important stress regulators that switch on defense pathways via the production of ROS, like ERD5, PRX34, and RBOH. Other genes encoding ROS-generating enzymes are down-regulated presumably to prevent excess ROS formation. The very different sets of regulated genes under high and low Fe suggest that ROS response patterns are adapted to these two opposing nutrient stress conditions. A single *PX* gene, *PER50*, is induced in a FIT-dependent manner only when FIT is present. Six *PX* genes, *APX1*, *GSTU10*, a *GRX*, and *TRX5*

AGI	symbol	annotation	
AT1G09090	RBOH	RBOHB	respiratory burst oxidase homolog B
AT5G47910	RBOH	RBOHD	respiratory burst oxidase homolog D
AT1G55020	LOX	LOX1	lipoxygenase"
AT1G72520	LOX		lipoxygenase"
AT3G45140	LOX	LOX2	lipoxygenase"
AT1G17420	LOX	LOX3	lipoxygenase"
AT1G67560	LOX		lipoxygenase"
AT2G28190	SOD	CSD2	superoxide dismutase
AT4G25100	SOD	FSD1	superoxide dismutase
AT1G20620	CAT	CAT3	catalase
AT1G05240	PX		peroxidase
AT1G33660	PX		peroxidase
AT1G44970	PX		peroxidase
AT1G49570	PX		peroxidase
AT2G18150	PX		peroxidase
AT2G37130	PX	PER21	peroxidase
AT3G17070	PX		peroxidase
AT1G30870	PX		peroxidase
AT1G49570	PX		peroxidase
AT2G18150	PX		peroxidase
AT2G22420	PX	PER17	peroxidase
AT2G34060	PX		peroxidase
AT2G35380	PX		peroxidase
AT2G38380	PX	PER22	peroxidase
AT2G38390	PX		peroxidase
AT2G39040	PX		peroxidase
AT2G43480	PX		peroxidase
AT3G01190	PX	PER27	peroxidase
AT3G21770	PX	PER30	peroxidase
AT3G49110		PERX33	peroxidase
AT3G49120	PX	PERX34	peroxidase
AT3G49960	PX		peroxidase
AT4G08770	PX		peroxidase
AT4G08780	PX		peroxidase
AT4G17690	PX		peroxidase
AT4G36430	PX	PX	peroxidase
AT4G37520	PX	PER50	peroxidase
AT4G37530			peroxidase
AT5G05340	PX		peroxidase
AT5G06730	PX		peroxidase
AT5G14130	PX		peroxidase
AT5G15180	PX		peroxidase
AT5G17820	PX	PER57	peroxidase
AT5G19880	PX		peroxidase
AT5G19890	PX		peroxidase
AT5G22410	PX		peroxidase
AT5G24070	PX		peroxidase
AT5G39580	PX		peroxidase
AT5G42180	PX	PER64	peroxidase
AT5G58390	PX		peroxidase
AT5G64100	PX		peroxidase
AT5G64110	PX		peroxidase
AT5G64120	PX		peroxidase
AT5G67400	PX	PER73	peroxidase

Fig. 7 Expression patterns of genes encoding ROS generation and scavenging enzymes affected by Fe supply. The comparative transcriptome datasets for high Fe, low Fe, and very low Fe (see Fig. 6) were scanned for differentially expressed genes coding for RBOH, LOX, SOD, CAT, and PX (see Fig. 5). Only differentially expressed genes are represented, up-regulated (dark gray), down-regulated (light gray)

AGI	symbol	annotation	
AT4G11600	GPX	GPX6	glutathione peroxidase
AT2G25080	GPX	GPX1	glutathione peroxidase
AT4G31870	GPX	GPX7	glutathione peroxidase
AT1G07890	APX	APX1	ascorbate peroxidase
AT4G09010	APX	APX4	ascorbate peroxidase
AT1G77490	APX	TAPX	ascorbate peroxidase
AT1G02920	GST	GSTF7	glutathionine-S-transferase
AT1G02930	GST	GSTF6	glutathionine-S-transferase
AT1G10360	GST	GSTU18	glutathionine-S-transferase
AT1G10370	GST	GSTU17	glutathionine-S-transferase
AT1G17170	GST	GSTU24	glutathionine-S-transferase
AT1G17180	GST	GSTU25	glutathionine-S-transferase
AT1G17190	GST	GSTU26	glutathionine-S-transferase
AT1G27140	GST	GSTU14	glutathionine-S-transferase
AT1G49860	GST	GSTF14	glutathionine-S-transferase
AT1G53680	GST	GSTU28	glutathionine-S-transferase
AT1G65820	GST		glutathionine-S-transferase
AT1G69920	GST	GSTU12	glutathionine-S-transferase
AT1G69930	GST	GSTU11	glutathionine-S-transferase
AT1G74590	GST	GSTU10	glutathionine-S-transferase
AT1G78320	GST	GSTU23	glutathionine-S-transferase
AT1G78360	GST	GSTU21	glutathionine-S-transferase
AT2G02390	GST	GSTZ1	glutathionine-S-transferase
AT2G02930	GST	GSTF3	glutathionine-S-transferase
AT2G29440	GST	GSTU6	glutathionine-S-transferase
AT2G29450	GST	GSTU5	glutathionine-S-transferase
AT2G29460	GST	GSTU4	glutathionine-S-transferase
AT2G29470	GST	GSTU3	glutathionine-S-transferase
AT3G03190	GST	GSTF11	glutathionine-S-transferase
AT3G62760	GST	GSTF13	glutathionine-S-transferase
AT4G02520	GST	GSTF2	glutathionine-S-transferase
AT5G17220	GST	GSTF12	glutathionine-S-transferase
AT1G60740	PRX		peroxiredoxin
AT5G06290	PRX		2-cys peroxiredoxin, chloroplast
AT3G26060	PRX		"peroxiredoxin Q
AT3G11630	PRX	BAS1	2-cys peroxiredoxin, chloroplast
AT1G48130	PRX	PER1	1-cys peroxiredoxin
AT4G33040	GRX		glutaredoxin
AT1G28480	GRX		glutaredoxin
AT2G47880	GRX		glutaredoxin
AT1G03850	GRX		glutaredoxin
AT5G63030	GRX		glutaredoxin
AT3G62960	GRX		glutaredoxin
AT2G20270	GRX	GRXS16	glutaredoxin
AT5G14070	GRX		glutaredoxin
AT2G30540	GRX		glutaredoxin
AT4G28730	GRX		glutaredoxin
AT3G62930	GRX		glutaredoxin
AT4G15670	GRX		glutaredoxin
AT4G15690	GRX		glutaredoxin
AT4G15680	GRX		glutaredoxin
AT1G03020	GRX	GRXS1	glutaredoxin
AT4G15700	GRX		glutaredoxin
AT4G15660	GRX	GRXS15	glutaredoxin
AT1G64500	GRX		glutaredoxin
AT5G18600	GRX		glutaredoxin
AT3G62950	GRX		glutaredoxin
AT1G69880	TRX	TH8	thioredoxin
AT1G45145	TRX	TRX5	thioredoxin
AT5G16400	TRX	TRXF2	thioredoxin
AT5G06690		TRXL5	thioredoxin
AT1G43560	TRX	ATY2	thioredoxin
AT1G50320	TRX	ATHX	thioredoxin
AT3G09940	MDHAR	MDAR3	monodehydroascorbate reductase
AT2G41680	TRXR		thioredoxin reductase
AT5G08410		FTRA2	ferredoxin(thioredoxin reductase

Fig. 8 Expression patterns of genes encoding ROS scavenging enzymes affected by Fe supply (**continued from** Fig. 7). The comparative transcriptome datasets for high Fe, low Fe, or very low Fe (see Fig. 6) were scanned for differentially expressed genes coding for GPX, APX, GST, PRX, GRX, TRX, MDHAR, DHAR, GR, and TRXR (see Fig. 5). Only differentially expressed genes are represented, up-regulated (dark gray), down-regulated (light gray)

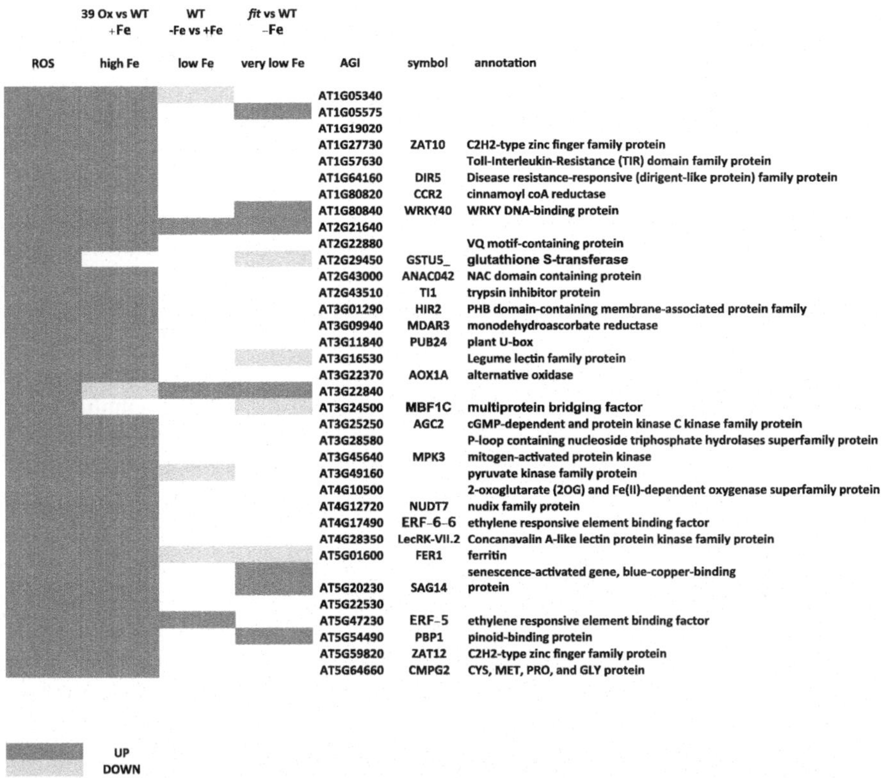

Fig. 9 Expression patterns of 35 Fe-regulated ROS signature genes. The comparative transcriptome datasets for high Fe, low Fe, or very low Fe (see Fig. 6) were scanned for differentially expressed genes from the list of 62 up-regulated ROS signature genes selected from Vaahtera et al. (2014; Table 2, genes listed under high light, O_3, H_2O_2, $1O_2$, general ROS, and Table 3 all listed genes). Only differentially expressed genes are represented, up-regulated (dark gray), down-regulated (light gray)

display partly criteria of a potential FIT target gene (up in 39Ox, down in *fit*); however, the genes are not induced at –versus +Fe in wild type when FIT is also active. *LOX2* and *GSTF12* are induced in all three Fe conditions, while none of the genes is repressed in all three. Four *PX* and six *GST* genes are up-regulated under high and very low Fe, but not under low Fe, while one *PX*, one *GPX* and two *GRX* genes are down-regulated under these same conditions. Overall, the high Fe condition is characterized by the strongest ROS metabolism response, followed by the very low and then the low Fe condition.

(b) Regulation of ROS signature genes

Out of the 35 Fe-regulated ROS signature genes, 32 genes were up-regulated under high Fe (Fig. 9). That was expected since the 39Ox plants showed signs of oxidative stress. Low Fe regulated only six genes in total, and very low Fe 11 genes, three of the low Fe-regulated genes and additional eight more genes.

Four genes, namely At1g05575 with unknown function, *WRKY40*, *SAG14*, and PINOID binding-encoding *PBP1*, were up-regulated by high and very low Fe. We had also previously reported that the ROS marker *ZAT12* that is up-regulated by high Fe, but does not appear regulated by low and very low Fe in Fig. 9, becomes up-regulated upon low Fe in a time-dependent manner under prolonged Fe deficiency stress beyond 8 days (Le et al. 2016). Prolonged Fe deficiency conditions may well result in stress and localized Fe mobilization caused by cell death and autophagy, which might again be promoted by ROS signals for induced cell death (see discussion later in this chapter).

FER1 displayed a unique regulatory pattern. This ROS marker was up-regulated by high Fe, but down-regulated under low and very low Fe. Ferritin is a multisubunit protein complex and storage place for numerous Fe ions in plastids and thereby helps to reduce oxidative stress. On the other side, its degradation is linked with the internal mobilization of Fe (Briat et al. 2010). Particular patterns are also displayed by *GSTU5* and *MBF1C* which are not induced by high Fe and down-regulated under very low Fe, as well as At3g22840 down-regulated under high Fe and induced by low and very low Fe.

Quite interestingly, we had reported that the *fer* mutant of tomato, deficient in the ortholog of the FIT transcription factor, develops brown necrotic areas on its leaf surface, potentially resulting from severe Fe deficiency-induced cell death (Ling et al. 2002).

Hence, ROS signature gene regulation in response to Fe indicates that high Fe results in the strongest response, followed by very low and then low Fe.

4 Examples for Interconnections of ROS and Fe Signaling During Development and External Stress

ROS and Fe signaling encompass several cellular compartments, opening up possibilities for ROS and Fe signaling to merge. We summarize some examples of recent investigations in Fe and ROS signaling.

(a) ROS generation coupled to NADPH oxidase signaling and heme breakdown

Fe is important for ROS generation and stress signaling. However, several stress situations cause a repression of Fe acquisition, e.g. high or low light, dehydration, and salt stress (Séguéla et al. 2008). Presumably, Fe is more deleterious in such stresses and an accumulation of Fe could contribute to additional ROS stress. For example, cold-induced ROS stress is reduced in the presence of glycine betaine, which is dependent on FRO2 in the root (Einset et al. 2008). Fe uptake might be linked with its potential to catalyze ROS generation.

ROS signaling is employed to communicate iteratively stress signals via ROS waves from cell to cell. This long-distance signaling involves the action of RBOH plasma membrane-bound oxidases generating $\cdot O_2-$ in the apoplast (Fig. 5), which is subsequently eliminated by SODs. H_2O_2 is then taken up by neighboring cells via aquaporin transporters where it can further elicit calcium increases that ultimately are sensed and transmitted by calcium-dependent CIPK protein kinases to activate RBOH (Zandalinas and Mittler 2017). *RBOHB* and *RBOHD* genes are Fe-regulated (Fig. 7). Conversely, Fe uptake by FRO2/IRT1 is under control of CIPK23 (Tian et al. 2016; Dubeaux et al. 2018), opening up the possibility that Fe information is passed on via ROS waves.

The liberation of internal ferrous Fe through heme degradation is coupled with a nutritional mobilization of Fe but also contributes to ROS production. Heme can be non-enyzmatically degraded through ROS resulting in an oxidation of Fe and attack of the tetrapyrrole rings (Nagababu and Rifkind 2004). Oftentimes, the degradation of heme into biliverdin, carbon monoxide, and free ferrous Fe is catalyzed by heme oxygenase (HO) as part of a nutritional response in plastids. HO is a heme enzyme and its activity is stimulated by nitrosyl-heme. HO1 is the main regulated HO enzyme in *Arabidopsis*. In one study, HO1 was found to promote nitric oxide (NO) signaling for Fe uptake, suggesting that liberation of Fe from heme reduces the sensitivity to Fe deficiency (Li et al. 2013). However, in none of the three Fe transcriptomics datasets, we could find the *HO1* gene to show differential expression (data not shown). It is still possible that positive HO1 protein control occurs at the level of a regulatory nitrosyl-heme complex as NO is a positive signal for Fe uptake (Li et al. 2013; Singh and Bhatla 2016).

(b) Glutathione-based signaling

There is a strong connection of GSH with Fe homeostasis in the cells and it was even proposed that the primary role of GSH is to regulate Fe metabolism (Berndt and Lillig 2017). GSH and redoxreactive proteins coordinate via their sulfur atoms Fe and NO. These complexes are termed dinitrosyl-Fe complexes (DNICs). Especially diglutathionyl-dinitrosyl-Fe complexes (Fig. 10) are abundant NO-bound derivatives in the cells that stabilize NO and may have NO signaling functions. However, it was discussed recently that due to their abundance DNICs may also be cytosolic transport forms for Fe delivery to target compartments (Berndt and Lillig 2017). In this respect, it is interesting to note that glutathione serves as substrate for a higher order thiol-rich polymer derivative containing besides Glu–Cys–Gly one or more additional stretches of Glu–Cys, whereby during synthesis Gly is cleaved by phytochelatin synthase. Phytochelatins have roles in heavy metal and arsenate detoxification, as well as in the regulation of secondary metabolism related to biotic stress (Kuhlenz et al. 2014; De Benedictis et al. 2018).

GSH is coupled to ROS scavenging through its role in the ascorbate-glutathione cycle where it serves to provide ascorbate to APX for H_2O_2 elimination (see Fig. 5). Monomeric GSH can be oxidized and form dimeric disulfide (GSSG) forms. GSH is oxidized to GSSG upon reduction of H_2O_2 by glutathione peroxidase (GPX) or upon reduction of other peroxides by peroxiredoxins (PRXs), thioredoxins (TRX), and glutaredoxins (GRX). The backwards reaction from GSSG to replenish GSH is catalyzed by glutathione reductase (GR), a FAD and NADPH-dependent enzyme.

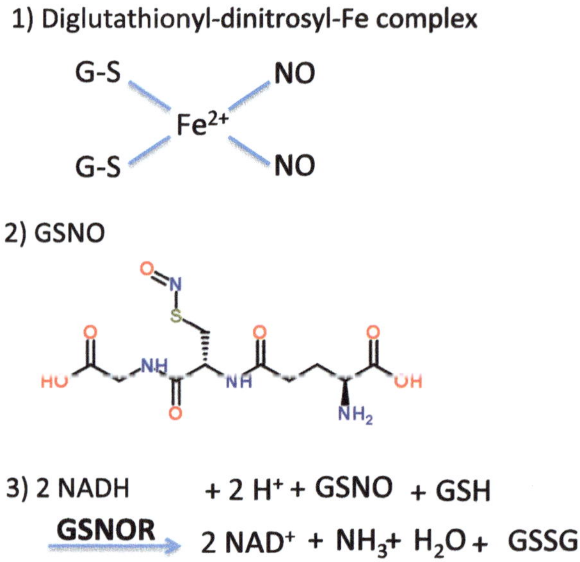

Fig. 10 Glutathione-Fe signaling. (1) Diglutathionyl-dinitrosyl-Fe complex as example of a dinitrosyl-Fe complex (DNIC); (2) S-nitrosoglutathione (GSNO); (3) Reaction of S-nitrosoglutathione reductase (GSNOR) enzyme

Particularly due to its relatively high concentrations in cells GSH/GSSG is considered a redox buffer in cells (Berndt and Lillig 2017).

GSSG is also generated under the action of S-nitrosoglutathione reductase (GSNOR), an enzyme with multiple regulatory cysteine residues that integrates NO and ROS signals (Fig. 10; Lindermayr 2017). From our survey of transcriptomics data, we could not find evidence for Fe regulation of GSNOR. ROS-mediated inhibition of GSNOR causes accumulation of GSNO. GSNO is an NO donor for nitrosylation reactions, resulting in enhanced protein S-nitrosylation, a protective mechanism for thiol groups and regulatory switches of target proteins (Lindermayr 2017). NO and GSNO are signals that positively affect Fe deficiency responses and enhance Fe acquisition (García et al. 2011; Garcia et al. 2013). Glutathione is also important for up-regulation of Fe acquisition and growth of plants under Fe-deficient conditions. Glutathione-deficient mutants have Fe deficiency leaf chlorosis phenotypes (Shanmugam et al. 2015). Recently, GSNO was found to act downstream of NO in regulating Fe deficiency responses and FIT (Kailasam et al. 2018). NO signaling is linked in multiple ways also via enzyme S-nitrosylation of hormone pathways such as ethylene to Fe acquisition (Freschi 2013). Such NO-linked regulatory mechanisms could also be relevant in long-distance Fe signaling (Ramirez et al. 2011; Garcia et al. 2013). On the other hand, NO might mediate ferritin regulation under high Fe. Application of the NO donor sodium nitroprusside (SNP) resulted in an enhancement of ferritin gene expression in *Arabidopsis* and soybean, perhaps linked with the regulation of a ferritin-linked proteasomal inhibitor (Arnaud et al. 2006; Huang et al. 2018a). SNP and GSNO indeed have different biological effects. SNP is a toxic cyanide-based compound while GSNO naturally occurs as a signal in plants. Whether and how GSH, NO, GSNO, ROS, and Fe act in concert to generate signals for communicating the Fe nutritional status of cells requires further experimental attention.

(c) Retrograde chloroplast signaling

Chloroplasts are a major compartment for ROS generation due to the electron transport activities and light reactions. At the same time, chloroplasts are a storage compartment for Fe, either in the form of ferritin, heme or as Fe–S cofactors in the electron transport chains. Ferritin is a ROS marker and prevents oxidative stress through forming a large protein complex for sequestering Fe. The levels of ferritin correlate with the presence of Fe (see Fig. 9). ROS participate in several retrograde signaling pathways that instruct the nucleus to regulate specific responses to ROS (Leister 2017). One of the retrograde signaling pathways links sulfur metabolism, Fe, and ethylene via the action of ROS: Under high light stress, SAL1 phosphatase is inactivated via ROS and GSSG, therefore no longer degrading 3′phosphoadenosine 5′-phosphate (PAP). PAP is normally a byproduct of sulfur metabolism and eliminated by SAL1. If PAP accumulates, it initiates a retrograde signal from chloroplast to instruct gene expression in the nucleus, which could possibly involve targeting of an exoribonuclease (XRN) (Estavillo et al. 2011). XRN is also relevant in ethylene signaling, and convergence of ethylene, high light, and Fe deficiency signaling could ensure acclimation (Olmedo et al. 2006; Lingam et al. 2011). Very interestingly in this respect is the finding that *EARLY LIGHT-INDUCED PROTEIN* (*ELIP2*) is a target of the retrograde signaling via the SAL1/PAP pathway (Estavillo et al. 2011). *ELIP2* is thus induced under high light where it down-regulates chlorophyll biosynthesis. At the same time, *ELIP2* is also induced in our datasets by low Fe, where leaf chlorosis occurs, but repressed under high Fe (data not shown). *ELIP2* induction upon Fe deficiency involves the ethylene transcription factors EIN3/EIL1 (Lingam et al. 2011). Thus, gene regulation confirms that Fe deficiency results in a photooxidative stress that may resemble a high light condition and hence could result in ROS generation requiring acclimation also via Fe acquisition.

(d) Fe–S cluster signaling

Fe–S clusters are synthesized and transferred to target proteins in mitochondria, plastids, and in the cytosol. Fe–S cluster de novo synthesis occurs in several steps initiated by the incorporation of Fe and sulfur on a scaffold protein, subsequently the detachment of Fe–S clusters and transfer to trafficking proteins. After an additional Fe–S cluster maturation step, the cofactors are finally delivered to Fe–S cluster-requiring target proteins (Balk and Pilon 2011; Braymer and Lill 2017; Przybyla-Toscano et al. 2018). The vulnerability of free Fe–S clusters to degradation under oxygen is presumably the reason for the need of protein iron–sulfur cluster (ISC) assembly machineries in mitochondria, plastids and in the cytosol.

Frataxin is a Fe-binding and Fe donor protein for Fe–S cluster and heme production in mitochondria and indirectly for this reason affects the regulation of Fe uptake and ROS signaling. Reduced expression of frataxin in certain organs occurring in genetically inherited Friedreich's ataxia causes iron overload and oxidative stress symptoms followed by cardio and neurodegenerative diseases (Chiang et al. 2016). In plants, frataxin proteins occur in mitochondria and plastids (Turowski et al. 2015). Full frataxin deficiency in plants is embryo-lethal (Vazzola et al. 2007). However, knockdown plants show increased NO and ROS signals and slightly elevated ferritin levels (Martin et al. 2009).

Interesting new avenues came from the studies of GRXs (Fig. 5). Some GRX genes are regulated by high or low Fe conditions in plants (Fig. 8). GRX proteins are divided in different classes. Dithiol GRXs can regulate the formation of disulfide bridges via their redox-active cysteines and catalyze thiol-disulfide exchange reactions in target proteins. Such a thiol regulation may serve post-translational control of target protein activities in response to a redox or ROS signal (Gutsche et al. 2015). GSH is required for Fe–S biogenesis in mitochondria (Braymer and Lill 2017). Monothiol class II GRX enzymes bind Fe–S clusters, are associated with the ISC assembly machineries and transfer Fe–S clusters to target proteins (Couturier et al. 2015; Braymer and Lill 2017). GRXS14, GRXS15, and GRXS16 are *Arabidopsis* plastidic and mitochondrial monothiol GRX and associated with Fe–S client proteins. For example, GRXS15 is an essential mitochondrial enzyme for Fe–S cluster biogenesis where it affects lipoic acid cofactor synthesis and arsenic toxin responses (Moseler et al. 2015; Stroher et al. 2016). Cytosolic monothiol GRX mediate Fe–S cluster transfer to Fe regulatory transcription factors and thereby regulate aspects of Fe deficiency responses in yeast, mammals, bacteria, and perhaps also plants (Couturier et al. 2015). In plants, haemerythrin domain-containing E3 ligases of the BRUTUS (BTS) type may act as Fe sensors and affect regulatory transcription factors (Kobayashi et al. 2013; Selote et al. 2015). GRXS17 is the only cytosolic monothiol GRX in *Arabidopsis* and associates with cytosolic Fe-S cluster assembly but is not essential in *Arabidopsis* (Inigo et al. 2016). GRXS17 is ubiquitinated by RGLG3 and RGLG4 (Durand et al. 2016). Interestingly, RGLG1 and RGLG2 play a role in Fe deficiency responses (Pan et al. 2015). RGLG2, RGLG3, and RGLG4 interact with the same E2 ubiquitine-conjugating enzyme UBC8, suggesting that perhaps there could be a role for GRXS17 also in Fe deficiency (Durand et al. 2016). Indeed, *grxs17*-deficient mutants show enhanced ROS and Fe reductase activity and an increased seed Fe content along with enhanced ROS levels in roots (Yu et al. 2017). GRXS17 could, therefore, have a negative role on Fe uptake.

(e) Ferroptosis

Different cell death mechanisms are distinguished in animal and fungal cells based on morphological features, regulatory features, and molecular markers. For plants, distinctive cell death mechanisms are still under debate. Cell death mostly involves the action of ROS and at least indirectly of Fe. Recently, a new form of cell death was defined in animal cells involving Fe and lipid peroxides, named "ferroptosis" (Dixon et al. 2012; Cao and Dixon 2016). Evidence for ferroptosis was found in plants exposed to a 55 °C heat shock (Distefano et al. 2017). Ferroptosis has been recently the subject of several reviews (Conlon and Dixon 2017; Doll and Conrad 2017; Stockwell et al. 2017). A hallmark of ferroptosis is that in the presence of a "labile Fe" pool and oxygen-specific lipids with polyunsaturated fatty acids (PUFAs; in mammals particularly a specific class of phosphatidylethanolamines) are transformed under the action of a lipooxygenase into lipid peroxides. The concrete signals and cellular regulation are still under examination, but it seems that there is a specificity requirement for lipid peroxide generation. At the same time, depletion of GSH causes inactivation of a cytosolic GPX so that lipid peroxides are not eliminated by GPX and instead cause cell death. Ferroptosis is triggered by chemicals that either deplete GSH, for example in mammals by inhibition of the transporter for cystine, a precursor for GSH synthesis, or by inhibition of GSH synthesis and inhibition of the GPX. On the other side, ferroptosis is prevented upon application of Fe chelators and compounds that eliminate lipid peroxides, such as tocopherols or prevents its formation. In plants, calcium was required for ferroptosis and EGTA could also prevent it (Distefano et al. 2017). In addition, ferroptosis is morphologically distinct from other cell death mechanisms because of cytoplasmic retraction and minimized mitochondria. The *KISS OF DEATH* (*KOD*) gene is a marker for heat stress-induced ferroptosis in *Arabidopsis* (Distefano et al. 2017). This gene was described earlier as 25-amino acid peptide inducer of programmed cell death (Blanvillain et al. 2011).

(f) Role of the ROS marker ZAT12 in Fe regulation

Zinc finger proteins (ZFPs) form a superfamily of regulators involved in plant development and resistance mechanisms toward biotic and abiotic stress. ZFPs vary widely in structure and function. The zinc finger domain is a protein structural motif stabilizing the protein by the coordination of a zinc ion. According to their zinc-binding topology and distribution of cysteine and histidine motifs, ZFPs are classified into several different types: TFIIIA type or classical C_2H_2, C_3H, C_2HC_5 (LIM-type), C_3HC_4 (termed RING finger), C_4HC_3, C_4 (Dof-type), C_6, C_8. These proteins play a pivotal role in many cellular functions through DNA binding (C_2H_2- and C_4-type), RNA binding and protein–protein interactions (LIM- and RING finger types) (Yanagisawa 2004; Gamsjaeger et al. 2007).

One of the most common types of ZFPs is the C_2H_2-type in which two cysteines and two histidines coordinate a single zinc atom to form a finger construct that can bind the major groove of DNA (Pavletich and Pabo 1991; Choo and Klug 1997). Two of the genes encoding this type of ZFPs are the ROS- and high Fe-responsive genes *ZAT10* and *ZAT12*.

ZAT10/STZ is induced in response to salinity, low temperature, and dehydration (Lippuner et al. 1996; Sakamoto et al. 2000; Gong et al. 2001; Sakamoto et al. 2004), affecting stress responses in a complex manner (Mittler et al. 2006). It was shown to serve as an active transcriptional repressor through its ethylene-responsive element-binding factor associated amphiphilic repression (EAR) motif (Ohta et al. 2001). Recently, ZAT10 was reported as a target of several factors involved in abiotic stress and metabolic adaptation. The calcium-dependent protein kinase CPK1 was described as a positive regulator of salt and drought stress responses in *Arabidopsis*, leading to the up-regulation of *ZAT10* and its downstream target *APX2* (Huang et al. 2018b). The seed-specific WRKY DNA-BINDING PROTEIN 43 (WRKY43) which is a modulator of fatty acid desaturation and seed filling, increasing abiotic stress tolerance, positively affected the ABA-induced regulation of ZAT10 (Geilen et al. 2017). *ZAT10* transcription was repressed by the NAD^+-dependent histone deacetylase AtSRT1 that regulates primary metabolism and stress response (Liu et al. 2017).

ZAT12 is one of the genes responding early to multiple stresses, classified as a core abiotic stress-responsive gene (Hahn et al. 2013). It is induced by ROS in the context of high light, salinity, cold, oxidative and osmotic stress (Rizhsky et al. 2004; Davletova et al. 2005; Vogel et al. 2005). ZAT12 is regulated by CIRCADIAN CLOCK ASSOCIATED 1 (CCA1) in correlation with ROS level changes (Lai et al. 2012). ZAT12 is the target of many transcriptional regulators. Chromatin immunoprecipitation sequencing (ChIP-seq) revealed that *ZAT12* is a candidate target of EIN3 (Chang et al. 2013). Ben Daniel et al. (2016) also identified 15 potential *ZAT12* transcriptional regulators. These included bZIP29, AtUSB1, DRM2, At3g03170, and LEA18, which interacted with either the ROS- or the salt-responsive segment of the *ZAT12* promoter. Together with ZAT10, it was recently shown to be the target of the DNA-binding with one finger (DOF)-type transcription factor CDF3. CDF3 was induced by drought, high temperature, and ABA, linking flowering time and abiotic stress tolerance partially through its effect on *ZAT10* and *ZAT12* (Corrales et al. 2017). Furthermore, *ZAT12* expression was reduced after treatment of wild-type plants with small IDA-like peptides IDL6 and IDL7. After wounding, *ZAT12* expression was suppressed by IDL7 and IDL6, suggesting that these peptides act as negative regulators of stress-induced ROS signaling (Vie et al. 2017).

Recently, a link between ZAT12 and *Arabidopsis* responses to prolonged Fe deficiency was described, which involves H_2O_2-mediated regulation of ZAT12 protein stability (Brumbarova et al. 2016; Le et al. 2016). *ZAT12* expression is increased after prolonged Fe deficiency. ZAT12 protein interacted with FIT. Under Fe-deficient conditions, H_2O_2 levels within the root were elevated in a FIT-dependent manner. FIT protein was stabilized by H_2O_2 in the presence of ZAT12. Since this led to an attenuated Fe deficiency response, ZAT12 functions as a negative regulator of Fe uptake forming an inactive H_2O_2-dependent complex with FIT. A role of this inhibition might be to limit the uncontrolled uptake of other, potentially harmful, elements, such as zinc and manganese. Thus, ZAT12 action might limit the risk of ROS generation under prolonged Fe deficiency (Le et al. 2016). At the same time, ZAT12 protein stability is also regulated by H_2O_2. The process is dependent on the ZAT12 EAR motif and differs between root cell types (Brumbarova et al. 2016; Le et al. 2016). The mechanism underlying this regulation remains unknown. However, the current data based on analysis of intact ZAT12 and a form lacking the EAR motif, suggest a hierarchical interaction mechanism involving an E3 ubiquitin ligase and two other, as yet unknown, proteins (Brumbarova et al. 2016). Thus, H_2O_2 regulates ZAT12 availability in the cell at both transcriptional and post-transcriptional level. The positive regulators of FIT activity EIN3 and EIL1 are also EAR motif-containing proteins. Therefore, it is possible that the outcome for FIT activity under certain stress conditions may depend on the competition between the positive and negative EAR-containing FIT interactors. The fact that *ZAT12* itself is a direct target of EIN3 (Chang et al. 2013) may allow fine-tuning of the FIT regulatory system at multiple levels.

Twenty-six *ZFP* genes are up-regulated in Indica rice in response to cold, dehydration, and salt (Agarwal et al. 2007), stress conditions that may interfere with Fe nutrition. One of these genes, ZOS3-22 (Os03g0820400, LOC_Os03g6057, ZFP37), a homolog of *ZAT12*, was also found consistently up-regulated in seedlings from 21 different rice genotypes treated with H_2O_2 and in the shoots of plants exposed to Fe deficiency and combined Fe and P deficiency (de Abreu Neto and Frei 2015). This suggests that the rice ZAT12 homolog may also be involved in the crosstalk between oxidative stress and Fe uptake regulation.

5 Concluding Remarks

The differentially expressed genes identified in the datasets reflecting high Fe, low Fe, and very low Fe are regulated in the order high Fe $\gg$ very low Fe $>$ low Fe, and this phenomenon was seen for ROS metabolism genes as well as for the ROS signature genes. One conclusion is that high Fe, as known from its role in ROS production, indeed is a signal for regulating ROS metabolism genes, but the same must be happening although at much lower level for very low Fe. The many examples of ROS and Fe signaling interference demonstrate that during the course of evolution organisms and particularly plants have used ROS as signal to regulate Fe homeostasis and vice versa.

References

Abadía J, Vázquez S, Rellán-Álvarez R, El-Jendoubi H, Abadía A, Alvarez-Fernández A, López-Millán AF (2011) Towards a knowledge-based correction of iron chlorosis. Plant Physiol Biochem 49:471–482

Agarwal S, Kapoor A, Lakshmi OS, Grover A (2007) Production and phenotypic analysis of rice transgenics with altered levels of pyruvate decarboxylase and alcohol dehydrogenase proteins. Plant Physiol Biochem 45:637–646

Arnaud N, Murgia I, Boucherez J, Briat JF, Cellier F, Gaymard F (2006) An iron-induced nitric oxide burst precedes ubiquitin-dependent protein degradation for *Arabidopsis AtFer1* ferritin gene expression. J Biol Chem 281:23579–23588

Balk J, Pilon M (2011) Ancient and essential: the assembly of iron-sulfur clusters in plants. Trends Plant Sci 16:218–226

Ben Daniel BH, Cattan E, Wachtel C, Avrahami D, Glick Y, Malichy A, Gerber D, Miller G (2016) Identification of novel transcriptional regulators of Zat12 using comprehensive yeast one-hybrid screens. Physiol Plant 157:422–441

Berndt C, Lillig CH (2017) Glutathione, glutaredoxins, and iron. Antioxid Redox Signal 27:1235–1251

Blanvillain R, Young B, Cai YM, Hecht V, Varoquaux F, Delorme V, Lancelin JM, Delseny M, Gallois P (2011) The Arabidopsis peptide kiss of death is an inducer of programmed cell death. EMBO J 30:1173–1183
Braymer JJ, Lill R (2017) Iron-sulfur cluster biogenesis and trafficking in mitochondria. J Biol Chem 292:12754–12763
Briat JF, Duc C, Ravet K, Gaymard F (2010) Ferritins and iron storage in plants. Biochim Biophys Acta 1800:806–814
Brumbarova T, Bauer P, Ivanov R (2015) Molecular mechanisms governing *Arabidopsis* iron uptake. Trends Plant Sci 20:124–133
Brumbarova T, Le CT, Ivanov R, Bauer P (2016) Regulation of ZAT12 protein stability: the role of hydrogen peroxide. Plant Signal Behav 11:e1137408
Cao JY, Dixon SJ (2016) Mechanisms of ferroptosis. Cell Mol Life Sci 73:2195–2209
Chang KN, Zhong S, Weirauch MT, Hon G, Pelizzola M, Li H, Huang SS, Schmitz RJ, Urich MA, Kuo D, Nery JR, Qiao H, Yang A, Jamali A, Chen H, Ideker T, Ren B, Bar-Joseph Z, Hughes TR, Ecker JR (2013) Temporal transcriptional response to ethylene gas drives growth hormone cross-regulation in *Arabidopsis*. Elife 2:e00675
Chiang S, Kovacevic Z, Sahni S, Lane DJ, Merlot AM, Kalinowski DS, Huang ML, Richardson DR (2016) Frataxin and the molecular mechanism of mitochondrial iron-loading in Friedreich's ataxia. Clin Sci (Lond) 130:853–870
Choo Y, Klug A (1997) Physical basis of a protein-DNA recognition code. Curr Opin Struct Biol 7:117–125
Choudhury FK, Rivero RM, Blumwald E, Mittler R (2017) Reactive oxygen species, abiotic stress and stress combination. Plant J 90:856–867
Conlon M, Dixon SJ (2017) Ferroptosis-like death in plant cells. Mol Cell Oncol 4:e1302906
Connorton JM, Balk J, Rodriguez-Celma J (2017) Iron homeostasis in plants—a brief overview. Metallomics 9:813–823
Corrales AR, Carrillo L, Lasierra P, Nebauer SG, Dominguez-Figueroa J, Renau-Morata B, Pollmann S, Granell A, Molina RV, Vicente-Carbajosa J, Medina J (2017) Multifaceted role of cycling DOF factor 3 (CDF3) in the regulation of flowering time and abiotic stress responses in *Arabidopsis*. Plant Cell Environ 40:748–764
Couturier J, Przybyla-Toscano J, Roret T, Didierjean C, Rouhier N (2015) The roles of glutaredoxins ligating Fe-S clusters: sensing, transfer or repair functions? Biochim Biophys Acta 1853:1513–1527
Czarnocka W, Karpinski S (2018) Friend or foe? Reactive oxygen species production, scavenging and signaling in plant response to environmental stresses. Free Radic Biol Med
Davletova S, Schlauch K, Coutu J, Mittler R (2005) The zinc-finger protein Zat12 plays a central role in reactive oxygen and abiotic stress signaling in *Arabidopsis*. Plant Physiol 139:847–856
De Abreu Neto JB, Frei M (2015) Microarray meta-analysis focused on the response of genes involved in Redox Homeostasis to Diverse Abiotic stresses in rice. Front Plant Sci 6:1260
De Benedictis M, Brunetti C, Brauer EK, Andreucci A, Popescu SC, Commisso M, Guzzo F, Sofo A, Castiglione MR, Vatamaniuk OK, Di Toppi LS (2018) The *Arabidopsis thaliana* knockout mutant for Phytochelatin Synthase1 (cad1-3) is defective in callose deposition, bacterial pathogen defense and auxin content, but shows an increased stem lignification. Front Plant Sci 9
Distefano AM, Martin MV, Cordoba JP, Bellido AM, D'ippolito S, Colman SL, Soto D, Roldan JA, Bartoli CG, Zabaleta EJ, Fiol DF, Stockwell BR, Dixon SJ, Pagnussat GC (2017) Heat stress induces ferroptosis-like cell death in plants. J Cell Biol 216:463–476
Dixon SJ, Lemberg KM, Lamprecht MR, Skouta R, Zaitsev EM, Gleason CE, Patel DN, Bauer AJ, Cantley AM, Yang WS, Morrison B 3rd, Stockwell BR (2012) Ferroptosis: an iron-dependent form of nonapoptotic cell death. Cell 149:1060–1072
Doll S, Conrad M (2017) Iron and ferroptosis: a still ill-defined liaison. IUBMB Life 69:423–434
Dubeaux G, Neveu J, Zelazny E, Vert G (2018) Metal sensing by the IRT1 transporter-receptor orchestrates its own degradation and plant metal nutrition. Mol Cell 69(953–964):e955

Durand AN, Inigo S, Ritter A, Iniesto E, De Clercq R, Staes A, Van Leene J, Rubio V, Gevaert K, De Jaeger G, Pauwels L, Goossens A (2016) The *Arabidopsis* iron-sulfur protein GRXS17 is a target of the ubiquitin E3 ligases RGLG3 and RGLG4. Plant Cell Physiol 57:1801–1813

Einset J, Winge P, Bones AM, Connolly EL (2008) The FRO2 ferric reductase is required for glycine betaine's effect on chilling tolerance in *Arabidopsis* roots. Physiol Plant 134:334–341

Estavillo GM, Crisp PA, Pornsiriwong W, Wirtz M, Collinge D, Carrie C, Giraud E, Whelan J, David P, Javot H, Brearley C, Hell R, Marin E, Pogson BJ (2011) Evidence for a SAL1-PAP chloroplast retrograde pathway that functions in drought and high light signaling in *Arabidopsis*. Plant Cell 23:3992–4012

Freschi L (2013) Nitric oxide and phytohormone interactions: current status and perspectives. Front Plant Sci 4:398

Gamsjaeger R, Liew CK, Loughlin FE, Crossley M, Mackay JP (2007) Sticky fingers: zinc-fingers as protein-recognition motifs. Trends Biochem Sci 32:63–70

Garcia MJ, Romera FJ, Stacey MG, Stacey G, Villar E, Alcantara E, Perez-Vicente R (2013) Shoot to root communication is necessary to control the expression of iron-acquisition genes in Strategy I plants. Planta 237:65–75

García MJ, Suárez V, Romera FJ, Alcántara E, Pérez-Vicente R (2011) A new model involving ethylene, nitric oxide and Fe to explain the regulation of Fe-acquisition genes in Strategy I plants. Plant Physiol Biochem 49:537–544

Geilen K, Heilmann M, Hillmer S, Bohmer M (2017) WRKY43 regulates polyunsaturated fatty acid content and seed germination under unfavourable growth conditions. Sci Rep 7:14235

Gong Z, Koiwa H, Cushman MA, Ray A, Bufford D, Kore-Eda S, Matsumoto TK, Zhu J, Cushman JC, Bressan RA, Hasegawa PM (2001) Genes that are uniquely stress regulated in salt overly sensitive (sos) mutants. Plant Physiol 126:363–375

Gutsche N, Thurow C, Zachgo S, Gatz C (2015) Plant-specific CC-type glutaredoxins: functions in developmental processes and stress responses. Biol Chem 396:495–509

Hahn A, Kilian J, Mohrholz A, Ladwig F, Peschke F, Dautel R, Harter K, Berendzen KW, Wanke D (2013) Plant core environmental stress response genes are systemically coordinated during abiotic stresses. Int J Mol Sci 14:7617–7641

Huang J, Wei H, Li L, Yu S (2018a) Transcriptome analysis of nitric oxide-responsive genes in upland cotton (*Gossypium hirsutum*). PLoS ONE 13:e0192367

Huang K, Peng L, Liu Y, Yao R, Liu Z, Li X, Yang Y, Wang J (2018b) *Arabidopsis* calcium-dependent protein kinase AtCPK1 plays a positive role in salt/drought-stress response. Biochem Biophys Res Commun 498:92–98

Inigo S, Durand AN, Ritter A, Le Gall S, Termathe M, Klassen R, Tohge T, De Coninck B, Van Leene J, De Clercq R, Cammue BPA, Fernie AR, Gevaert K, De Jaeger G, Leidel SA, Schaffrath R, Van Lijsebettens M, Pauwels L, Goossens A (2016) Glutaredoxin GRXS17 associates with the cytosolic iron-sulfur cluster assembly pathway. Plant Physiol 172:858–873

Jakoby M, Wang HY, Reidt W, Weisshaar B, Bauer P (2004) *FRU* (*BHLH029*) is required for induction of iron mobilization genes in *Arabidopsis thaliana*. FEBS Lett 577:528–534

Jena NR (2012) DNA damage by reactive species: mechanisms, mutation and repair. J Biosci 37:503–517

Kailasam S, Wang Y, Lo JC, Chang HF, Yeh KC (2018) S-Nitrosoglutathione works downstream of nitric oxide to mediate iron-deficiency signaling in *Arabidopsis*. Plant J 94:157–168

Kobayashi T, Nagasaka S, Senoura T, Itai RN, Nakanishi H, Nishizawa NK (2013) Iron-binding haemerythrin RING ubiquitin ligases regulate plant iron responses and accumulation. Nat Commun 4:2792

Kuhlenz T, Schmidt H, Uraguchi S, Clemens S (2014) *Arabidopsis thaliana* phytochelatin synthase 2 is constitutively active in vivo and can rescue the growth defect of the PCS1-deficient cad1-3 mutant on Cd-contaminated soil. J Exp Bot 65:4241–4253

Lai AG, Doherty CJ, Mueller-Roeber B, Kay SA, Schippers JH, Dijkwel PP (2012) CIRCADIAN CLOCK-ASSOCIATED 1 regulates ROS homeostasis and oxidative stress responses. Proc Natl Acad Sci USA 109:17129–17134

Le CT, Brumbarova T, Ivanov R, Stoof C, Weber E, Mohrbacher J, Fink-Straube C, Bauer P (2016) Zinc finger of *Arabidopsis thaliana*12 (ZAT12) interacts with FER-like iron deficiency-induced transcription factor (FIT) linking iron deficiency and oxidative stress responses. Plant Physiol 170:540–557

Leister D (2017) Piecing the puzzle together: the central role of reactive oxygen species and redox hubs in chloroplast retrograde signaling. Antioxid Redox Signal

Li H, Song JB, Zhao WT, Yang ZM (2013) AtHO1 is involved in iron homeostasis in an NO-dependent manner. Plant Cell Physiol 54:1105–1117

Lindermayr C (2017) Crosstalk between reactive oxygen species and nitric oxide in plants: key role of S-nitrosoglutathione reductase. Free Radic Biol Med S0891–5849(0817):31231–31235

Ling HQ, Bauer P, Bereczky Z, Keller B, Ganal M (2002) The tomato *fer* gene encoding a bHLH protein controls iron-uptake responses in roots. Proc Natl Acad Sci USA 99:13938–13943

Lingam S, Mohrbacher J, Brumbarova T, Potuschak T, Fink-Straube C, Blondet E, Genschik P, Bauer P (2011) Interaction between the bHLH transcription factor fit and ethylene insensitive3/ethylene insensitive3-like1 reveals molecular linkage between the regulation of iron acquisition and ethylene signaling in *Arabidopsis*. Plant Cell 23:1815–1929

Lippuner V, Cyert MS, Gasser CS (1996) Two classes of plant cDNA clones differentially complement yeast calcineurin mutants and increase salt tolerance of wild-type yeast. J Biol Chem 271:12859–12866

Liu X, Wei W, Zhu W, Su L, Xiong Z, Zhou M, Zheng Y, Zhou DX (2017) Histone deacetylase AtSRT1 links metabolic flux and stress response in *Arabidopsis*. Mol Plant 10:1510–1522

Mai HJ, Pateyron S, Bauer P (2016) Iron homeostasis in *Arabidopsis thaliana*: transcriptomic analyses reveal novel FIT-regulated genes, iron deficiency marker genes and functional gene networks. BMC Plant Biol 16:211

Martin M, Colman MJR, Gomez-Casati DF, Lamattina L, Zabaleta EJ (2009) Nitric oxide accumulation is required to protect against iron-mediated oxidative stress in frataxin-deficient *Arabidopsis* plants. FEBS Lett 583:542–548

Mittler R, Kim Y, Song L, Coutu J, Coutu A, Ciftci-Yilmaz S, Lee H, Stevenson B, Zhu JK (2006) Gain- and loss-of-function mutations in Zat10 enhance the tolerance of plants to abiotic stress. FEBS Lett 580:6537–6542

Moseler A, Aller I, Wagner S, Nietzel T, Przybyla-Toscano J, Muhlenhoff U, Lill R, Berndt C, Rouhier N, Schwarzlander M, Meyer AJ (2015) The mitochondrial monothiol glutaredoxin S15 is essential for iron-sulfur protein maturation in *Arabidopsis thaliana*. Proc Natl Acad Sci USA 112:13735–13740

Nagababu E, Rifkind JM (2004) Heme degradation by reactive oxygen species. Antioxid Redox Signal 6:967–978

Naranjo Arcos MA, Maurer F, Meiser J, Pateyron S, Fink-Straube C, Bauer P (2017) Dissection of iron signaling and iron accumulation by overexpression of subgroup Ib bHLH039 protein. Sci Rep 7:10911

Niki E (2009) Lipid peroxidation: physiological levels and dual biological effects. Free Radic Biol Med 47:469–484

Ohta M, Matsui K, Hiratsu K, Shinshi H, Ohme-Takagi M (2001) Repression domains of class II ERF transcriptional repressors share an essential motif for active repression. Plant Cell 13:1959–1968

Olmedo G, Guo HW, Gregory BD, Nourizadeh SD, Aguilar-Henonin L, Li HJ, An FY, Guzman P, Ecker JR (2006) ETHYLENE-INSENSITIVE5 encodes a $5' \rightarrow 3'$ exoribonuclease required for regulation of the EIN3-targeting F-box proteins EBF1/2. Proc Natl Acad Sci USA 103:13286–13293

Pan IC, Tsai HH, Cheng YT, Wen TN, Buckhout TJ, Schmidt W (2015) Post-Transcriptional coordination of the Arabidopsis iron deficiency response is partially dependent on the E3 ligases ring domain ligase1 (RGLG1) and RING DOMAIN LIGASE2 (RGLG2). Mol Cell Proteomics 14:2733–2752

Pavletich NP, Pabo CO (1991) Zinc finger-DNA recognition: crystal structure of a Zif268-DNA complex at 2.1 A. Science 252:809–817

Pospisil P (2016) Production of reactive oxygen species by photosystem II as a response to light and temperature stress. Front Plant Sci 7:1950
Przybyla-Toscano J, Roland M, Gaymard F, Couturier J, Rouhier N (2018) Roles and maturation of iron-sulfur proteins in plastids. J Biol Inorg Chem
Ramirez L, Simontacchi M, Murgia I, Zabaleta E, Lamattina L (2011) Nitric oxide, nitrosyl iron complexes, ferritin and frataxin: a well equipped team to preserve plant iron homeostasis. Plant Sci 181:582–592
Rizhsky L, Davletova S, Liang H, Mittler R (2004) The zinc finger protein Zat12 is required for cytosolic ascorbate peroxidase 1 expression during oxidative stress in *Arabidopsis*. J Biol Chem 279:11736–11743
Sakamoto H, Araki T, Meshi T, Iwabuchi M (2000) Expression of a subset of the *Arabidopsis* Cys(2)/His(2)-type zinc-finger protein gene family under water stress. Gene 248:23–32
Sakamoto H, Maruyama K, Sakuma Y, Meshi T, Iwabuchi M, Shinozaki K, Yamaguchi-Shinozaki K (2004) Arabidopsis Cys2/His2-type zinc-finger proteins function as transcription repressors under drought, cold, and high-salinity stress conditions. Plant Physiol 136:2734–2746
Séguéla M, Briat JF, Vert G, Curie C (2008) Cytokinins negatively regulate the root iron uptake machinery in *Arabidopsis* through a growth-dependent pathway. Plant J 55:289–300
Selote D, Samira R, Matthiadis A, Gillikin JW, Long TA (2015) Iron-binding E3 ligase mediates iron response in plants by targeting basic helix-loop-helix transcription factors. Plant Physiol 167:273–286
Shahid M, Pourrut B, Dumat C, Nadeem M, Aslam M, Pinelli E (2014) Heavy-metal-induced reactive oxygen species: phytotoxicity and physicochemical changes in plants. Rev Environ Contam Toxicol 232:1–44
Shanmugam V, Wang YW, Tsednee M, Karunakaran K, Yeh KC (2015) Glutathione plays an essential role in nitric oxide-mediated iron-deficiency signaling and iron-deficiency tolerance in Arabidopsis. Plant J 84:464–477
Singh N, Bhatla SC (2016) Nitric oxide and iron modulate heme oxygenase activity as a long distance signaling response to salt stress in sunflower seedling cotyledons. Nitric Oxide-Biol Chem 53:54–64
Stadtman ER, Levine RL (2003) Free radical-mediated oxidation of free amino acids and amino acid residues in proteins. Amino Acids 25:207–218
Stockwell BR, Friedmann Angeli JP, Bayir H, Bush AI, Conrad M, Dixon SJ, Fulda S, Gascon S, Hatzios SK, Kagan VE, Noel K, Jiang X, Linkermann A, Murphy ME, Overholtzer M, Oyagi A, Pagnussat GC, Park J, Ran Q, Rosenfeld CS, Salnikow K, Tang D, Torti FM, Torti SV, Toyokuni S, Woerpel KA, Zhang DD (2017) Ferroptosis: a regulated cell death nexus linking metabolism, redox biology, and disease. Cell 171:273–285
Stroher E, Grassl J, Carrie C, Fenske R, Whelan J, Millar AH (2016) Glutaredoxin S15 is involved in Fe-S cluster transfer in mitochondria influencing lipoic acid-dependent enzymes, plant growth, and arsenic tolerance in *Arabidopsis*. Plant Physiol 170:1284–1299
Tian Q, Zhang X, Yang A, Wang T, Zhang WH (2016) CIPK23 is involved in iron acquisition of *Arabidopsis* by affecting ferric chelate reductase activity. Plant Sci 246:70–79
Turowski VR, Aknin C, Maliandi MV, Buchensky C, Leaden L, Peralta DA, Busi MV, Araya A, and Gomez-Casati DF (2015) Frataxin is localized to both the chloroplast and mitochondrion and is involved in chloroplast fe-s protein function in *Arabidopsis*. Plos One 10
Vaahtera L, Brosche M, Wrzaczek M, Kangasjarvi J (2014) Specificity in ROS signaling and transcript signatures. Antioxid Redox Signal 21:1422–1441
Vazzola V, Losa A, Soave C, Murgia I (2007) Knockout of frataxin gene causes embryo lethality in *Arabidopsis*. FEBS Lett 581:667–672
Vie AK, Najafi J, Winge P, Cattan E, Wrzaczek M, Kangasjarvi J, Miller G, Brembu T, Bones AM (2017) The IDA-LIKE peptides IDL6 and IDL7 are negative modulators of stress responses in *Arabidopsis thaliana*. J Exp Bot 68:3557–3571

Vogel JT, Zarka DG, Van Buskirk HA, Fowler SG, Thomashow MF (2005) Roles of the CBF2 and ZAT12 transcription factors in configuring the low temperature transcriptome of *Arabidopsis*. Plant J 41:195–211

Wang HY, Klatte M, Jakoby M, Baumlein H, Weisshaar B, Bauer P (2007) Iron deficiency-mediated stress regulation of four subgroup Ib *BHLH* genes in *Arabidopsis thaliana*. Planta 226:897–908

Wu LB, Ueda Y, Lai SK, Frei M (2017) Shoot tolerance mechanisms to iron toxicity in rice (Oryza sativa L.). Plant Cell Environ 40:570–584

Yanagisawa S (2004) Dof domain proteins: plant-specific transcription factors associated with diverse phenomena unique to plants. Plant Cell Physiol 45:386–391

Yu H, Yang J, Shi YF, Donelson J, Thompson SM, Sprague S, Roshan T, Wang DL, Liu JZ, Park S, Nakata PA, Connolly EL, Hirschi KD, Grusak MA, and Cheng NH (2017) *Arabidopsis* Glutaredoxin S17 Contributes to Vegetative Growth, Mineral Accumulation, and Redox Balance during Iron Deficiency. *Frontiers in Plant Science* 8

Yuan Y, Wu H, Wang N, Li J, Zhao W, Du J, Wang D, Ling HQ (2008) FIT interacts with AtbHLH38 and AtbHLH39 in regulating iron uptake gene expression for iron homeostasis in *Arabidopsis*. Cell Res 18:385–397

Zandalinas SI, Mittler R (2017) ROS-induced ROS release in plant and animal cells. Free Radic Biol Med

Interactions Between Circadian Rhythms, ROS and Redox

Noriane M. L. Simon, Suzanne Litthauer, Matthew A. Jones and Antony N. Dodd

Abstract Circadian rhythms are endogenous biological cycles with a period of about 24 h. In plants, circadian rhythms have a pervasive influence upon metabolism, physiology and development, yet we are still discovering how these rhythms interact with reactive oxygen species (ROS) generation and signalling. Recent work has identified circadian rhythms of ROS generation and ROS-scavenging enzymes, and there are also circadian rhythms of ROS-generating photosynthesis. Here, we summarise our current understanding of the relationship between the circadian system and ROS, and suggest roles for ROS in circadian signalling between organelles and the circadian regulation of guard cell function. There are circadian rhythms of peroxiredoxin oxidation state that occur in the absence of transcription and translation. It seems that there could be multiple levels of integration between ROS, redox and circadian regulation.

1 Introduction

The rotation of the Earth on its axis exposes the majority of species on the planet to daily oscillations in the environment. In plants and algae, the onset of photosynthesis after dawn triggers the accumulation of photosynthetically derived metabolites, including the production of reactive oxygen species (ROS) (Dodd et al. 2015; Jones 2017). These daily environmental and metabolic fluctuations are thought to have guided the evolution of a prognostic mechanism, known as the circadian system, which enables the anticipation of these environmental and metabolic transitions (Hut and Beersma 2011). The benefits of an endogenous circadian timing mechanism are numerous, enabling many physiological, biochemical and metabolic processes to be restricted to appropriate times of the day while also providing an internal timing reference for developmental processes such as flowering (de Montaigu et al. 2010;

N. M. L. Simon · A. N. Dodd (✉)
School of Biological Sciences, University of Bristol, 24 Tyndall Avenue, Bristol, UK
e-mail: antony.dodd@bristol.ac.uk

S. Litthauer · M. A. Jones
School of Biological Sciences, University of Essex, Colchester, UK

S. K. Panda and Y. Y. Yamamoto (eds.), *Redox Homeostasis in Plants*,
Signaling and Communication in Plants, https://doi.org/10.1007/978-3-319-95315-1_4

Millar 2016; Jones 2017). Plants with impaired endogenous rhythms are less fit than wild-type plants and have altered susceptibility to biotic and abiotic stresses (Green et al. 2002; Dodd et al. 2005a; Nakamichi et al. 2009; Goodspeed et al. 2012). Despite the importance of the circadian system, comparatively little is known about how ROS signalling interacts with circadian timing mechanisms in plants (Spoel and van Ooijen 2014). Such an interaction would be advantageous to ensure that energy-demanding ROS-scavenging mechanisms are activated only when needed, and would also provide the ability to anticipate oxidative stress occurring at specific times of day (Lai et al. 2012a). Here, we will review how ROS signalling is modulated by the circadian system as well as highlighting how ROS feed back to modulate circadian rhythms.

2 Generation of Circadian Rhythms

The circadian system has a number of properties that enable it to function as a reliable timekeeper within a fluctuating environment (Fig. 1) (McClung 2006). First, the circadian system becomes synchronised with the phase of the environment through the process of entrainment, which causes changes in the circadian phase in response to light, temperature and metabolic cues (Fig. 1) (Hsu and Harmer 2014). This mechanism also allows adaptation of the circadian oscillator to seasonal variability and geographical latitude (Jones 2009). The circadian oscillator underlies temporal changes in the sensitivity of cellular responses to environmental stimuli, which is termed "circadian gating". There is circadian gating of environmental inputs that entrain the circadian oscillator (Fig. 1) and also circadian gating of a variety of other signalling pathways including cellular responses to temperature and light (Salter et al. 2003; Fowler et al. 2005; Allen et al. 2006; Dodd et al. 2006; Nusinow et al. 2011a; Noordally et al. 2013; Greenham and McClung 2015). This circadian gating of environmental signals and physiological processes leads to circadian rhythms having extensive consequences for plant responses to the environment. Finally, the circadian oscillator is temperature compensated. This allows it to provide consistent timing across a range of physiologically relevant temperatures (Gould et al. 2006), although transcriptional rhythms of many oscillator components stop under cold temperature conditions (Bieniawska et al. 2008).

The circadian system is comprised of transcription/translation feedback loops and biochemical components that generate self-sustaining endogenous rhythms. In the nucleus, successive waves of transcriptional repressors generate a transcriptional rhythm with a period of about 24 h. CCA1 and LHY are morning-phased transcriptional repressors that are induced by light (Wang et al. 1997; Alabadi et al. 2001; Kim et al. 2003). These transcription factors are accompanied by the accumulation of PSEUDO-RESPONSE REGULATOR9 (PRR9) and its homologues PRR7, PRR5 and TOC1/PRR1, which generate a successive wave of transcriptional repression. In the evening, complex (comprising EARLY FLOWERING 3 (ELF3), ELF4 and LUX ARRHYTHMO (LUX)) limits transcription during the early part of the night

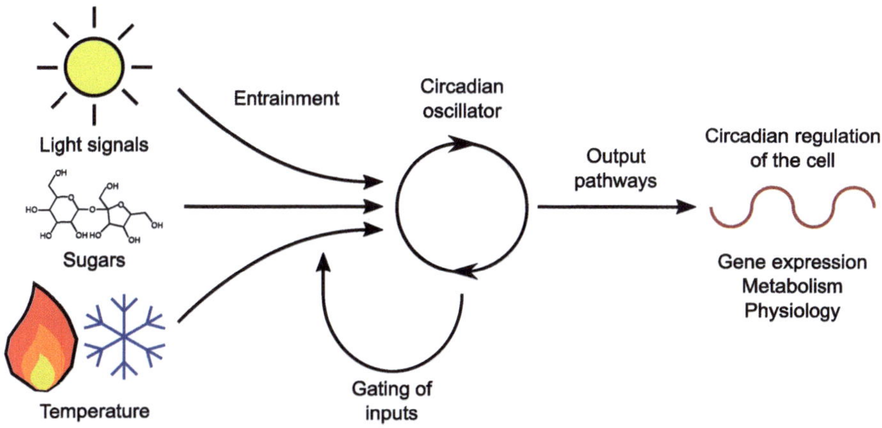

Fig. 1 Conceptual illustration of general features of plant circadian system. Light signals detected by photoreceptors and also photosynthetic sugars and temperature fluctuations act as zeitgebers (time givers) that adjust the phase of the circadian oscillator to match the daily phase of the environment. The canonical circadian oscillator comprises nuclear-encoded transcription-translation feedback loops with repressive steps. An emergent property of this network is that it completes a cycle approximately every 24 h. The circadian oscillator gates its sensitivity to entrainment signals, so that zeitgebers cause larger phase alterations at certain times of day. Components of the circadian oscillator and their interactors are transcription factors, expressed at various times of day, that interact with circadian-regulated promoter motifs within the genome. This leads to circadian regulation of the transcriptome and downstream biochemistry and physiology, including rhythms of ROS production and scavenging

(Matsushika et al. 2000; Nakamichi et al. 2010; Nusinow et al. 2011b; Gendron et al. 2012; Huang et al. 2012). In concert with these repressive elements, transcriptional activators including NIGHT LIGHT-INDUCIBLE AND CLOCK REGULATED 1 (LNK1), LNK2 and REVEILLE8 (RVE8) modulate and maintain these transcriptional rhythms (Hsu et al. 2013; Rugnone et al. 2013; Xie et al. 2014). These transcriptional rhythms are coupled with oscillations of post-transcriptional and post-translational regulation (Seo and Mas 2014). Degradation of ELF3 is mediated by the E3 ubiquitin ligase CONSTITUTIVELY PHOTOMORPHOGENIC1 (COP1) (Yu et al. 2008), while the loss of protein deubiquitinases can alter the circadian phenotype (Cui et al. 2013). Similarly, there are rhythms of protein phosphorylation and oxidation (Edgar et al. 2012; Choudhary et al. 2015), which have various consequences including altered dimerisation (Daniel et al. 2004) and protein stability (Fujiwara et al. 2008; Choudhary et al. 2015). This multi-tiered regulation of transcription and translation is thought to ensure that the circadian system serves as a reliable timekeeper.

3 Rhythms of ROS Production and Scavenging

Plants, like other aerobic organisms, continuously generate ROS as by-products of normal oxygen metabolism (Apel and Hirt 2004). The various types of ROS differ in stability and cytotoxicity, and include hydrogen peroxide (H_2O_2, the most stable and most common ROS), singlet oxygen, superoxide and hydroxyl radicals (Apel and Hirt 2004; de Souza et al. 2017). In plants, ROS are produced mainly in chloroplasts (via light capturing, photosynthetic electron transport and photochemistry), mitochondria (through respiratory electron transport) and peroxisomes (as a by-product of photorespiration), as well as in the cytosol and cellular membrane through the action of peroxidases (Apel and Hirt 2004). Accumulating ROS can cause oxidative damage to cellular components such as DNA, proteins and lipids which, if not limited, can lead to further ROS production, irreparable cellular damage and cell death (Apel and Hirt 2004; de Souza et al. 2017). Therefore, plants have evolved a range of enzymatic (including peroxidases, catalase and superoxide dismutase) and non-enzymatic (involving glutathione, ascorbate, flavonoids, alkaloids and carotenoids) ROS-scavenging antioxidant mechanisms that maintain ROS homeostasis (Apel and Hirt 2004; de Souza et al. 2017). Chloroplasts, mitochondria, peroxisomes, vacuoles, plasma membranes and cell walls all contribute to maintaining ROS homeostasis through redox mechanisms that are localised in particular cellular compartments (Apel and Hirt 2004). Despite these mechanisms, intracellular ROS levels can rise rapidly in response to environmental stresses, disrupting ROS homeostasis (Apel and Hirt 2004; de Souza et al. 2017). The "oxidative burst" in response to biotic stress occurs mainly from the activity of NADPH-dependent oxidases, while abiotic stresses can result in the production of ROS by photodynamically active molecules and over-reduction of the photosynthetic electron transport chain (Apel and Hirt 2004; de Souza et al. 2017).

Despite the negative consequences of excessive ROS accumulation, the rapid nature of ROS production in response to stimuli combined with high ROS turnover rates and the specificity of ROS production sites make ROS ideal candidates for signalling molecules (Apel and Hirt 2004; de Souza et al. 2017). It has been proposed that the specific activity of ROS in different cellular compartments, or the specific activity of secondary signals generated from ROS-oxidised molecules, could provide the cellular precision required for decoding complex environmental signals (de Souza et al. 2017). Given the sensitivity of the circadian system to environmental changes (Jones 2009; Hsu et al. 2013; Millar 2016), it is likely that changes in ROS accumulation inform circadian timing, and also that the circadian system regulates ROS signalling pathways.

In *Arabidopsis thaliana* (Arabidopsis), ROS accumulation and scavenging have daily oscillations under light/dark cycles (Lai et al. 2012a), and 34% of transcripts within various ROS GO categories are circadian regulated (Covington et al. 2008). The circadian oscillator also gates the oxidative burst that occurs following plant exposure to pathogen-associated molecular patterns (PAMPs) such as flg22, because a greater oxidative burst in response to flg22 occurs shortly after dawn compared with the end of the day and this rhythm is absent in circadian-arrhythmic *lux* (*pcl1*) mutants (Korneli et al. 2014). In plants grown under 12-h light/12-h dark cycles, H_2O_2 and catalase levels peak at noon and reach the lowest levels at midnight (Lai et al. 2012a), and 140 of 167 transcripts involved in ROS production and scavenging have similar daily rhythms in abundance (Lai et al. 2012b). Interestingly, these oscillations of H_2O_2 and ROS-related transcript accumulation persist in constant light, albeit with much reduced amplitude, suggesting the ROS-scavenging network is under circadian control.

Efforts have been made to determine how the circadian system regulates ROS accumulation (Lai et al. 2012a). Expression of the three catalases *CAT1*, *CAT2* and *CAT3* in response to oxidative stress is altered upon overexpression of *CCA1* or mutations in *CCA1* or *LHY*, and becomes arrhythmic in an *elf3* mutant (Lai et al. 2012b). Evidence suggests that *CCA1* regulates the transcription of ROS genes even in the absence of oxidative stress. For example, there are changes in the daily patterns of accumulation of the ROS-related transcripts *APX4*, *HSP182*, *PAL1*, *HSFA4A*, *MYB9* and the peroxidase *At2g22420* in both a *CCA1* overexpresser and *cca1-1/lhy-11* and *elf3-1* mutants, pointing further to a role for the circadian oscillator in regulating ROS homeostasis (Lai et al. 2012a).

In Arabidopsis, the family of ascorbate peroxidase (APX) enzymes that scavenge hydrogen peroxide contains members that are cytosolic, chloroplast targeted and microsomal. Four of these have circadian oscillations in transcript abundance, including a thylakoid lumen-localised APX (*APX4*, At4g09010) and a stromal APX (*SAPX*, At4g08390). There is evidence that alterations in *APX* gene transcription can alter scavenging capacity, because decreased *APX2* transcript accumulation is associated with increased H_2O_2 accumulation (Fryer et al. 2003). Therefore, these circadian changes in *APX4* and *SAPX* transcript abundance might cause oscillations in H_2O_2 scavenging capacity.

Examination of the sensitivity of plants to ROS provides information about connections between ROS signalling and the circadian system. Plants with mutations in core circadian oscillator genes including *CCA1*, *LHY*, *ELF3*, *ELF4*, *PRR5* and *LUX* are hypersensitive to methyl viologen (MV) treatment, while *CCA1* overexpression is linked to MV hyposensitivity in both light/dark cycles and constant light (Lai et al. 2012b). Similarly, mutations within the oscillator component *GIGANTEA* lead to greater tolerance to the superoxide-generating herbicide paraquat (Kurepa et al. 1998). Conversely, changes in ROS scavenging also contribute to circadian timing. Alterations in ROS scavenging in ascorbic acid *vtc* mutants lead to alterations in circadian oscillator and flowering time transcript accumulation, although the effect of *vtc* mutants upon flowering time appears to be independent from these changes in gene expression (Kotchoni et al. 2009). Such data demonstrate how the circadian system contributes to ROS scavenging while also being responsive to ROS production.

4 Circadian Rhythms of Peroxiredoxin Redox State

A remarkable emerging mechanism involving circadian regulation and peroxiredoxin sheds light upon both circadian rhythms and redox regulation, and also the evolutionary conservation of circadian rhythms (Edgar et al. 2012). Peroxiredoxins (PRX) are antioxidant enzymes that scavenge ROS. Their activity depends on a key cysteine residue in their active site, which is highly conserved with neighbouring proline and threonine/serine residues in a Pxx(T/S)xxC pattern (Edgar et al. 2012). This catalytic cysteine undergoes oxidation when ROS accumulate, forming a sulphenic acid derivative (Dietz 2011). An intermolecular disulphide bridge subsequently forms with another cysteine residue from a different PRX molecule (Dietz 2011). At this stage, the dimer can either return to its active monomeric form through thioredoxin-mediated reduction or be over- or hyperoxidised to form decamers and higher multimers (Dietz 2011). These are catalytically inactive but play a role as molecular chaperones or in ROS signalling, and can be rescued by sulphiredoxin activity (Dietz 2011). Higher plant genomes commonly encode the 2-Cys PRX, 1-Cys PRX, PrxQ and PrxII sub-types of peroxiredoxins (Dietz 2011). In plants, the 2-Cys PRX is targeted to chloroplasts and protects chloroplast membranes against oxidative damage resulting from photosynthetic ROS generation (O'Neill et al. 2011).

The oxidation of PRX proteins can be examined using an antiserum directed against their oxidised active site, known as the PRX-$SO_{2/3}$ antiserum. Initial studies in the picoeukaryotic alga *Ostreococcus tauri* found that PRXs undergo 24-hour redox cycles under conditions where the transcription and translation feedback loops of the circadian oscillator (TTFL) do not cycle (constant darkness), and also in the presence of inhibitors of transcription and translation (O'Neill et al. 2011). This remarkable discovery demonstrated that nuclear transcription is not essential for the generation of certain circadian rhythms (O'Neill and Reddy 2011). Comparable oscillations in PRX redox state also occur in anucleate red blood cells (O'Neill and Reddy 2011). These findings challenged the view that eukaryotic circadian rhythms are based entirely upon TTFL (van Ooijen and Millar 2012).

All species tested, including Arabidopsis, have PRX redox cycles with an oxidation peak occurring at subjective dawn, leading the authors to conclude that this phenomenon forms a universal or conserved biomarker for circadian rhythms (Edgar et al. 2012). This suggests that while the genetic TTFL may have evolved independently on several occasions within the kingdoms of life, there might be highly conserved features of circadian regulation that incorporate or are reported by oscillations in PRX redox state. One possibility is that this oscillation in PRX redox state is conserved across life because it derives from a single ancestral proto-clock (van Ooijen and Millar 2012).

The relationship between this non-transcriptional oscillator (NTO) and the TTFL system remains tentative in plants. Work using mutant Arabidopsis and the alga *O. tauri* has started to clarify this relationship (Edgar et al. 2012). When the TTFL system was disrupted in each of these species by overexpression of *TOC1*, PRX redox state oscillated with a slightly altered period in Arabidopsis, whereas alterations in

PRX oscillations in TOC1-ox in *O. tauri* were less evident (Edgar et al. 2012). In Arabidopsis, a double knockout of 2-Cys PRX changes the phase of circadian oscillations of delayed chlorophyll fluorescence, a reporter of the plant circadian rhythms (Gould et al. 2009), demonstrating that 2-Cys PRX is required for correct circadian oscillations of delayed fluorescence (Edgar et al. 2012). This could reflect a change in the functioning of the TTFL in response to the PRX mutations, or alternatively an alteration in the photosynthetic processes that cause circadian oscillations of delayed fluorescence. Overall, these results suggest that there might be coupling between the NTO and TTFL (O'Neill and Reddy 2011; O'Neill et al. 2011; van Ooijen and Millar 2012). Therefore, the discovery of PRX redox circadian rhythms has not only created a new tool for studying NTOs, but has also transformed our view of circadian oscillator architecture and of how circadian timekeeping evolved.

5 ROS and Redox in the Circadian Regulation of Chloroplasts

There are circadian rhythms in photosynthesis that include circadian rhythms of net CO_2 uptake (Hennessey and Field 1991; Dodd et al. 2005b) and rhythms of several different measures of chlorophyll fluorescence (Gould et al. 2009; Litthauer et al. 2015; Dakhiya et al. 2017). Given that changes in the rate of photosynthesis lead to alterations in the production of singlet oxygen, circadian regulation of photosynthesis might give rise to rhythms of singlet oxygen production.

The nuclear-encoded circadian oscillator is important for the circadian regulation of photosynthesis because mutations that alter properties of the circadian oscillator, such as its period and phase, cause equivalent changes in the properties of circadian oscillations of photosynthesis (Dodd et al. 2004, 2005a; Gould et al. 2009; Litthauer et al. 2016). This suggests that anterograde signalling pathways communicate circadian timing information from the nuclear-encoded circadian oscillator to chloroplasts, in order to regulate photosynthesis. Circadian oscillations in the abundance of many nuclear-encoded photosynthesis-related transcripts (Harmer et al. 2000) have the potential to contribute to the circadian rhythms of photosynthesis, although the degree to which these transcriptional oscillations give rise to circadian changes in photosynthetic protein abundance or activity within chloroplasts is unclear. For example, a study in *O. tauri* reported circadian oscillations in the abundance of chloroplast-localised granule-bound starch synthase I protein, but no significant oscillations in proteins involved directly in photosynthesis (Le Bihan et al. 2011). In comparison, there are circadian oscillations of phosphopeptides derived from organellar and photosynthetic complex proteins (Choudhary et al. 2015), suggesting that rhythms of regulatory phosphorylation might be important for the circadian regulation of photosynthetic processes.

In addition to the global regulation of the transcriptome, proteome and protein phosphorylation, signalling mechanisms might participate in the circadian regulation of ROS-generating photosynthesis. One such mechanism involves signalling mechanisms that regulate chloroplast genome transcription. Plastid genes are transcribed by two types of RNA polymerase, plastid-encoded plastid RNA polymerase (PEP) and nuclear-encoded plastid RNA polymerase (NEP). PEP is a multi-subunit enzyme homologous to bacterial RNA polymerase. Although PEP is encoded by the plastid genome, its σ-like subunit (sigma factor) is encoded by the nuclear genome in plants and algae (Liu and Troxler 1996; Tanaka et al. 1996, 1997; Kanamaru et al. 1999; Fujiwara et al. 2000). Sigma factors are required for DNA binding and transcription initiation, providing the nucleus with a mechanism to control the specificity of plastid transcription (Tanaka et al. 1996; Privat et al. 2003; Kanamaru and Tanaka 2004; Loschelder et al. 2006; Schweer et al. 2009; Hanaoka et al. 2012). The Arabidopsis nuclear genome encodes six sigma factors that are thought to regulate the transcription of overlapping sets of chloroplast genes in a partially redundant manner. Circadian oscillations in nuclear-encoded SIGMA FACTOR5 (SIG5) drive circadian oscillations of transcription from a promoter of chloroplast-encoded *psbD*, which encodes the D2 protein of PSII (Noordally et al. 2013). As with peroxiredoxin mutations (Edgar et al. 2012), *sig5* mutants influence the phase of the circadian rhythm of delayed chlorophyll fluorescence (Noordally et al. 2013). This potentially indicates a contribution by SIG5 to the circadian timing of photosynthetic light harvesting and photosynthetic ROS generation.

Nuclear-encoded sigma factors are also involved in a signalling pathway that modulates energy balance between PSII and PSI. Because the pigments associated with PSII and PSI absorb slightly different wavelengths of light, changes in light quality will alter the excitation balance between PSII and PSI (Chow et al. 1990). To acclimate to changes in relative PSII and PSI excitation, plants alter the stoichiometry of PSII and PSI complexes to prevent an electron transfer imbalance (Chow et al. 1990). This is thought to occur through the sensing of plastoquinone redox state by chloroplast-localised CHLOROPLAST SENSOR KINASE (CSK), which alters the stoichiometry of PSII and PSI by indirectly regulating chloroplast gene expression (Puthiyaveetil et al. 2008, 2010). This is hypothesised to occur through changes in the specificity of chloroplast transcription by SIG1, caused by changes in SIG1 phosphorylation state in response to the light conditions (Shimizu et al. 2010). Interestingly, transcripts encoding both CSK and SIG1 are circadian regulated and phased to the subjective morning, and also have rhythms of abundance under light/dark cycles (Harmer et al. 2000; Bläsing et al. 2005; Edwards et al. 2006; Mockler et al. 2007). This raises the possibility that circadian regulation influences the redox balance between PSII and PSI. We speculate that this might establish a chloroplast transcriptional programme that adapts plants to changing light quantity or quality during the course of the day.

Plants also regulate photosynthesis through changes in chloroplast position in response to ambient light conditions. Under excessive light, the chloroplast avoidance response reduces light absorption to minimise photodamage, presumably by limiting accumulation of ROS (Kasahara et al. 2002). As the chloroplast avoidance pathway is primarily induced by the phototropin family of photoreceptors in higher plants (Sakai et al. 2001; Luesse et al. 2010), it is expected that the loss of phototropin activity would increase the accumulation of ROS. In agreement with this hypothesis, loss of *PHOT1A* in rice induces the accumulation of H_2O_2 (Goh et al. 2009).

Loss of the phototropin blue light photoreceptors also reduces the amplitude of circadian oscillations of the operating efficiency of PSII (F_q'/F_m') (Litthauer et al. 2015). The phototropins appear to act redundantly in their contribution to circadian oscillations of F_q'/F_m' because single phototropin mutants are without effect on circadian oscillations of F_q'/F_m', whereas a double mutant reduces the amplitude of the oscillation or abolishes the oscillation, depending on light conditions (Litthauer et al. 2015). Since mutations in phototropins do not alter the functioning of the nuclear circadian oscillator (Litthauer et al. 2015, 2016), one possibility is that the loss of the chloroplast avoidance response in *phot1 phot2* lines induces the generation of ROS that ultimately disrupts the rhythms of F_q'/F_m'.

6 Retrograde Signals Linked to ROS Production and ROS Scavenging

ROS-related signalling contributes to the retrograde signals that are thought to communicate information to the nucleus from organelles such as chloroplasts and mitochondria (Nott et al. 2006). Chloroplasts and mitochondria serve as complex metabolic hubs, which makes them particularly vulnerable to environmental stress. Sub-optimal environmental conditions perturb metabolism in these organelles, causing alterations in metabolite accumulation (e.g. sugars and co-factors) and also increased ROS accumulation (Chan et al. 2016; de Souza et al. 2017). Although there is increasing evidence for mechanisms that move ROS such as H_2O_2 directly from the chloroplast to nucleus (Exposito-Rodriguez et al. 2017), most studies have examined the role of oxidised proteins and fatty acids, as well as other redox-induced intermediates, as part of a signalling cascade initiated by the accumulation of short-lived ROS (Møller and Sweetlove 2010; Chan et al. 2016; de Souza et al. 2017). ROS (specifically $^1O_2$) also cause lipid peroxidation, which results in the production of free fatty acids and derivatives that act as signals in response to abiotic and biotic stress (Mène-Saffrané et al. 2009).

Redox-induced intermediates include β-cyclocitral, 3′-phosphoadenosine 5′-phosphate (PAP) and 2-C-methyl-D-erythritol 2,4-cyclodiphosphate (MEcPP) (reviewed in (Chan et al. 2016; de Souza et al. 2017)). Analyses of these pathways are complicated by the pleiotropic consequences of manipulating these intermediates, with multiple hormone and metabolic pathways perturbed by their accumulation (Ramel et al. 2013; Bjornson et al. 2017; Pornsiriwong et al. 2017). β-cyclocitral accumulates in response to high light stress and has been proposed as an intermediate in a $^1O_2$ retrograde signalling pathway (Ramel et al. 2013; Lv et al. 2015), while PAP accumulates in response to oxidative stress in the chloroplast (Estavillo

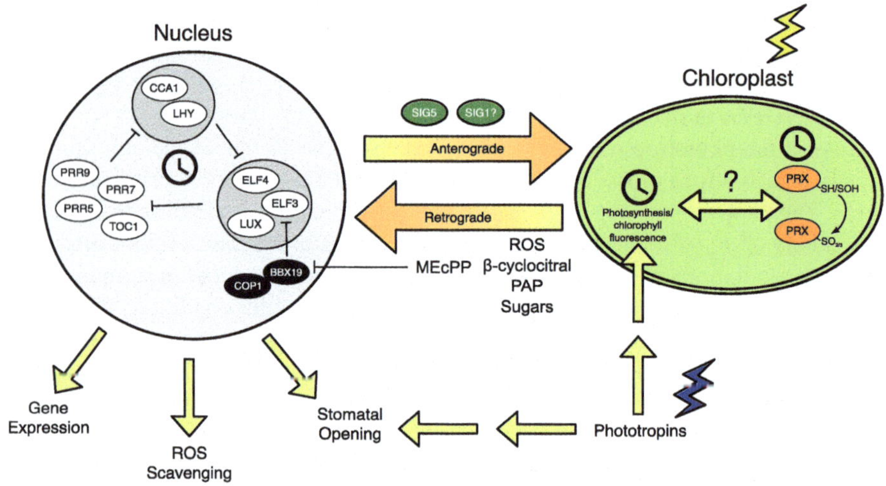

Fig. 2 Interactions between the circadian system and redox signalling. The nuclear-encoded circadian system is well understood and simplified within the light great circle in this diagram (see Hsu and Harmer 2014; Millar 2016 for a more detailed description). Signals derived from the nuclear-encoded oscillator are linked to rhythms of chlorophyll fluorescence and photosynthesis in the chloroplast in part via SIG5. There are circadian rhythms of chloroplast PRX oxidation. Rhythms of photosynthesis (driven by either the circadian system or daily environmental rhythms) induce oscillations of generation of ROS, photosynthetically derived sugars and other metabolites that feed back to influence nuclear gene expression. These retrograde signalling pathways continue to be elucidated, but a role for MEcPP in regulating ELF3 accumulation through BBX19- and COP1-mediated ubiquitination has been identified. These interlinked oscillations between the nucleus and chloroplast have a variety of consequences for gene expression, ROS scavenging and stomatal opening. Phototropins optimise stomatal opening and chloroplast positioning in response to ambient light conditions. Transcriptional repressors are shown in white, with dark grey circles surrounding physically interacting proteins. Black ovals represent components necessary for post-translational regulation of ELF3 accumulation. Proteins that convey timing information from the nuclear-encoded circadian oscillator to chloroplasts are indicated in green. Oxidation states of PRX within chloroplasts are shown in orange. Abbreviations: BBX19, B-BOX19; CCA1, CIRCADIAN CLOCK ASSOCIATED1; COP1, CONSTITUTIVELY PHOTOMORPHOGENIC1; CSK, CHLOROPLAST SENSOR KINASE; ELF, EARLY FLOWERING; LHY, LATE ELONGATED HYPOCOTYL; LUX, LUX ARRHYTHMO; MEcPP, methylerythritol cyclodiphosphate; PAP, 3′-phosphoadenosine 5′-phosphate; PRR, PSEUDO-RESPONSE REGULATOR; PRX, peroxiredoxin; SIGx, SIGMA FACTORx; TOC1, TIMING OF CAB2 EXPRESSION1

et al. 2011). Although each of these signalling pathways induces changes in nuclear transcript accumulation, a direct link with circadian timing components has yet to be demonstrated. In contrast, a role for MEcPP in the modulation of circadian timing is becoming apparent (Fig. 2).

MEcPP is a small metabolite that acts as a precursor of isoprenoids in the plastid methylerythritol phosphate (MEP) pathway (Xiao et al. 2012). MEcPP accumulates in response to a range of stresses, including high light stress, oxidative stress and wounding (to a lesser degree), and an increase in vitro MEcPP correlates with the upregulation of specific nuclear stress-responsive genes (Wang et al. 2015). Elevated levels of endogenous MEcPP accumulate in *constitutively expressing HPL* (*ceh1*) plants, with these having a dwarf phenotype and flowering early (Wang and Dehesh 2015). The elevated levels of MEcPP repress accumulation of *B-BOX DOMAIN PROTEIN 19* (*BBX19*) transcripts, a positive regulator of growth and a negative regulator of flowering time (Wang et al. 2014; Wang and Dehesh 2015). BBX19 promotes hypocotyl growth by recruiting ELF3 for degradation by COP1, preventing repression of *PIF4* and *PIF5* expression by ELF3 (Wang et al. 2015). Although the circadian phenotype of *ceh1* or *BBX19* over-expressing lines has yet to be studied, it is plausible that MEcPP could alter circadian pace through modulation of the evening complex of the circadian oscillator, as well as acting downstream of the circadian oscillator to regulate growth and flowering time (Wang et al. 2014, 2015).

7 Circadian Regulation and ROS Signalling in Stomatal Guard Cells

There are circadian rhythms of stomatal opening that are regulated by the circadian oscillator (Hennessey and Field 1991; Somers et al. 1998; Dodd et al. 2004, 2005b). The circadian oscillator influences the daily patterns of stomatal opening under cycles of light and dark. For example, transpiration rates are greater in CCA1-ox arrhythmic plants compared with the wild type (Dodd et al. 2005a). This suggests that circadian regulation could be an important contributor to water use efficiency. Mechanistic interactions between components of the circadian oscillator and stomatal regulation are emerging (Kinoshita et al. 2011; Legnaioli et al. 2009; Liu et al. 2013; Hassidim et al. 2017), and circadian regulation appears to influence canopy gas exchange (Resco de Dios et al. 2016a, b), yet much important work remains in this area. It is possible that circadian oscillator output pathways interface with known signal transduction pathways that control stomatal aperture, either by regulating stomatal aperture directly or through the circadian gating of the responses of guard cells to the environmental signals that control stomatal aperture (Webb 2003; Dodd et al. 2006; Legnaioli et al. 2009; Joo et al. 2017).

ROS signals have an important role in the regulation of stomatal aperture. ROS signalling has been studied extensively within guard cells, particularly within the context of the regulation of stomatal closure (Song et al. 2014; Singh et al. 2017). For example, the phytohormone abscisic acid (ABA) induces H_2O_2 production in guard cells via NADPH oxidases, and the resultant increase in ROS signals this information to a variety of transcription factors and ion channels, ultimately causing stomatal closure (Chen and Gallie 2004; Song et al. 2014; Singh et al. 2017). The overarching consensus is that guard cells strictly regulate the concentration and types of ROS that are generated, by balancing ROS production and their scavenging, to achieve signalling specificity within the control of stomatal dynamics (Song et al. 2014; Singh et al. 2017). Interestingly, the levels of ROS and its scavengers are regulated in daily opposing manners. In guard cells, H_2O_2 levels peak in the afternoon, while the redox form of the H_2O_2 scavenger ascorbic acid decreases twofold (Chen and Gallie 2004). Transgenic plants containing high levels of the ascorbic acid had a higher rate of transpiration and were less responsive to ABA signalling (Chen and Gallie 2004), which implies that this diel rhythmicity of ROS production may be important for the guard cell signalling networks. This could suggest further crosstalk between ROS and the circadian oscillator, supporting the findings of Lai et al. (2012a). Therefore, since ROS are central players in guard cell signalling, ROS are generated with diel rhythms that affect stomatal aperture, and ROS interact with the circadian oscillator (Lai et al. 2012b), it is possible that the circadian regulation of ROS generation within guard cells might contribute to the circadian regulation of stomatal aperture. It has even been suggested that signals from ROS could drive circadian clocks that are specific to each tissue type, or may possibly synchronise different tissue-specific clocks (Schippers et al. 2013). Despite the recent advances in this area, a role for ROS in the circadian regulation of stomatal aperture remains hypothetical and represents a potentially valuable area for future investigation.

8 Concluding Remarks

There is a growing body of evidence for roles for ROS and redox regulation within both the circadian oscillator and circadian-regulated processes (Fig. 2). Understanding the mechanistic basis for circadian oscillations in peroxiredoxin redox state will be key to identifying novel and conserved features of circadian oscillators. In addition, ROS and redox signalling might reside within the processes underlying the circadian regulation of photosynthesis and circadian regulation of stomatal opening. Given the importance of the circadian system in regulating plant function, it is likely that signalling information derived from ROS generation informs circadian timing, although additional work will be required to understand these signalling pathways.

Acknowledgements We thank Dr Zeenat Noordally and Dr Dora Cano-Ramirez for helpful discussion.

References

Alabadi D, Oyama T, Yanovsky M, Harmon F, Mas P, Kay S (2001) Reciprocal regulation between TOC1 and LHY/CCA1 within the Arabidopsis circadian clock. Science 293:880–883

Allen T, Koustenis A, Theodorou G, Somers DE, Kay SA, Whitelam GC, Devlin PF (2006) Arabidopsis FHY3 specifically gates phytochrome signaling to the circadian clock. Plant Cell 18:2506–2516

Apel K, Hirt H (2004) Reactive Oxygen Species: metabolism, oxidative stress, and signal transduction. Annu Rev Plant Biol 55:373–399

Bieniawska Z, Espinoza C, Schlereth A, Sulpice R, Hincha DK, Hannah MA (2008) Disruption of the Arabidopsis circadian clock is responsible for extensive variation in the cold-responsive transcriptome. Plant Physiol 147:263–279

Bjornson M, Balcke GU, Xiao Y, de Souza A, Wang J-Z, Zhabinskaya D, Tagkopoulos I, Tissier A, Dehesh K (2017) Integrated omics analyses of retrograde signaling mutant delineate interrelated stress-response strata. Plant J 91:70–84

Bläsing OE, Gibon Y, Günther M, Höhne M, Morcuende R, Osuna D, Thimm O, Usadel B, Scheible W-R, Stitt M (2005) Sugars and circadian regulation make major contributions to the global regulation of diurnal gene expression in Arabidopsis. Plant Cell 17:3257

Chan KX, Phua SY, Crisp P, McQuinn R, Pogson BJ (2016) Learning the languages of the chloroplast: Retrograde signaling and beyond. Annu Rev Plant Biol 67:25–53

Chen Z, Gallie DR (2004) The ascorbic acid redox state controls guard cell signaling and stomatal movement. Plant Cell 16:1143

Choudhary MK, Nomura Y, Wang L, Nakagami H, Somers DE (2015) Quantitative circadian phosphoproteomic analysis of Arabidopsis reveals extensive clock control of key components in physiological, metabolic, and signaling pathways. Mol Cell Proteomics 14:2243–2260

Chow WS, Melis A, Anderson JM (1990) Adjustments of photosystem stoichiometry in chloroplasts improve the quantum efficiency of photosynthesis. Proc Natl Acad Sci 87:7502–7506

Covington MF, Maloof JN, Straume M, Kay SA, Harmer SL (2008) Global transcriptome analysis reveals circadian regulation of key pathways in plant growth and development. Genome Biol 9:R130

Cui X, Lu F, Li Y, Xue Y, Kang Y, Zhang S, Qiu Q, Cui X, Zheng S, Liu B, Xu X, Cao X (2013) Ubiquitin-specific proteases UBP12 and UBP13 act in circadian clock and photoperiodic flowering regulation in Arabidopsis. Plant Physiol 162:897–906

Dakhiya Y, Hussien D, Fridman E, Kiflawi M, Green R (2017) Correlations between circadian rhythms and growth in challenging environments. Plant Physiol 173:1724–1734

Daniel X, Sugano S, Tobin E (2004) CK2 phosphorylation of CCA1 is necessary for its circadian oscillator function in Arabidopsis. Proc Natl Acad Sci U S A 101:3292–3297

de Montaigu A, Tóth R, Coupland G (2010) Plant development goes like clockwork. Trends Genet 26:296–306

de Souza A, Wang J-Z, Dehesh K (2017) Retrograde signals: integrators of interorganellar communication and orchestrators of plant development. Annu Rev Plant Biol 68:85–108

Dietz K-J (2011) Peroxiredoxins in plants and cyanobacteria. Antioxid Redox Signal 15:1129–1159

Dodd AN, Belbin FE, Frank A, Webb AAR (2015) Interactions between circadian clocks and photosynthesis for the temporal and spatial coordination of metabolism. Front Plant Sci 6:245

Dodd AN, Jakobsen MK, Baker AJ, Telzerow A, Hou S-W, Laplaze L, Barrot L, Scott Poethig R, Haseloff J, Webb AAR (2006) Time of day modulates low-temperature Ca2 + signals in Arabidopsis. Plant J 48:962–973

Dodd AN, Parkinson K, Webb AAR (2004) Independent circadian regulation of assimilation and stomatal conductance in the *ztl*-1 mutant of Arabidopsis. New Phytol 162:63–70

Dodd AN, Salathia N, Hall A, Kevei E, Toth R, Nagy F, Hibberd JM, Millar AJ, Webb AA (2005a) Plant circadian clocks increase photosynthesis, growth, survival, and competitive advantage. Science 309(5734):630–633

Dodd AN, Salathia N, Hall A, Kevei E, Tóth R, Nagy F, Hibberd JM, Millar AJ, Webb AAR (2005b) Plant circadian clocks increase photosynthesis, growth, survival, and competitive advantage. Science 309:630–633

Edgar RS, Green EW, Zhao Y, van Ooijen G, Olmedo M, Qin X, Xu Y, Pan M, Valekunja UK, Feeney KA, Maywood ES, Hastings MH, Baliga NS, Merrow M, Millar AJ, Johnson CH, Kyriacou CP, O'Neill JS, Reddy AB (2012) Peroxiredoxins are conserved markers of circadian rhythms. Nature 485:459–464

Edwards KD, Anderson PE, Hall A, Salathia NS, Locke JC, Lynn JR, Straume M, Smith JQ, Millar AJ (2006) FLOWERING LOCUS C mediates natural variation in the high-temperature response of the Arabidopsis circadian clock. Plant Cell 18(3):639–650

Estavillo GM, Crisp PA, Pornsiriwong W, Wirtz M, Collinge D, Carrie C, Giraud E, Whelan J, David P, Javot H, Brearley C, Hell R, Marin E, Pogson BJ (2011) Evidence for a SAL1-PAP chloroplast retrograde pathway that functions in drought and high light signaling in Arabidopsis. Plant Cell 23:3992–4012

Exposito-Rodriguez M, Laissue PP, Yvon-Durocher G, Smirnoff N, Mullineaux PM (2017) Photosynthesis-dependent H_2O_2 transfer from chloroplasts to nuclei provides a high-light signalling mechanism. Nat Commun 8:49

Fowler SG, Cook D, Thomashow MF (2005) Low temperature induction of Arabidopsis CBF1, 2, and 3 is gated by the circadian clock. Plant Physiol 137:961–968

Fryer MJ, Ball L, Oxborough K, Karpinski S, Mullineaux PM, Baker NR (2003) Control of ascorbate peroxidase 2 expression by hydrogen peroxide and leaf water status during excess light stress reveals a functional organisation of Arabidopsis leaves. Plant J 33:691–705

Fujiwara M, Nagashima A, Kanamaru K, Tanaka K, Takahashi H (2000) Three new nuclear genes, *sigD*, *sigE* and *sigF*, encoding putative plastid RNA polymerase σ factors in Arabidopsis thaliana. FEBS Lett 481:47–52

Fujiwara S, Wang L, Han L, Suh S, Salome P, McClung C, Somers D (2008) Post-translational regulation of the Arabidopsis circadian clock through selective proteolysis and phosphorylation of pseudo-response regulator proteins. J Biol Chem 283:23073–23083

Gendron JM, Pruneda-Paz JL, Doherty CJ, Gross AM, Kang SE, Kay SA (2012) Arabidopsis circadian clock protein, TOC1, is a DNA-binding transcription factor. Proc Natl Acad Sci USA 109:3167–3172

Goh C-H, Jang S, Jung S, Kim H-S, Kang H-G, Y-i Park, Bae H-J, Lee C-H, An G (2009) Rice *phot1a* mutation reduces plant growth by affecting photosynthetic responses to light during early seedling growth. Plant Mol Biol 69:605–619

Goodspeed D, Chehab EW, Min-Venditti A, Braam J, Covington MF (2012) Arabidopsis synchronizes jasmonate-mediated defense with insect circadian behavior. PNAS 109:4674–4677

Gould PD, Diaz P, Hogben C, Kusakina J, Salem R, Hartwell J, Hall A (2009) Delayed fluorescence as a universal tool for the measurement of circadian rhythms in higher plants. Plant J 58:893–901

Gould PD, Locke JCW, Larue C, Southern MM, Davis SJ, Hanano S, Moyle R, Milich R, Putterill J, Millar AJ, Hall A (2006) The molecular basis of temperature compensation in the Arabidopsis circadian clock. Plant Cell 18:1177–1187

Green RM, Tingay S, Wang Z-Y, Tobin EM (2002) Circadian rhythms confer a higher level of fitness to Arabidopsis plants. Plant Physiol 129:576–584

Greenham K, McClung CR (2015) Integrating circadian dynamics with physiological processes in plants. Nat Rev Genet 16:598–610

Hanaoka M, Kato M, Anma M, Tanaka K (2012) SIG1, a sigma factor for the chloroplast RNA polymerase, differently associates with multiple DNA regions in the chloroplast chromosomes *in vivo*. Int J Mol Sci 13:12182–12194

Harmer SL, Hogenesch JB, Straume M, Chang HS, Han B, Zhu T, Wang X, Kreps JA, Kay SA (2000) Orchestrated transcription of key pathways in Arabidopsis by the circadian clock. Science 290(5499):2110–2113

Hassidim M, Dakhiya Y, Turjeman A, Hussien D, Shor E, Anidjar A, Goldberg K, Green RM (2017) CIRCADIAN CLOCK ASSOCIATED 1 (CCA1) and the circadian control of stomatal aperture. Plant Phys

Hennessey TL, Field CB (1991) Circadian rhythms in photosynthesis: oscillations in carbon assimilation and stomatal conductance under constant conditions. Plant Physiol 96:831–836

Hsu PY, Devisetty UK, Harmer SL (2013) Accurate timekeeping is controlled by a cycling activator in Arabidopsis. Elife 2:e00473

Hsu PY, Harmer SL (2014) Wheels within wheels: the plant circadian system. Trends Plant Sci 19:240–249

Huang W, Perez-Garcia P, Pokhilko A, Millar AJ, Antoshechkin I, Riechmann JL, Mas P (2012) Mapping the core of the Arabidopsis circadian clock defines the network structure of the oscillator. Science 336:75–79

Hut RA, Beersma DGM (2011) Evolution of time-keeping mechanisms: early emergence and adaptation to photoperiod. Phil Trans R Soc Lond B Biol Sci 366:2141–2154

Jones M (2009) Entrainment of the Arabidopsis circadian clock. J Plant Biol 52:202–209

Jones M (2017) Interplay of circadian rhythms and light in the regulation of photosynthesis-derived metabolism. vol 79. https://link.springer.com/chapter/10.1007/124_2017_2. In press

Joo Y, Fragoso V, Yon F, Baldwin IT, Kim S-G (2017) Circadian clock component, LHY, tells a plant when to respond photosynthetically to light in nature. J Integr Plant Biol 59:572–587

Kanamaru K, Fujiwara M, Seki M, Katagiri T, Nakamura M, Mochizuki N, Nagatani A, Shinozaki K, Tanaka K, Takahashi H (1999) Plastidic RNA polymerase sigma factors in Arabidopsis. Plant Cell Physiol 40:832–842

Kanamaru K, Tanaka K (2004) Roles of chloroplast rna polymerase sigma factors in chloroplast development and stress response in higher plants. Biosci Biotechnol Biochem 68:2215–2223

Kasahara M, Kagawa T, Oikawa K, Suetsugu N, Miyao M, Wada M (2002) Chloroplast avoidance movement reduces photodamage in plants. Nature 420(6917):829

Kim J-Y, Song H-R, Taylor BL, Carré IA (2003) Light-regulated translation mediates gated induction of the Arabidopsis clock protein LHY. The EMBO journal 22:935–944

Kinoshita T, Ono N, Hayashi Y, Morimoto S, Nakamura S, Soda M, Kato Y, Ohnishi M, Nakano T, Inoue S-I, Shimazaki K-I (2011) FLOWERING LOCUS T regulates stomatal opening. Current Biology 21(14):1232–1238

Korneli C, Danisman S, Staiger D (2014) Differential control of pre-invasive and post-invasive antibacterial defense by the Arabidopsis circadian clock. Plant Cell Physiol 55:1613–1622

Kotchoni SO, Larrimore KE, Mukherjee M, Kempinski CF, Barth C (2009) Alterations in the endogenous ascorbic acid content affect flowering time in Arabidopsis. Plant Physiol 149:803

Kurepa J, Smalle J, Va M, Montagu N, Inzé D (1998) Oxidative stress tolerance and longevity in Arabidopsis: the late-flowering mutant gigantea is tolerant to paraquat. Plant J 14:759–764

Lai AG, Doherty CJ, Mueller-Roeber B, Kay SA, Schippers JHM, Dijkwel PP (2012a) CIRCADIAN CLOCK-ASSOCIATED 1 regulates ROS homeostasis and oxidative stress responses. Proc Natl Acad Sci 109:17129–17134

Lai AG, Doherty CJ, Mueller-Roeber B, Kay SA, Schippers JHM, Dijkwel PP (2012b) CIRCADIAN CLOCK-ASSOCIATED 1 regulates ROS homeostasis and oxidative stress responses. Proc Natl Acad Sci USA 109:17129–17134

Le Bihan T, Martin SF, Chirnside ES, van Ooijen G, Barrios-Llerena ME, O'Neill JS, Shliaha PV, Kerr LE, Millar AJ (2011) Shotgun proteomic analysis of the unicellular alga Ostreococcus tauri. J Proteomics 74:2060–2070

Legnaioli T, Cuevas J, Mas P (2009) TOC1 functions as a molecular switch connecting the circadian clock with plant responses to drought. EMBO J 28:3745–3757

Litthauer S, Battle MW, Jones MA (2016) Phototropins do not alter accumulation of evening-phased circadian transcripts under blue light. Plant Signal Beh 11:e1126029

Litthauer S, Battle MW, Lawson T, Jones MA (2015) Phototropins maintain robust circadian oscillation of PSII operating efficiency under blue light. Plant J 83:1034–1045

Liu T, Carlsson J, Takeuchi T, Newton L, Farré EM (2013) Direct regulation of abiotic responses by the Arabidopsis circadian clock component PRR7. Plant J 76:101–114

Liu B, Troxler RF (1996) Molecular characterization of a positively photoregulated nuclear gene for a chloroplast RNA polymerase sigma factor in *Cyanidium caldarium*. Proc Natl Acad Sci USA 93:3313–3318

Loschelder H, Schweer J, Link B, Link G (2006) Dual temporal role of plastid sigma factor 6 in Arabidopsis development. Plant Physiol 142:642

Luesse DR, DeBlasio SL, Hangarter RP (2010) Integration of phot1, phot2, and PhyB signalling in light-induced chloroplast movements. J Exp Bot 61:4387–4397

Lv F, Zhou J, Zeng L, Xing D (2015) β-cyclocitral upregulates salicylic acid signalling to enhance excess light acclimation in Arabidopsis. J Exp Bot 66:4719–4732

Matsushika A, Makino S, Kojima M, Mizuno T (2000) Circadian waves of expression of the APRR1/TOC1 family of pseudo-response regulators in *Arabidopsis thaliana*: insight into the plant circadian clock. Plant Cell Physiol 41:1002–1012
McClung CR (2006) Plant circadian rhythms. Plant Cell 18:792–803
Mène-Saffrané L, Dubugnon L, Chételat A, Stolz S, Gouhier-Darimont C, Farmer EE (2009) Nonenzymatic oxidation of trienoic fatty acids contributes to reactive oxygen species management in Arabidopsis. J Biol Chem 284:1702–1708
Millar AJ (2016) The intracellular dynamics of circadian clocks reach for the light of ecology and evolution. Annu Rev Plant Biol 67:595–618
Mockler TC, Michael TP, Priest HD, Shen R, Sullivan CM, Givan SA, McEntee C, Kay S, Chory J (2007) THE DIURNAL PROJECT: Diurnal and circadian expression profiling, model-based pattern matching and promoter analysis. Cold Spring Harb Symp Quant Biol 72:353–363
Møller IM, Sweetlove LJ (2010) ROS signalling-specificity is required. Trends Plant Sci 15:370–374
Nakamichi N, Fukushima A, Kusano M, Sakakibara H, Mizuno T, Saito K (2009) Linkage between circadian clock and tricarboxylic acid cycle in Arabidopsis. Plant Signal Beh 4:660–662
Nakamichi N, Kiba T, Henriques R, Mizuno T, Chua N-H, Sakakibara H (2010) PSEUDO-RESPONSE REGULATORS 9, 7, and 5 are transcriptional repressors in the Arabidopsis circadian clock. Plant Cell 22:594–605
Noordally ZB, Ishii K, Atkins KA, Wetherill SJ, Kusakina J, Walton EJ, Kato M, Azuma M, Tanaka K, Hanaoka M, Dodd AN (2013) Circadian control of chloroplast transcription by a nuclear-encoded timing signal. Science 339:1316–1319
Nott A, Jung HS, Koussevitzky S, Chory J (2006) Plastid-to-nucleus retrograde signaling. Annu Rev Plant Biol 57:739–759
Nusinow DA, Helfer A, Hamilton EE, King JJ, Imaizumi T, Schultz TF, Farre EM, Kay SA (2011a) The ELF4-ELF3-LUX complex links the circadian clock to diurnal control of hypocotyl growth. Nature 475:398–402
Nusinow DA, Helfer A, Hamilton EE, King JJ, Imaizumi T, Schultz TF, Farré EM, Kay SA (2011b) The ELF4-ELF3-LUX complex links the circadian clock to diurnal control of hypocotyl growth. Nature 475:398–402
O'Neill JS, Reddy AB (2011) Circadian clocks in human red blood cells. Nature 469:498–503
O'Neill JS, van Ooijen G, Dixon LE, Troein C, Corellou F, Bouget F-Y, Reddy AB, Millar AJ (2011) Circadian rhythms persist without transcription in a eukaryote. Nature 469:554–558
Pornsiriwong W, Estavillo GM, Chan KX, Tee EE, Ganguly D, Crisp PA, Phua SY, Zhao C, Qiu J, Park J, Yong MT, Nisar N, Yadav AK, Schwessinger B, Rathjen J, Cazzonelli CI, Wilson PB, Gilliham M, Chen Z-H, Pogson BJ (2017) A chloroplast retrograde signal, 3′-phosphoadenosine 5′-phosphate, acts as a secondary messenger in abscisic acid signaling in stomatal closure and germination. Elife 6:e23361
Privat I, Hakami MA, Buhot L, Favory JJ, Lerbs-Mache S (2003) Characterization of Arabidopsis plastid sigma-like transcription factors SIG1, SIG2 and SIG3. Plant Mol Biol 51:385–399
Puthiyaveetil S, Ibrahim IM, Jeličić B, Tomašić A, Fulgosi H, Allen JF (2010) Transcriptional control of photosynthesis genes: the evolutionarily conserved regulatory mechanism in plastid genome function. Genome Biology Evol 2:888–896
Puthiyaveetil S, Kavanagh TA, Cain P, Sullivan JA, Newell CA, Gray JC, Robinson C, van der Giezen M, Rogers MB, Allen JF (2008) The ancestral symbiont sensor kinase CSK links photosynthesis with gene expression in chloroplasts. Proc Natl Acad Sci 105:10061–10066
Ramel F, Ksas B, Akkari E, Mialoundama AS, Monnet F, Krieger-Liszkay A, Ravanat J-L, Mueller MJ, Bouvier F, Havaux M (2013) Light-induced acclimation of the Arabidopsis chlorina1 mutant to singlet oxygen. Plant Cell 25:1445–1462
Resco de Dios V, Gessler A, Ferrio JP, Alday JG, Bahn M, del Castillo J, Devidal S, García-Muñoz S, Kayler Z, Landais D, Martín-Gómez P, Milcu A, Piel C, Pirhofer-Walzl K, Ravel O, Salekin S, Tissue DT, Tjoelker MG, Voltas J, Roy J (2016a) Circadian rhythms have significant effects on leaf-to-canopy scale gas exchange under field conditions. GigaScience 5:43

Resco de Dios V, Loik ME, Smith R, Aspinwall MJ, Tissue DT (2016b) Genetic variation in circadian regulation of nocturnal stomatal conductance enhances carbon assimilation and growth. Plant, Cell Environ 39:3–11
Rugnone ML, Faigón Soverna A, Sanchez SE, Schlaen RG, Hernando CE, Seymour DK, Mancini E, Chernomoretz A, Weigel D, Más P, Yanovsky MJ (2013) LNK genes integrate light and clock signaling networks at the core of the Arabidopsis oscillator. Proc Natl Acad Sci USA 110:12120–12125
Sakai T, Kagawa T, Kasahara M, Swartz TE, Christie JM, Briggs WR, Wada M, Okada K (2001) Arabidopsis *nph1* and *npl1*: Blue light receptors that mediate both phototropism and chloroplast relocation. Proc Natl Acad Sci 98:6969–6974
Salter MG, Franklin KA, Whitelam GC (2003) Gating of the rapid shade-avoidance response by the circadian clock in plants. Nature 426:680
Schippers JHM, Lai AG, Mueller-Roeber B, Dijkwel PP (2013) Could ROS signals drive tissue-specific clocks? Transcription 4:206–208
Schweer J, Geimer S, Meurer J, Link G (2009) Arabidopsis mutants carrying chimeric sigma factor genes reveal regulatory determinants for plastid gene expression. Plant Cell Physiol 50:1382–1386
Seo PJ, Mas P (2014) Multiple layers of posttranslational regulation refine circadian clock activity in Arabidopsis. Plant Cell 26:79
Shimizu M, Kato H, Ogawa T, Kurachi A, Nakagawa Y, Kobayashi H (2010) Sigma factor phosphorylation in the photosynthetic control of photosystem stoichiometry. Proc Natl Acad Sci 107:10760–10764
Singh R, Parihar P, Singh S, Mishra RK, Singh VP, Prasad SM (2017) Reactive oxygen species signaling and stomatal movement: current updates and future perspectives. Redox Biology 11:213–218
Somers DE, Webb AA, Pearson M, Kay SA (1998) The short-period mutant, *toc1*-1, alters circadian clock regulation of multiple outputs throughout development in Arabidopsis thaliana. Development 125:485
Song Y, Miao Y, Song C-P (2014) Behind the scenes: the roles of reactive oxygen species in guard cells. New Phytol 201:1121–1140
Spoel SH, van Ooijen G (2014) Circadian redox signaling in plant immunity and abiotic stress. Antioxid Redox Signal 20:3024–3039
Tanaka K, Oikawa K, Ohta N, Kuroiwa H, Kuroiwa T, Takahashi H (1996) Nuclear encoding of a chloroplast RNA polymerase sigma subunit in a red alga. Science 272:1932
Tanaka K, Tozawa Y, Mochizuki N, Shinozaki K, Nagatani A, Wakasa K, Takahashi H (1997) Characterization of three cDNA species encoding plastid RNA polymerase sigma factors in *Arabidopsis thaliana*: evidence for the sigma factor heterogeneity in higher plant plastids. FEBS Lett 413:309–313
van Ooijen G, Millar AJ (2012) Non-transcriptional oscillators in circadian timekeeping. Trends Biochem Sci 37:484–492
Wang C, Dehesh K (2015) From retrograde signaling to flowering time. Plant Signal Beh 10:e1022012
Wang C-Q, Guthrie C, Sarmast MK, Dehesh K (2014) BBX19 interacts with CONSTANS to repress FLOWERING LOCUS T transcription, defining a flowering time checkpoint in Arabidopsis. Plant Cell 26:3589–3602
Wang ZY, Kenigsbuch D, Sun L, Harel E, Ong MS, Tobin EM (1997) A Myb-related transcription factor is involved in the phytochrome regulation of an Arabidopsis Lhcb gene. Plant Cell 9:491–507
Wang C-Q, Sarmast MK, Jiang J, Dehesh K (2015) The Transcriptional regulator BBX19 promotes hypocotyl growth by facilitating COP1-mediated EARLY FLOWERING3 degradation in Arabidopsis. Plant Cell 27:1128–1139
Webb AAR (2003) The physiology of circadian rhythms in plants. New Phytol 160:281–303

Xiao Y, Savchenko T, Baidoo EEK, Chehab WE, Hayden DM, Tolstikov V, Corwin JA, Kliebenstein DJ, Keasling JD, Dehesh K (2012) Retrograde signaling by the plastidial metabolite MEcPP regulates expression of nuclear stress-response genes. Cell 149:1525–1535
Xie Q, Wang P, Liu X, Yuan L, Wang L, Zhang C, Li Y, Xing H, Zhi L, Yue Z, Zhao C, McClung CR, Xu X (2014) LNK1 and LNK2 are transcriptional coactivators in the Arabidopsis circadian oscillator. Plant Cell 26:2843–2857
Yu J, Rubio V, Lee N, Bai S, Lee S, Kim S, Liu L, Zhang Y, Irigoyen M, Sullivan J, Lee I, Xie Q, Paek N, Deng X (2008) COP1 and ELF3 control circadian function and photoperiodic flowering by regulating GI stability. Mol Cell 32:617–630

Ascorbate Peroxidases: Scavengers or Sensors of Hydrogen Peroxide Signaling?

Andréia Caverzan, Douglas Jardim-Messeder, Ana Luiza Paiva and Marcia Margis-Pinheiro

Abstract Environmental factors can trigger the accumulation of reactive oxygen species (ROS) in plant cells. These molecules at a basal level are associated with essential cellular functions such as signaling for both developmental and defense responses in plants. On the other hand, ROS at high concentrations are significantly detrimental, leading to oxidative damage in different cell biomolecules. Plants have efficient antioxidant systems to neutralize toxic levels of ROS, which include various enzymatic and non-enzymatic components. Ascorbate peroxidase (APX) is a key enzyme in this context, acting in the control of toxic ROS levels in different subcellular compartments. In several species, APX expression is modulated in some developmental stages, and under biotic and abiotic stresses, indicating the importance of APX activity in controlling hydrogen peroxide (H_2O_2) content in intracellular compartments. The genetic manipulation of *APX* gene expression in diverse plant models has been found to trigger differential responses to stress conditions and affects the growth and development of the plant, indicating that these enzymes can play a role as H_2O_2 scavengers and also as sensors of redox alteration inside plant cells.

Andréia Caverzan, Douglas Jardim-Messeder and Ana Luiza Paiva—These authors contributed equally.

A. Caverzan
Faculdade de Agronomia e Medicina Veterinária, Programa de Pós-Graduação em Agronomia, Universidade de Passo Fundo, Passo Fundo, RS 99052-900, Brazil
e-mail: acaverzan@hotmail.com

D. Jardim-Messeder
Departamento de Genética, Universidade Federal do Rio de Janeiro, Rio de Janeiro, RJ 21941-590, Brazil
e-mail: douglas.messeder@gmail.com

A. L. Paiva · M. Margis-Pinheiro (✉)
Departamento de Genética, Universidade Federal do Rio Grande do Sul, Avenida Bento Gonçalves, 9500, Porto Alegre, RS 91501-970, Brazil
e-mail: marcia.margis@ufrgs.br; marcia.p.margis@gmail.com

A. L. Paiva
e-mail: ana_lulux@hotmail.com

S. K. Panda and Y. Y. Yamamoto (eds.), *Redox Homeostasis in Plants*, Signaling and Communication in Plants, https://doi.org/10.1007/978-3-319-95315-1_5

1 Introduction

Due to their sessile lifestyle, plants are susceptible to a significant variety of environmental stimuli, including biotic and abiotic stresses. These stimuli can affect whole plant development and physiology. At cellular levels, the primary response to stress is the production of reactive oxygen species (ROS), such as singlet oxygen ($^1O_2$), superoxide anion (O_2^-), hydrogen peroxide (H_2O_2), and hydroxyl radical (OH).

At the cellular level, ROS act as signaling molecules or secondary messenger participating in stress perception, gene expression regulation, and multiple stress response pathways. Many researchers have focused on H_2O_2 generation as a signaling molecule implicated in the regulation of a variety of biological processes. Thereby, evidence to support the function of ROS as signaling molecules and their physiological significance in plant stress responses has accumulated in recent decades. There are many site- and kind-specific pathways to ROS production that allows plant stress response flexibility and efficient completion of various developmental events (Mittler 2017).

Increase in H_2O_2 levels can be rapid and transient and are thought to constitute a general perception and signaling of cellular stress. H_2O_2 levels increase in plant cells after exposure to various environmental conditions, performing a central role in plant stress response. In fact, plants treated with low concentrations of H_2O_2 can develop differential stress responses, such as an increased tolerance to chilling (Wang et al. 2017) and oxidative stress induced by chilling (Prasad et al. 1994).

ROS operating as signaling molecules are able to modulate not only stress responses, but also plant growth and development, and the specificities of site and kind of ROS production pathways have been intensely studied to understand the ROS signaling process (Gadjev et al. 2006; Vaahtera et al. 2014; Shigeoka and Maruta 2014). Although these pathways remain largely unclear, the integration and cross talk between multiple pathways are associated with a finely tuned stress response in plants. Therefore, the antioxidant networks located in different cellular compartments may play a central role in the integration and cross talk processes through the spatiotemporal regulation of ROS signal availability in the cell.

Due to their high reactivity, ROS are able to cause oxidative damage in different cell biomolecules such as proteins, lipids, and nucleic acid, which are all able to impair cell metabolism. Since ROS can be cytotoxic molecules, the development of efficient antioxidant system(s) during evolution must have been essential for plant survival under changes in environmental conditions. Thus, oxidative stress occurs when ROS generation exceeds the pace of these antioxidant systems (Scandalios 2005; Sharma et al. 2012).

The list of ROS-scavenging enzymes includes superoxide dismutase (SOD; EC 1.15.1.1), catalases (CAT; EC 1.11.1.6), ascorbate peroxidases (APX; EC 1.11.1.11), glutathione/thioredoxin peroxidases (GPX; EC 1.11.1.9), guaiacol peroxidases (POX; EC 1.11.1.7), various types of peroxiredoxins (PRX; EC 1.11.1.15), and glutathione-S-transferases (GST; EC 2.5.1.18). Despite preventing stress-induced ROS, these antioxidant enzymes maintain the cellular concentration of ROS to a level necessary for normal plant growth and development (Asada 1999; Mittler 2002; Mittler et al. 2004).

In photosynthetic organisms, APX is the major component of enzymatic antioxidant pathways. APX can occur in chloroplasts, cytosol, peroxisomes, and mitochondria and efficiently eliminate even very low levels of H_2O_2 using ascorbate (AsA) as the electron donor. Various studies have demonstrated that APX expression is highly modulated in some developmental stages, stress conditions, and pathogen attack, indicating the importance of APX activity in the control of H_2O_2 levels, and in intracellular signaling processes (Zhang et al. 1997; Yoshimura et al. 2000; Sato et al. 2001; Agrawal et al. 2003; Fryer et al. 2003; Menezes-Benavente et al. 2004; Teixeira et al. 2006; Rosa et al. 2010; Gill and Tuteja 2010; Bonifacio et al. 2011; Caverzan et al. 2012, 2014; Wang et al. 2015). In addition, the genetic manipulation of APX gene expression affects plant development and alters plant responses to stress conditions (Teixeira et al. 2006; Rosa et al. 2010; Bonifacio et al. 2011, 2016; Caverzan et al. 2012; Jiang et al. 2016a; Ribeiro et al. 2017). These studies have provided insights into the role of APX as signaling modulators through the spatiotemporal regulation of ROS levels.

In this chapter, the focus will be on molecular and physiological findings regarding APX in photosynthetic organisms such as higher plants and eukaryotic and prokaryotic algal cells. In fact, the elucidation of the regulation mechanism of APX in response to development and environmental stress is a subject of great interest and can help clarify the signaling process and oxidative stress responses in plants.

2 Enzymatic Properties of APX Isoenzymes

Ascorbate peroxidases are found in plastid-containing organisms, such as prokaryotic algae and most eukaryotes including green algae and higher plants (Groden and Beck 1979; Kelly and Latzko 1979; Nakano and Asada 1981; De Leonardis et al. 2000; Battistuzzi et al. 2001; Sharma and Dubey 2004; Yadav et al. 2014). APX has also been found in other organisms, such as the protozoan *Trypanosoma cruzi* and the bovine eye (Boveris et al. 1980; Wada et al. 1998). These findings suggest that APX also contributes to the detoxification of ROS in non-photosynthetic organisms.

Ascorbate (AsA)-dependent peroxidase was first reported in 1979 in pea leaves (Kelly and Latzko 1979) and in lamellae isolated from spinach chloroplasts (Groden and Beck 1979). It has since been purified and characterized in several other plant species including tea, pumpkin, cotton, cucumber, tobacco, rice, olive, and berry (Chen and Asada 1989; Yamaguchi et al. 1995a, b; Bunkelmann and Trelease 1996; Battistuzzi et al. 2001; Madhusudhan et al. 2003; Sharma and Dubey 2004; López-Huertas and del Río 2014; Yadav et al. 2014).

Plant APXs are found in several cellular compartments including cytosol, chloroplasts, mitochondria, and microbodies such as peroxisomes. In plant cells, cytosolic (cAPX), chloroplastic (chlAPX), mitochondrial (mitAPX), and microbody (mAPX) isoforms of APX have been identified, and all these function as H_2O_2 scavengers, generated continuously in cells, allowing ROS produced in organelles to be efficiently scavenged by the organelles themselves (Miyake and Asada 1996). These different APX isoforms have distinct properties. For example, while cAPX is a dimer consisting of identical subunits, the chlAPX isoenzymes exist in a monomeric form (Mittler and Zilinskas 1991; Miyake et al. 1993).

APX isoenzymes are unstable in the absence of AsA. Concentrations of AsA lower than 20 μM result in rapid loss of APX activity. It is estimated that the half-inactivation time of chlAPX and mitAPX is less than 30 s. On the other hand, the half-inactivation time of cAPX and mAPX is approximately 1 h or more (Chen and Asada 1989; Miyake et al. 1993; Ishikawa et al. 1996; Yoshimura et al. 1998; De Leonardis et al. 2000). The instability of APX isoenzymes seems to be one of the reasons for the difficulty in obtaining large amounts of highly purified APX isoenzymes. This highly fragile nature of APXs could limit the cellular antioxidant defense, affecting the capacity of plants to tolerate stress.

APX isoenzymes have high specificity for AsA as the electron donor, which is particularly the case for the chlAPX and mitAPX isoenzymes (Yoshimura et al. 1998; Asada 1999; De Leonardis et al. 2000). In addition to oxidizing AsA, cAPX, mAPX of higher plants and algal APX can also oxidize artificial electron donors such as pyrogallol or guaiacol at appreciable rates (Chen and Asada 1989; Ishikawa et al. 1996; Yoshimura et al. 1998; Asada 1999).

APX is a heme-containing enzyme, the prosthetic group of which is a protoporphyrin. In fact, cyanide and azide-mediated inhibition in the activities of all APX isoforms indicate the heme peroxidase nature of APX (Mittler and Zilinskas 1991). Thus, iron plays an important role in the catalytic site of APX and, in fact, iron deficiency results in lowering cAPX activity by half (Zaharieva and Abadía 2003). In addition, the inhibition of APX by thiol-modifying reagents such as Ellman's reagent (5,5′-dithiobis-(2-nitrobenzoic acid; DTNB) and *p*-chloromercuribenzoate indicates that the thiol group participates at the enzyme's active center (Mandelman et al. 1998a, b). Suicide inhibitors such as hydroxylamine, *p*-aminophenol, and hydroxyurea also inhibit APX (Chen and Asada 1989).

3 Appearance of Different APX Isoenzymes During Plant Evolution

APXs are heme peroxidases and members of Class I non-animal peroxidases, which also include cytochrome C peroxidases (CCPs) and bacterial catalase peroxidases (CPs) (Welinder 1992; Passardi et al. 2007). CCPs are present in photosynthetic and non-photosynthetic eukaryotes, whereas APXs are only found in plastid-containing organisms, with some exceptions (Teixeira et al. 2004; Nedelcu et al. 2008). While cyanobacteria have no APX gene, most eukaryotic algae possess more than one APX gene (Passardi et al. 2007).

Analysis using a large-scale phylogenetic tree of the whole non-animal peroxidase superfamily suggests that monofunctional APXs are evolutionary descendants of atypical APX–CCP hybrid peroxidases, which are found in non-photosynthetic kinetoplastids, such as Trypanosoma and Leishmania (Zamocky et al. 2014). On the other hand, recent work suggests that a catalase-peroxidase gene (katG) present in an ancestral bacterium was transmitted to eukaryotic cells through the first endosymbiosis event, which allowed the emergence of mitochondria and originates the CCP gene. Posteriorly in the second endosymbiosis event, where an ancient eukaryote engulfed a Cyanobacterium, giving rise to chloroplast, the cyanobacteria KatG gene was transferred to the ancient eukaryotic genome and diverged into an ancestral APX gene (Lazzarotto et al. 2015). Thus, this latter gene then likely diverged into the current APX, and consequently, plants' APX evolved in the absence of CCP genes, while APX sequences in non-photosynthetic kinetoplastids may have hybridized with the CCP gene.

Phylogenetic analyses showed that the diversification of cAPX and chlAPX was the first step in the evolution of APX. These groups resulted from duplication events in the basal *Viridiplantae,* and it has been proposed that the organelle isoforms probably originated from cytosolic isoforms that gained an exon encoding a transmembrane domain (Shigeoka et al. 2002; Teixeira et al. 2004; Passardi et al. 2007).

The chlAPX can be divided into two subclasses: one located in the thylakoid membrane (tAPX) and other one soluble in stroma (sAPX), and it is recognized that the appearance of tAPX may have been important for the transition of charophytes to the first land plants in order to cope with the harsh conditions of terrestrial environments, such as stresses associated with drought and high light conditions (Maruta et al. 2012). Among the APXs in the chlAPX group, many isoforms are targeted solely to mitochondria or dual targeted to chloroplasts and mitochondria (Xu et al. 2013). In fact, there is a high similarity of chloroplastic and mitochondrial APXs; consequently, in the literature the chlAPX group includes not only chloroplastic APXs, but also mitochondrial and dual-targeted enzymes. Thus, the evolution of mitochondrial APXs is not well established. Finally, a more recent duplication event in the higher plants may have originated the mAPX, and the presence of mAPX isoforms only in land plants reinforces this idea (Shigeoka et al. 2002, Passardi et al. 2007).

The differences in subcellular localization and biochemical properties of antioxidant enzymes and the distinct responses in gene expression, in addition to the presence of non-enzymatic mechanisms, result in a versatile and flexible antioxidant system able to control the optimum ROS levels. *Arabidopsis thaliana* has been reported to have eight APX isozymes (soluble cytosolic APXs: *AtAPX1*, *AtAPX2*, *AtAPX6*; peroxisome membrane-bound APXs: *AtAPX3*, *AtAPX4*, *AtAPX5*; and chloroplastic *AtAPXs*: *sAPX* and *tAPX*) (Jespersen et al. 1997; Panchuk et al. 2002). Recently, it was demonstrated that *AtAPX6* and *AtAPX4* do not encode classical APXs, but different classes of enzymes called APX-R and APX-L, respectively, both target chloroplasts, which are present in different plant species (Lazzarotto et al. 2015). In the rice genome, eight *APX* genes have been found: two cytosolic (*OsAPX1* and *OsAPX2*), two peroxisomal isoforms (*OsAPX3* and *OsAPX4*), one mitochondrial

Table 1 Distribution of APX isoforms in distinct subcellular compartments in different species

Ascorbate peroxidase						
Species	cAPX	chlAPX	mitAPX	mAPX	Unknown	Total
Arabidopsis thaliana	3	2		3		8
Eucalyptus grandis	3	2		1		6
Vigna unguiculata	4	4		4		12
Spinacia oleracea	1	2		1	1	5
Solanum lycopersicum	3	2		2		7
Oryza sativa	2	3[a]	2[a]	2		8

[a]The *OsAPX5* is found in mitochondria and in the stroma of chloroplast

isoform (*OsAPX6*), two chloroplastic ones (*OsAPX7* and *OsAPX8*), and one isoform present in both mitochondria and chloroplasts (*OsAPX5*) (Teixeira et al. 2004). Table 1 illustrates the distribution of APX isoforms in distinct subcellular compartments in different species.

4 Post-translational Redox Regulation of APX

Multiple post-translational regulations of APX occur in plants, modulating plant development, and response to stress. A proteomic approach showed that cAPX is a target of thioredoxins (TRX), which drastically inhibits its activity thought reversible reduction of Cys residues (Marchand et al. 2004; Yamazaki et al. 2004; Gelhaye et al. 2005, 2006). In addition, the carbonylation of these residues leads to irreversible inhibition of APX activity, as demonstrated during *Antiaris toxicaria* seed desiccation (Bai et al. 2011).

Another important post-translational method of regulation of APX is phosphorylation. Under pathogen attack, the chloroplastic protein kinase Wheat Kinase Start 1.1 (WKS1.1) increases cellular H_2O_2 levels in order to activate defense responses by phosphorylation and inactivation of tAPX (Gou et al. 2015). APX activity is also modulated by NO, acting as a redox sensor of oxidative stress and participating in signaling pathways, as, for example, nitric oxide (NO) regulation of lateral root development. Several post-translational modifications mediated by NO have been reported to influence APX activity. In tobacco, NO is able to bind to the heme prosthetic group leading to inhibition of APX activity (Clark et al. 2000). Studies using a peroxynitrite donor showed that a potential redox modification in APX is the nitration of tyrosine residues in *Arabidopsis* (Lozano-Juste et al. 2011). In citrus plants subjected to increased levels of salinity, this modification is found in roots but not in shoots, and studies in pea showed that this modification occurs in Tyr5 and Tyr235, causing an irreversible inhibition of APX activity (Tanou et al. 2012; Begara-Morales et al. 2013; Begara-Morales et al. 2014).

Another important post-translational modification of APX is *S*-nitrosylation. Various studies have demonstrated distinct effects of this modification in APX activity. During seed germination or in salt-stressed plants, *S*-nitrosylation of cAPX enhances its activity (Bai et al. 2011). The cysteine residue Cys32, part of the ascorbate bidding site, is the main target of *S*-nitrosylation (Gelhaye et al. 2006; Begara-Morales et al. 2013; Fares et al. 2011; Correa-Aragunde et al. 2013). A mutation of this residue can result in the loss of at least two-thirds of APX activity (Mandelman et al. 1998a). On the other hand, *S*-nitrosylation of APX also occurs under heat shock conditions during programmed cell death (PCD) in tobacco BY2 cells, leading to inhibition of APX activity (de Pinto et al. 2013). In this experimental model, the inhibition of APX activity by *S*-nitrosylation, ubiquitination, and degradation appears to lead to PCD under stress conditions.

Therefore, during conditions of strong oxidative stress, such as heat, drought, salt stress, and H_2O_2 treatment, the post-translational modification of APX can have a pivotal role. Biotic and abiotic stresses induce an increase of ROS production, which is accompanied by the increase of NO and other reactive nitrogen species (RNS), which induce simultaneous carbonylation, nitration and *S*-nitrosylation of APX, irreversibly affecting its activity. In addition, this modification can induce polyubiquitination and finally degradation of APX by the proteasome (Bai et al. 2011; de Pinto et al. 2013; Tanou et al. 2012).

5 Antioxidant Role of Different APX Isoenzymes

5.1 Chloroplast APX—chlAPX

Chloroplasts are organelles responsible for photosynthesis, which generates molecular oxygen and reducing power from the splitting of water under illumination, thereby making them significant source of ROS. In chloroplasts, H_2O_2 is produced in high amounts as a consequence of the highly energetic reactions taking place during photosynthetic process. The over-reduction of the photosynthetic electron transport (PET) chain occurs under high light intensity due to the exhaustion of NADP+, a terminal electron acceptor, improving ROS production. This also occurs under conditions in which the Calvin cycle is inhibited (e.g., drought, salinity and low temperatures), which also limits NADP+ production (Collen et al. 1995; Asada 1999).

Since CATs are not present in chloroplasts, APX has a critical antioxidant role in this organelle. In all chloroplastic isoforms, a 19-residue transit peptide at the N-terminus has been identified that is processed in mature proteins (Madhusudhan et al. 2003). In addition, two signatures clearly identified the chloroplastic isoforms in higher plants (Teixeira et al. 2006). The first signature consists of seven amino acid residues (KNIEEWP) with the second having 16 amino acids (ETKYTKDGPGAPGGQS). Figure 1 shows the alignment of cAPXs and chlAPXs, highlighting the signatures of chlAPXs and the transmembrane domain.

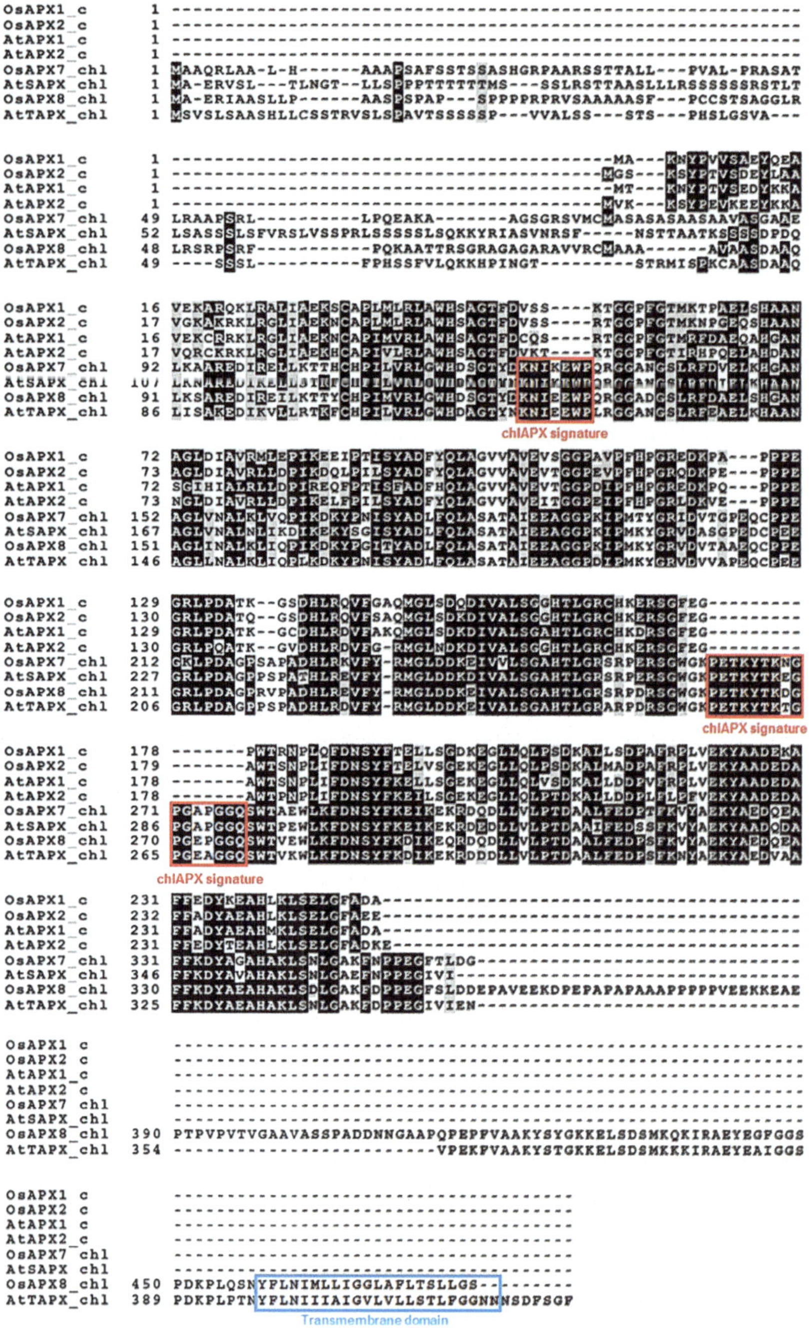

Fig. 1 Alignment of cAPXs and chlAPX from *Oryza sativa* and *Arabidopsis thaliana*. The red boxes indicate the signatures of chlAPXs, and the blue box indicates the transmembrane domain of thylakoidal isoforms as described by Madhusudhan et al. (2003) and Teixeira et al. (2006)

Both chlAPXs are able to scavenge the H_2O_2 produced in chloroplasts (Chen and Asada 1989; Ishikawa et al. 1996; Danna et al. 2003). In some plants, such as *Arabidopsis*, tomato and rice, two genes encoding APX isoforms (a thylakoid-bound isoform and a stromal isoform) have been reported (Chew et al. 2003; Teixeira et al. 2004; Davletova et al. 2005). In contrast, in other plant species including spinach, tobacco, pumpkin, and ice plant, stromal and thylakoid-bound isoforms are generated by alternative splicing of a single gene (Mano et al. 1997).

The amino acid sequences of tAPX and sAPX contain common chloroplast transit peptides (Ishikawa et al. 1996) and were identical, except for the C-terminal of tAPX, which corresponds to the thylakoid membrane domain and is 50 amino acids longer than that of sAPX (Dong et al. 2011). Thus, genes encoding isoenzymes of chlAPX are categorized into two groups: (1) through alternative splicing, a single gene encodes two enzymes and (2) individual genes coding *tAPX* and *sAPX* (reviewed by Caverzan et al. 2012).

In chloroplasts, APX participates in the AsA–GSH cycle, or Foyer–Halliwell–Asada pathway. APX utilizes AsA as its specific electron donor to reduce H_2O_2 to water with the concomitant generation of monodehydroascorbate (MDHA), which is reduced to AsA spontaneously or by the enzyme monodehydroascorbate reductase (MDHAR). MDHA is also spontaneously converted to dehydroascorbate (DHA), which is reduced back to AsA by the enzyme dehydroascorbate reductase (DHAR), which uses GSH as the reducing substrate, which is converted into oxidized glutathione (GSSG). Finally, GSSG is reduced back to GSH by glutathione reductase (GR). In addition to the AsA–GSH cycle, the water–water cycle, which involves the photoreduction of oxygen to water by the electrons derived from the water in photosystem II (PSII), participates in the ROS detoxification and dissipation of the energy of excess photons. In the water–water cycle, electrons excised from water at PSII are transferred to oxygen by PSI, resulting in the formation of O_2^-, which is subsequently converted into H_2O_2 by membrane-attached copper/zinc superoxide dismutase (Cu/Zn-SOD). Thylakoid membrane-bound ascorbate peroxidase reduces H_2O_2 back into water using ascorbate as an electron donor. The reactions of these enzymes act as the first layer of ROS scavenging, followed by their removal by iron SOD (Fe-SOD) and stromal APX as the second layer in stroma. The oxidized form of ascorbate, which is generated by the APX reaction, is reduced by ferredoxin-, glutathione-, and NAD(P)H-dependent pathways. Since electrons are consumed at many steps, including the photoreduction of oxygen and recycling of reductants, the water–water cycle acts not only as an antioxidant system, but also as a system for dissipating excess electrons from PET (Asada 1999). The AsA–GSH and water–water cycles are shown in Fig. 2.

The rates of O_2^- and H_2O_2 formation within intact chloroplasts under optimal conditions are estimated to be 240 and 120 mM per second, respectively (Asada and Takahashi 1987). To control the ROS levels, SOD and APX scavenging enzymes are preferentially located near the site of ROS production (Miyake and Asada 1992; Ogawa et al. 1995). Consequently, it is estimated that, under normal conditions, the H_2O_2 concentration is less than 8×10^{-7} M (Asada 1999).

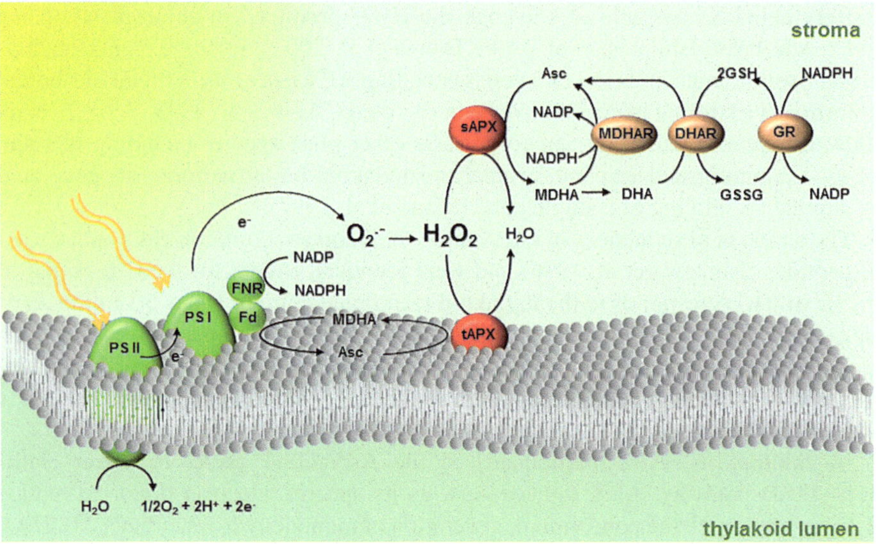

Fig. 2 AsA–GSH and water–water cycles act coordinately to remove hydrogen peroxide in chloroplasts. Legend: PS II: photosystem II, PS I: photosystem I, Fd: ferrodoxin, FNR: ferredoxin-NADP + reductase, tAPX: thylakoidal ascorbate peroxidase, sAPX: stromal ascorbate peroxidase, Asc: Ascorbate, MDHA: monodehydroascorbate, MDHAR: monodehydroascorbate reductase, DHA: dehydroascorbate, DHAR: dehydroascorbate reductase, GR: glutathione reductase

In higher plants, the activity of chlAPXs is rapidly inhibited under photooxidative stress (Yabuta et al. 2002), while algal APXs are considerably stable, and it is recognized that the fragile nature of chlAPX was acquired during plant evolution (Miyake et al. 1991). This sensitivity of APX to oxidative stress provides the "floodgate" hypothesis that chlAPX maintain cellular H_2O_2 at very low levels under favorable conditions, but permit higher levels under stressful conditions, allowing for the flexible use of H_2O_2 as a signaling molecule in plants (Wood et al. 2003). These may be linked to the current concept for the evolution of ROS signaling functions in which plants first acquired ROS-scavenging mechanisms, developed the ability to control intracellular levels of ROS, and then started to use these molecules for signaling purposes (Gill and Tuteja 2010; Mittler et al. 2011).

In wheat plants, a reduction in tAPX activity results in a lowered photosynthetic carbon assimilation, reduced growth rate, and seed production (Danna et al. 2003). The characterization of *tAPX* and *sAPX* single mutants in *Arabidopsis* and spinach plants show that, under exposure to high light or methyl viologen (MV) treatment, oxidized proteins were accumulated, and the expression of H_2O_2-responsive genes were drastically suppressed. These data suggest that chlAPXs contribute not only to photoprotection but also to the regulation of H_2O_2-responsive genes in response to photooxidative stress (Mano et al. 2001; Maruta et al. 2010).

In rice, a double knockdown of *sAPX* and *tAPX* genes showed that the chlAPXs are essential for the protection of PSII under MV-induced oxidative stress. These plants were affected more harshly than the non-transformed (NT) plants in terms of the activity and structure of PSII and CO_2 assimilation in the presence of MV. However, under high light-induced photooxidative stress, these proteins were less important to the PSII and overall photosynthesis in rice (Caverzan et al. 2014). In addition, the silencing of chlAPXs affected the expression levels of proteins involved with the photosynthetic process and oxidative metabolism under normal growth conditions, evidencing the important role played by these enzymes (Caverzan et al. 2014).

The tAPX and sAPX functional specificities in different plant species and stress conditions, and particularly their functional overlaps with other cellular antioxidant agents, remain poorly understood. The interaction between ROS and antioxidant enzymes seems to be more complex than we expected. In *Arabidopsis*, depending on the plant developmental stage, one isoform can be more effective in photoprotection than the other one. The sAPX is particularly important for photoprotection during the early greening process, while both isoforms are functionally redundant and crucial under oxidative stress in mature leaves (Kangasjärvi et al. 2008).

Due to the lability of chlAPX in the absence of AsA, a high level of endogenous AsA is necessary for successfully maintaining the antioxidant system protecting plants from oxidative damage caused by biotic and abiotic stresses (Shigeoka et al. 2002; Miyake et al. 2006; Ishikawa and Shigeoka 2008). For example, Yoshimura et al. (2000) demonstrated that in spinach leaves, the chlAPXs were inactivated by the AsA redox status change under photooxidative stress. Another particularly relevant report was described for tobacco, which showed that chlAPXs were significantly inactivated upon the depletion of AsA (Shikanai et al. 1998; Miyagawa et al. 2000). In tomato seedlings, chlAPXs were also inactivated due to a decrease in AsA rate, increasing H_2O_2 accumulation and a decrease in Rubisco activity (Liu et al. 2008).

5.2 *Mitochondrial APX—mitAPX*

Plant mitochondria are responsible for energy production during respiration reactions, providing the energy needed to drive metabolic and transport processes in the cells. Until recently, the central role of mitochondria as a source of ROS production in plant cells was largely unexplored. However, it is now widely accepted that ROS produced during respiration make an important contribution to oxidative metabolism in plant cells. In mitochondria, ROS are regularly produced in the mitochondrial electron transporter chain reactions, and respiratory complexes I, II, and III are reported to be major sites of mitochondrial superoxide radical production (Møller 2001; Jardim-Messeder et al. 2015). It is well established that superoxide formed via leakage of electrons from complexes I, II, and III is rapidly disputed to H_2O_2 by matrix Mn-SOD (Bowler et al. 1991).

In mitochondria, H_2O_2 is a potent inhibitor of different metabolic pathways, such as the tricarboxylic acid (TCA) cycle, due to oxidative damage (Verniquet et al. 1991;

Tretter and Adam-Vizi 2000; Nulton-Persson and Szweda 2001). Thus, H_2O_2 production need to be highly regulated by an efficient antioxidant system. Furthermore, H_2O_2 removal is required to prevent formation of the highly toxic hydroxyl radical, which is the most reactive and toxic ROS.

The mechanisms of H_2O_2 removal in plant mitochondria are still not well characterized. In plant cells, H_2O_2 is scavenged mainly by APX and CAT (Noctor and Foyer 1998). However, CAT is largely restricted to the peroxisomes, while the APX family maintains H_2O_2 levels at non-toxic concentrations in all of the other compartments of the cell, including mitochondria. In fact, a complete ascorbate–glutathione enzyme cycle and membrane-bound APX isoforms have been detected in plant mitochondria, where mitAPX can be found bound to mitochondrial membranes or as soluble in the matrix (Jimenez et al. 1997)

Studies regarding mitAPX isoforms are currently scarce. However, some works have already shown the important role of this enzyme when plants are submitted to some stress conditions. In *Oryza sativa*, for example, the mitAPX isoform was induced after salt exposure (Lázaro et al. 2013), and in a salt-tolerant wheat cultivar, mitochondrial APX enzyme activity was higher than in sensitive plants (Sairam and Srivastava 2002).

The role of mitAPX in salt stress was also verified in pea plants. In this plant, mitAPX and MDHAR activities were increased at an early stage under mild salt stress and progressively increased under high salt concentrations (Hernández et al. 2000). A decreased oxidative stress in salt-tolerant tomato plants was also attributed to induced activities of Mn-SOD and mitAPX, suggesting that these enzymes have a role in the alleviation of salt-induced oxidative stress (Mittova et al. 2003, 2004a, b). Similar results were also found in *Robinia pseudoacacia* L., which showed an increase of mitAPX under NaCl treatment (Luo et al. 2017).

Biotic stresses can also specifically induce the expression of mitAPX. In cucumber and tomato leaves, cauliflower mosaic virus (CMV) infection is also able to induce mitAPX expression (Song et al. 2009). On the other hand, a continuous decrease in mitAPX activity is observed in tomato leaves infected with the necrotrophic fungus *Botrytis cinerea* (Kuźniak and Skłodowska 2004).

The location of mitAPX in mitochondria appears to have a pivotal role in the control of H_2O_2 levels and in plant stress response patterns. In tomatoes, for example, previous work has explored the metabolic basis for salt tolerance in the wild salt-tolerant *Lycopersicon pennellii* (Lpa), compared with the relative salt sensitivity of the cultivated tomato, *Lycopersicon esculentum* (Lem) (Mittova et al. 2004a, b). In sensitive species, only 16% of APX activity is soluble, and the ratio of soluble to membrane-bound APX activity was changed in leaves of plants that had been exposed to salt stress. In this case, the proportion of soluble APX activity was significantly increased in relation to the membrane-bound form. On the other hand, the mitochondrial distribution of APX was different in tolerant species, where APX was largely soluble (61%); the activities of both soluble and membrane-bound APX isoforms increased significantly in response to salt stress, maintaining the proportion of soluble and membrane-bound APX activity.

5.3 Microsomal APX—mAPX

In plant cells, microbodies, as peroxisomes and glyoxysomes, are probably the major sites of intracellular ROS production. The main metabolic process source of ROS production are the photorespiratory glycolate oxidase reaction, the β-oxidation of fatty acids, the enzymatic reaction of flavin oxidase, xanthine oxidase and NADH oxidase, and a small electron transport chain, composed of monodehydroascorbate reductase (MDAR), a cytochrome b, and a peroxisomal NADPH: cytochrome P450 reductase (Huang et al. 1983; Baker and Graham 2002; del Río et al. 2002; Foyer and Noctor 2003; López-Huertas et al. 1999; Corpas 2015).

Under normal conditions, ROS production is controlled by the antioxidant systems present in these organelles, as CAT and mAPX, mainly peroxisomal APX (pAPX). On the other hand, in most biotic and abiotic stress conditions, an overproduction of ROS in peroxisomes has been demonstrated, which contributes to oxidative damage associated with plant stress (Dat et al. 2000; Mittler 2002).

Although CAT is capable of scavenging large concentrations of H_2O_2, its localization in the peroxisomal/glyoxysomal matrix along with its low affinity for H_2O_2 limits its ability to keep H_2O_2 concentrations low enough to prevent it from diffusing into other subcellular compartments (Karyotou and Donaldson 2005). In this context, due to its high affinity to H_2O_2, it is recognized that APX prevents the H_2O_2 leaking out from peroxisomes. Peroxisomal APX preferentially accumulates in spongy parenchyma rather than palisade parenchyma and can also be found in large amounts near the central vascular bundles (Pereira et al. 2005).

In microbodies, the N-terminal active domain of the peroxisomal APX enzyme faces the cytosol, and its C-terminal domain is anchored in the membrane (Lisenbee et al. 2003). Thus, APX has a central role scavenging the H_2O_2 leaking from microbodies (Yamaguchi et al. 1995a; Ishikawa et al. 1997).

The peroxisomal targeting signal comprises a COOH-terminal transmembrane domain rich in valine and alanine, followed by a positively charged domain containing five amino acid residues (Mullen and Trelease 2000). mAPX was also reported within a subdomain of rough endoplasmic reticulum (RER), suggesting that this organelle can serve as a constitutive sorting compartment likely to be involved in the posttranslational routing of constitutively synthesized pAPX (Lisenbee et al. 2003; Teixeira et al. 2006).

The role of pAPX in stress response has already been explored. In Arabidopsis, the expression of pAPX increases in response to cold (Zhang et al. 1997); in *Avicennia marina,* it was induced in response to salt, high light, H_2O_2 treatment, and excess iron (Kavitha et al. 2008); also, in pea leaves, cadmium stress leads to an increase of pAPX (Romero-Puertas et al. 1999). In rice, the expression of *pAPX* (*OsAPX3* and *OsAPX4*) genes increased under drought stress and application with exogenous H_2O_2 (Rosa et al. 2010), high salinity and mannitol (Cunha et al. 2016).

Some publications have also explored the possible role of peroxisomal APX under biotic stress, and it was confirmed that the *pAPX* gene was differentially expressed in resistant and sensitive wheat after powdery mildew infection (Chen et al. 2006).

Furthermore, it has been demonstrated that pAPX exerts a pivotal role in the senescence process. It is known that during the onset of senescence, there is a decrease of antioxidant enzymes, such as CAT and mAPX. In fact, the decrease in transcripts levels of pAPX was observed at the beginning of senescence in *Arabidopsis* (Panchuk et al. 2005) and in rice plants (Li et al. 2015). In addition, the silencing of the pAPX (*OsAPX4*) gene led to early senescence in rice (Ribeiro et al. 2017), suggesting that *OsAPX4* plays an important role in this process.

5.4 Cytosolic APX—cAPX

Cytosolic APX is the most studied isoform of this gene and, consequently, these genes have more information published. These isoforms have been implicated in many defense processes, being induced under many stresses, and being involved in plant homeostasis redox regulation. The cAPX isoforms are considered much more resistant to ascorbate depletion than the other isoforms, which are inactivated rapidly in the absence of ascorbate (Miyake and Asada 1996). This fact is likely to be associated with their more reported functions under stress conditions.

Many examples of the role of cAPX under abiotic stresses have already been reported. The *apx1*-deficient *Arabidopsis* plants presented delayed development, late flowering, and perturbed stomatal responses (Pnueli et al. 2003). In transgenic tobacco, the down-regulation of cAPX increased susceptibility to ozone stress (Orvar and Ellis 1997), whereas the overexpression of cAPX conferred protection against MV and UV-C-induced oxidative stress (Pitcher et al. 1994; Webb and Allen 1996; Saxena et al. 2011). Transgenic tomato overexpressing the cAPX gene from pea ameliorated oxidative injury induced by chilling and salt stress (Wang et al. 2005).

The reduced expression of the two cAPX (*OsAPX1* and *OsAPX2*) genes in rice altered expression of several genes, particularly those associated with photosynthesis and antioxidant defense (Ribeiro et al. 2012). In addition, the induction of the genes involved in the antioxidant defense suggests that these genes were induced to compensate for the reduction of cAPX and to promote a decrease in the amount of H_2O_2 accumulated in rice plants. Moreover, c*APX* knockout mutants showed increased sensitivity to drought, salt, and cold. Furthermore, the *OsAPX2* knockout mutants exhibited a pleiotropic phenotype including semidwarf, severe leaf mimic lesion, and male sterility, demonstrating that *OsAPX2* is important to growth and development in rice (Zhang et al. 2013). These results corroborate in part with those obtained by Rosa et al. (2010) in rice plants, where it was demonstrated that the silencing of *OsAPX1* and *OsAPX2* genes harmed plant development, producing a semidwarf phenotype. On the other hand, the double silencing of the cAPX genes (APx1/2s) produced plants with a normal phenotype and development (Rosa et al. 2010). Curiously, deficiency in both *cAPX* (APx1/2s) genes was not crucial for oxidative protection in cytosol of rice roots submitted to salt and osmotic treatment (Cunha et al. 2016), once these enzymes act to scavenge and maintain homeostasis of H_2O_2 in foliar tissues of several species.

In *Arabidopsis*, a key role of the cytosolic *APX1* gene protecting chloroplasts during light stress has been described (Davletova et al. 2005). In this study, the absence of *APX1* provoked the entire chloroplastic H_2O_2-scavenging system to collapse, with increases in H_2O_2 and protein oxidation also being observed. These results indicate that *APX1* plays a key role in cross-compartment protection of chloroplast functions.

APX1-deficient mutant (*apx1*) plants were more sensitive when exposed to heat stress combined with drought (Koussevitzky et al. 2008). In addition, cAPX seems to be more versatile under heat stress than the other isoforms, with its activity enhanced after moderate heat shock (HS) and reduced during plant cell death (PCD) induction (Locato et al. 2009).

The role of cAPX under biotic stresses has been shown to be important to plant defense against bacterial and viral infections. The resistant *Theobroma cacao* plants, for example, show an increase in cAPX activity after *Moniliophthora perniciosa* infection (Camillo et al. 2013), and cAPX overexpression in broccoli increases resistance to downy mildew and heat stress (Jiang et al. 2016b). In tobacco, the overexpression of cAPX alone or in combination with Cu/Zn-superoxide dismutase increased resistance to bacterial wildfire caused by *Pseudomonas syringae* pv. Tabaci (Faize et al. 2012).

The role of cAPX during *B. cinerea* infection in tomato leaves has also been documented (Kuźniak and Skłodowska 1999, 2001). In rice plants, cAPX isoforms (*OsAPX1* and *OsAPX2*) have an important role during pathogen response pathways, growth, and reproduction (Agrawal et al. 2003). In addition, *OsAPX1* gene expression is induced in leaves inoculated with *Magnaporthe oryzae* and *Xanthomonas oryzae* pv. oryzae (Wang et al. 2015).

Cytosolic APX also has a central role in chloroplast protection. Although water–water and ascorbate–glutathione cycles are thought to be sufficient for proper H_2O_2 scavenging in thylakoid and stroma, respectively, the cytosolic ascorbate–glutathione cycle is also required for removal of H_2O_2 leaked during photosynthesis. In fact, under high-intensity light, the chloroplast ROS-scavenging pathway can be insufficient in the face of the higher ROS production, indicating that cAPX may be essential for chloroplast protection during stress conditions. Thus, it is suggested that the cytosolic pathway could even compensate or protect the cell when the stromal pathway is absent or insufficient (Davletova et al. 2005; Pnueli et al. 2003). In fact, tobacco plants overexpressing cAPX have better performance under drought stress and have improved photosynthetic rate, confirming the role of APX protecting chloroplasts (Faize et al. 2011; Shrivastava et al. 2015). Since H_2O_2 is transported across biological membranes, this type of cross-compartment protection may also be found with other organelles, such as peroxisomes or mitochondria.

Other publications have also shown that the lack of chloroplastic APX triggers a specific signal, inducing heat tolerance, while a deficiency in cytosolic APX triggers a different signal, which results in reduced growth and increased oxidative stress sensitivity. However, when both chlAPX and cAPX enzymes are suppressed in a cell, new responses are detected, such as late flowering, low protein oxidation during light stress, and enhanced accumulation of anthocyanins, which all suggest the complex integration of the two different signals (Miller et al. 2007). These reports indicate that there is a complex plasticity in plant ROS signaling and the existence of redundant pathways for ROS protection that compensate for the lack of classical ROS removal enzymes such as chloroplastic APXs.

Thus, the notion that each individual compartment is protected by its set of antioxidant pathways has been reevaluated, and a new view of the ROS network has been adopted, in which the cellular redox requires the coordinated function of antioxidant pathways from different cellular compartments to modulate ROS levels, preventing oxidative damage, and controlling signaling pathways (Maruta et al. 2012). In addition, the cytosol is a cellular compartment for communicating signals from each organelle to the nucleus; thus, cAPXs have a critical function in controlling the H_2O_2 concentration in signaling pathways, and their localization in the cytosol has strong relevance to gene expression regulation and therefore also to the mechanism of response to oxidative stress (Shigeoka et al. 2002). These signaling pathways seem to be very complex, and further studies are required to improve our understanding of how different organelles can act individually and in concert with others to promote redox homeostasis.

6 The Role of Ascorbate Peroxidase in the Abiotic and Biotic Stresses Responses

Plants are continuously exposed to abiotic and biotic stress conditions, which seriously reduce their productivity. Abiotic stresses are originated from the surrounding environment and non-living factors (i.e., cold, heat, salinity, drought, nutritional deficiency, heavy metal toxicity, flooding, herbicides, and radiation). Biotic stresses are induced by living organisms, such as bacteria, fungi, viruses, nematodes, and insects that can cause disease or damage. As sessile organisms, plants developed via evolution efficient mechanisms that allowed them to cope and avoid the negative effects induced by these challenge conditions.

As already discussed in the previous section, one of the most common negative effects of many stresses is oxidative stress, which is induced by a disequilibrium between ROS production and scavengers, leading to ROS accumulation in plant cells. It is known that ROS are naturally produced during normal plant development due to aerobic metabolism and that, in low concentrations, ROS are very important as signaling compounds, activating different signaling cascades, and modulating transduction pathways through mitogen-activated protein kinases (MAPK) and transcription factors (Lázaro et al. 2013). However, during non-optimal conditions, an overproduction of these molecules can occur, disturbing cell homeostasis. In excess, these molecules can induce significant damage in plant cells, such as oxidation of proteins, nucleic acids, lipids, and others, leading to cell death (Das and Roychoudhury 2014; Kapoor et al. 2015). In this context, antioxidant enzymes, such as APX, have a key role in maintaining an equilibrium in redox cell homeostasis, removing the excess, and avoiding toxic levels of ROS. In general, a plant's response to stress will depend on the plant's species, tissue, developmental stage, exposure time, intensity and kind of stress, genotype, among many other factors.

APX gene expression is differentially regulated during plant development and also under stress conditions, and it has been reported as being one of the most regulated enzymes during H_2O_2 detoxification (Saxena et al. 2011). APX expression can be modulated by different environmental conditions, such as heat, chilling, high light, drought, salinity, heavy metal toxicity, methyl viologen, and H_2O_2 treatment (Zhang et al. 1997; Yoshimura et al. 2000; Sato et al. 2001; Agrawal et al. 2003; Fryer et al. 2003; Menezes-Benavente et al. 2004; Teixeira et al. 2006; Rosa et al. 2010; Lázaro et al. 2013; Bonifacio et al. 2011; Caverzan et al. 2012, 2014) and after pathogenic attack (Agrawal et al. 2003; Wang et al. 2015). In these contexts, antioxidant enzymes could act not just as ROS scavengers, but also as redox sensor proteins, interacting with regulatory proteins and modulating redox signal transduction (Passaia and Margis-Pinheiro 2015).

Depending on the APX isoform and its respective subcellular localization, different pathways can be activated, triggering various specific responses under different stress conditions (Shigeoka et al. 2002; Pandey et al. 2017). Table 2 illustrates some examples of different APX isoforms from the model plants *Arabidopsis* and rice under different stresses. Although the major part of the literature has reported the importance of APX enzymes under different stresses without indicating the specific isoforms (Table 2), some reports permit us to discuss specific roles of the different isoforms encoded by these genes. cAPX, chAPX, and mitAPX are mostly involved with drought, cold, salt, high light, pathogens, wounding, photooxidative, and oxidative stresses, while pAPX is associated with heavy metal and salt stresses.

As expected, the genetic modification of ROS-scavenging systems can lead to considerable changes in oxidative stress tolerance. The combined results of studies of APX isoenzymes, together with advances in knowledge of AsA metabolism, have provided a greater understanding of the physiological functions of APX in plant stress responses. Changes in *APX* gene expression due to biotic and abiotic stresses suggest that *APX* genes may function in stress tolerance in plants. The modulation of *APX* genes under several stress conditions and the involvement of APX in plant tolerance to environmental stresses have been reported in different plant species (Caverzan et al. 2012, 2016). Besides APX, other key antioxidant enzymes such as SOD, CAT, and GPX have been manipulated to provide enhanced ROS-scavenging capacity and protection against oxidative stress. Often, transgenic plants overexpressing APX plus other antioxidant enzymes confer greater tolerance to certain stress conditions (Caverzan et al. 2016).

Overall, plants genetically engineered using APX enzyme have higher H_2O_2-scavenging APX activity, consequently decreasing oxidative damage and improving tolerance to a stress condition. Several studies have shown that the overexpression of a particular APX in a subcellular compartment render transgenic plants tolerant to certain stresses. Table 3 illustrates some examples of the overexpression of *APX* genes providing tolerance to salt, cold, drought, heat, photooxidative, flooding, chilling, heavy metal, and oxidative stress in different species.

Table 2 Different APX isoforms have various roles under stress

Gene	Stress related	References
Arabidopsis thaliana		
Cytosolic		
AtAPX1	Drought and heat stress	Panchuk et al. (2002), Koussevitzky et al. (2008)
AtAPX2	High light, heat stress	Panchuk et al. (2002). Fryer et al. (2003), Rossel et al. (2006)
AtAPX6	Oxidative stress	Chen et al. (2014)
Microsomal		
AtAPX3	Not respond	Narendra et al. (2006)
AtAPX4	Not described	–
AtAPX5	Not described	–
Chloroplastic		
AtsAPX	Photooxidative stress	Maruta et al. (2010)
AttAPX	Photooxidative stress and cell death induced by nitric oxide	Murgia et al. (2004), Maruta et al. (2010)
Oryza sativa L.		
Cytosolic		
OsAPX1 and *2*	Wounding, blast pathogen, oxidative stress, drought, salinity, cold	Agrawal et al. (2003), Morita et al. (2011), Zhang et al. (2013)
Microsomal		
OsAPX3	Salinity	Vighi et al. (2017)
OsAPX4	Salinity, heavy metals	Guan et al. (2010)
Mitochondrial		
OsAPX5[a]	Oxidative stress, salinity, drought	Rosa et al. (2010)
OsAPX6	Not described	Rosa et al. (2010)
Chloroplastic		
OsAPX7 and *8*	Oxidative stress, salinity, drought	Teixeira et al. (2006), Rosa et al. (2010)

[a]The *OsAPX5* is also found in the stroma of chloroplast

Along with stress tolerance, other traits have also been observed in these plants. The overexpression of *Jatropha curcas APX* in *Arabidopsis* was found to increase germination rate, number of leaves, and rosette area (Chen et al. 2015). In addition, the transgenic plants under salt stress conditions had longer primary root and greater number of lateral roots, higher total chlorophyll content, higher total APX activity, and lower H_2O_2 content. When tobacco plants were transformed to constitutively overexpress the *Arabidopsis* gene for peroxisomal *APX3* and subject to repeated

Table 3 Transgenic plants with potential stress tolerance expressing *APX* genes

Gene	Native species	Target species	Stress tolerance	Reference
cAPX	*Arabidopsis thaliana*	*Nicotiana tabacum*	Salt, drought, PEG, MV, Na_2SO_3	Badawi et al. (2004)
	Pisum sativum	*Lycopersicon esculentum*	Salt, chilling	Wang et al. (2005)
	Oryza sativa	*Arabidopsis thaliana*	Salt	Lu et al. (2007)
	Oryza sativa	*Oryza sativa*	Cold	Sato et al. (2011)
	Oryza sativa	*Oryza sativa*	Salt, cold, drought	Zhang et al. (2013)
	Oryza sativa	*Medicago sativa*	Salt	Zhang et al. (2014)
	Lycium chinense	*Nicotiana tabacum*	Salt	Wu et al. (2014)
	Solanum melongena	*Oryza sativa*	Flooding	Chiang et al. (2015a)
	Populus tomentosa	*Nicotiana tabacum*	Drought, salt, Oxidative stress	Cao et al. (2017)
	Panax ginseng	*Arabidopsis thaliana*	Salt	Sukweenadhi et al. (2017)
chlAPX	*Spinacia oleracea*	*Nicotiana tabacum*	MV, chilling	Yabuta et al. (2002)
	Arabidopsis thaliana	*Arabidopsis thaliana*	MV	Murgia et al. (2004)
	Cyanidioschyzon merolae	*Arabidopsis thaliana*	MV, heat	Hirooka et al. (2009)
	Lycopersicon esculentum	*Lycopersicon esculentum*	Chilling	Duan et al. (2012)
	Jatropha curcas	*Nicotiana tabacum*	Salt	Liu et al. (2014)
pAPX	*Hordeum vulgare*	*Arabidopsis thaliana*	Heat	Shi et al. (2001)
	Hordeum vulgare	*Arabidopsis thaliana*	Zinc, cadmium	Xu et al. (2008)
	Populus tomentosa	*Nicotiana tabacum*	MV, salt, drought	Li et al. (2009)
	Salicornia brachiata	*Nicotiana tabacum*	Salt, drought	Singh et al. (2014)
	Puccinellia tenuiflora	*Arabidopsis thaliana*	Salt	Guan et al. (2015)

APX—ascorbate peroxidase, cAPX—cytosolic APX, chlAPX—chloroplastic APX, pAPX—peroxisomal APX, MV—methyl viologen, Na_2SO_3—sodium sulfite water-deficit cycles, the fruit number and seed mass of transgenic plants were higher compared to non-transgenic plants (Yan et al. 2003), demonstrating the important role played by pAPX.

Transgenic tobacco plants overexpressing *Populus tomentosa* cAPX under normal conditions showed slightly increased APX and SOD activity. However, when these plants were submitted to stress conditions, they exhibited enhanced tolerance to drought, salt, and oxidative stresses, as demonstrated by their decreased levels of malondialdehyde and increased levels of chlorophyll (Cao et al. 2017). Also in tobacco, transgenic plants expressing the *Lycium chinense cAPX* gene exhibited a lower amount of H_2O_2 and relatively higher APX activity, proline content, and net photosynthetic rate (PN) under salt stress (Wu et al. 2014). These results emphasize that metabolites such as proline favor tolerance to salt and drought stress by permitting osmotic adjustment in order to avoid water losses. In addition, proline is considered a powerful antioxidant (Gill and Tuteja 2010), due to its ability to interact with other chemical compounds, such as phytohormones, and contribute to plant stress tolerance (Per et al. 2017).

Moreover, the contents of chlorophyll and proline, and APX activity were higher in transgenic plants expressing *O. sativa cAPX* gene in *Medicago sativa* under salt and drought stresses (Zhang et al. 2014). Higher chlorophyll content, relative water content, total APX activity, proline content, and lower H_2O_2 accumulation were shown in transgenic plants expressing *Panax ginseng APX* gene in *Arabidopsis* under salt stress (Sukweenadhi et al. 2017). These results suggest that expression of the *APX* gene could decrease ROS production caused by salt and drought stresses, protecting plants from oxidative stress. Salt and drought stresses are factors that affect crops worldwide, causing several disturbances such as reduction of cell/leaf expansion, cell and metabolic activities, stomatal closure, photosynthetic inhibition, leaf abscission, alteration in the carbon partition, destabilization of membranes and proteins, ROS production, ionic cytotoxicity, and cell death (Taiz and Zeiger 2013). These effects of salt and drought stress lead to significant loss of crop productivity.

Transgenic rice seedlings expressing *Solanum melongena APX* exhibited tolerance to flood due to higher APX activity and chlorophyll content (Chiang et al. 2015a). In addition, when rice seeds were anaerobically germinated under water for five days, root (ranging from 1.7 to 6.5 mm) and coleoptile (ranging from 2.74 to 3.41 cm) lengths of all transgenic lines grew more quickly than non-transformed seedlings. It is known that under flooding some processes are triggered, such as reduction of respiration, inadequate ATP production, stomatal closure, and ROS production (Taiz and Zeiger 2013), and photosynthetic pigments may be destroyed, leading to photosynthetic capacity losses. Thus, higher APX activity may contribute to tolerance to flood stress.

High temperatures cause destabilization of membranes and proteins, and photosynthetic and respiratory inhibition, ROS production and cell death (Taiz and Zeiger 2013). The overexpression of APX is able to confer tolerance to heat stress. Thus, the overexpression of the *Brassica campestris APX* gene in transgenic *Arabidopsis* enhanced heat tolerance. In addition, these plants showed higher APX activity, chlorophyll content, and germination rate. Furthermore, the transgenic lines subjected to heat stress presented lower malondialdehyde content (Chiang et al. 2015b). Tolerance to heat stress was also achieved by overexpression of the *pAPX* gene of barley and *chAPX* of *Cyanidioschyzon merolae* in *Arabidopsis* plants (Shi et al. 2001; Hirooka et al. 2009).

Cold stress causes many effects in plants, such as membrane dysfunction and physical destruction of the cell (Taiz and Zeiger 2013). In rice plants, enhanced chilling tolerance at the booting stage was achieved by overexpression of the rice *cAPX* gene (*OsAPX1*) (Sato et al. 2011). In this study, higher APX activity enhanced H_2O_2-scavenging capacity and protected spikelets from lipid peroxidation, thus increasing spikelet fertility under cold stress. Moreover, rice plants overexpressing the cytosolic *OsAPX2* gene showed increased APX activity and stress tolerance (Zhang et al. 2013). In these plants, spikelet fertilities were higher under drought, cold, and salt treatments. In addition, the plants were more tolerant to drought stress at the booting stage (Zhang et al. 2013).

Transgenic tobacco plants overexpressing a *chlAPX* gene target to the thylakoid showed increased tolerance to both photooxidative stress, MV (50 mM) or chilling under light intensity (Yabuta et al. 2002). Further, overexpression of *tAPX* in *Arabidopsis* plants increased resistance to photooxidative stress (Murgia et al. 2004). These results evidence the role of chlAPX enzymes against photooxidative stress, which is mostly induced by the absorption of excess excitation energy leading to over-reduction of the electron transport chains generating ROS.

Heavy metals occur naturally in the earth's crust at various levels, but they become harmful when released in excess, causing severe symptoms of plant toxicity. Several studies have already emphasized the important role of APX enzymes in H_2O_2 scavenging in plants growing under toxic metal levels and have shown that the expression of APX was altered both with respect to mRNA levels and enzymatic activity (Caverzan et al. 2012). Moreover, *Arabidopsis* plants overexpressing peroxisomal barley *APX1* gene showed tolerance to zinc and cadmium and higher APX activity and lower H_2O_2 and malondialdehyde contents (Xu et al. 2008). On the other hand, Rosa et al. (2010) demonstrated that rice plants double silenced for cytosolic APX genes, *OsAPX1* and *OsAPX2,* exhibited normal development and enhanced tolerance to a toxic concentration of aluminum. Besides, the simultaneous overexpression of both CuZnSOD and APX in transgenic tall fescue plants conferred increased tolerance against copper, cadmium, and arsenic due to depressed oxidative stress (Lee et al. 2007).

Considering the findings as summarized above, the results of the overexpression of APX and other enzymes in different plant species clearly demonstrate that increased expression of these antioxidant enzymes provides plants with increased abiotic stress tolerance. The rapid scavenging of toxic levels of ROS in the cells and the restoration of the redox homeostasis is crucial to plants exposed to stress. Therefore, a high level of antioxidant enzymes will help the plant to protect itself against oxidative damage and consequently tolerate a stress condition more effectively, reducing losses. Nevertheless, under field conditions the transgenic plant performance still needs to be better studied and proved to be useful in crop breeding programs targeting stress tolerance.

7 Concluding Remarks

The discovery of APX enzymes over more than two decades has contributed to a better understanding of plant antioxidant systems. The activity of all APXs in ROS detoxification, allowing ROS produced in the organelles to be efficiently scavenged by the organelles themselves, and thus contributing to homeostasis redox, is of vital importance to cellular life. Advances in the understanding of the biochemical, molecular, and physiological mechanisms of APX have allowed elucidation of the role of these enzymes in processes such as growth, development, and plant senescence. Moreover, the responses of APX to adverse conditions have demonstrated that the expression of *APX* encoding genes is differentially modulated by environmental stresses in different plant species, and that APX may function in stress tolerance in plants. Meantime, many issues still need to be elucidated regarding APX enzymes and more studies are needed for determining the role of APX as signaling regulators, as well as APX interactions with other proteins. In this sense, many challenges remain and others likely will appear.

References

Agrawal GK, Jwa NS, Iwahashi H, Rakwal R (2003) Importance of ascorbate peroxidases *OsAPX1* and *OsAPX2* in the rice pathogen response pathways and growth and reproduction revealed by their transcriptional profiling. Gene 11:93–103

Asada K, Takahashi M (1987) Production and scavenging of active oxygen in photosynthesis. In: Kyle DL, Osmond CB, Arntzen CJ (eds) Photoinhibition. Elsevier Scientific Publishers, Amsterdam, pp 227–287

Asada K (1999) The water–water cycle in chloroplasts, scavenging of active oxygen and dissipation of excess photons. Annu Rev Plant Physiol Plant Mol Biol 50:601–639

Badawi GH, Kawano N, Yamauchi Y, Shimada E, Sasaki R, Kubo A, Tanaka K (2004) Over-expression of ascorbate peroxidase in tobacco chloroplasts enhances the tolerance to salt stress and water deficit. Physiol Plant 121:231–238

Bai X, Yang L, Tian M, Chen J, Shi J, Yang Y, Hu X (2011) Nitric oxide enhances desiccation tolerance of recalcitrant *Antiaris toxicaria* seeds via protein *S*-nitrosylation and carbonylation. PLoS ONE 6:e20714

Baker A, Graham I (2002) Plant peroxisomes. Biochemistry, cell biology and biotechnological applications. Kluwer Academic Publishers, Dordrecht

Begara-Morales JC, Chaki M, Sánchez-Calvo B, Mata-Pérez C, Leterrier M, Palma JM, Barroso JB, Corpas FJ (2013) Protein tyrosine nitration in pea roots during development and senescence. J Exp Bot 64:1121–1134

Begara-Morales JC, Sanchez-Calvo B, Chaki M, Valderrama R, Mata-Pérez C, López-Jaramillo J, Padilha MN, Carreras A, Corpas FJ, Barroso JB (2014) Dual regulation of cytosolic ascorbate peroxidase (APX) by tyrosine nitration and *S*-nitrosylation. J Exp Bot 65:527–538

Bonifacio A, Martins MO, Ribeiro CW, Fontenele AV, Carvalho FE, Margis-Pinheiro M, Silveira JA (2011) Role of peroxidases in the compensation of cytosolic ascorbate peroxidase knockdown in rice plants under abiotic stress. Plant, Cell Environ 34:1705–1722

Bonifacio A, Carvalho FEL, Martins MO, Lima Neto MC, Cunha JR, Ribeiro CW, Margis-Pinheiro M, Silveira JAG (2016) Silenced rice in both cytosolic ascorbate peroxidases displays pre-acclimation to cope with oxidative stress induced by 3-aminotriazole-inhibited catalase. J Plant Physiol 201:17–27

Battistuzzi G, D'Onofrio M, Loschi L, Sola M (2001) Isolation and characterization of two peroxidases from *Cucumis sativus*. Arch Biochem Biophys 388:100–112

Boveris AH, Sies H, Martino EE, Decampo R, Turreus JF, Stoppani AOM (1980) Deficient metabolic utilization of hydrogen peroxide in *Trypanosoma cruzi*. Biochem J 188:643–648

Bowler C, Slooten L, Vandenbranden S, De Rycke R, Botterman J, Sybesma C, Van Montagu M, Inzé D (1991) Manganese superoxide dismutase can reduce cellular damage mediated by oxygen radicals in transgenic plants. EMBO J 10:1723–1732

Bunkelmann JR, Trelease RN (1996) Ascorbate peroxidase, a prominent membrane protein in oilseed glyoxysomes. Plant Physiol 110:589–598

Camillo LR, Filadelfo CR, Monzani PS, Corrêa RX, Gramacho KP, Micheli F, Pirovani CP (2013) Tc-cAPX, a cytosolic ascorbate peroxidase of *Theobroma cacao* L. engaged in the interaction with *Moniliophthora perniciosa*, the causing agent of witches' broom disease. Plant Physiol Biochem 73:254–265

Cao S, Du XH, Li LH, Liu YD, Zhang L, Pan X, Li Y, Li H, Lu H (2017) Overexpression of *Populus tomentosa* cytosolic ascorbate peroxidase enhances abiotic stress tolerance in tobacco plants. Russ J Plant Physiol 64:224–234

Caverzan A, Passaia G, Rosa SB, Ribeiro CW, Lazzarotto F, Margis-Pinheiro M (2012) Plant responses to stresses: role of ascorbate peroxidase in the antioxidant protection. Genet Mol Biol 35:1011–1019

Caverzan A, Bonifacio A, Carvalho FEL, Andrade CMB, Passaia G, Schünemann M, Maraschin FS, Martins MO, Teixeira FK, Rauber R, Margis R, Silveira JAG, Margis-Pinheiro M (2014) The knockdown of chloroplastic ascorbate peroxidases reveals its regulatory role in the photosynthesis and protection under photo-oxidative stress in rice. Plant Sci 214:74–87

Caverzan A, Casassola A, Brammer AP (2016) Reactive oxygen species and antioxidant enzymes. In: Shanker AK, Shanker C (eds) Abiotic and biotic stress in plants—recent advances and future perspectives involved in plant tolerance to stress. Intech, Croatia, pp 463–480

Chen GX, Asada K (1989) APX in tea leaves, occurrence of two isoenzymes, the differences in their enzymatic and molecular properties. Plant Cell Physiol 30:987–998

Chen Y, Wang H, Wang X, Cao A, Chen P (2006) Cloning and expression of peroxisomal ascorbate peroxidase gene from wheat. Mol Biol Rep 33:207–213

Chen C, Letnik I, Hacham Y, Dobrev P, Ben-Daniel BH, Vanková R, Amir R, Miller G (2014) Ascorbate peroxidase 6 protects *Arabidopsis* desiccating and germinating seeds from stress and mediates cross talk between reactive oxygen species, abscisic acid, and auxin. Plant Physiol 166:370–383

Chen Y, Cai J, Yang FX, Zhou B, Zhou LR (2015) Ascorbate peroxidase from *Jatropha curcas* enhances salt tolerance in transgenic *Arabidopsis*. Genet Mol Res 14:4879–4889

Chew O, Whelan J, Millar H (2003) Molecular definition of the ascorbate-glutathione cycle in *Arabidopsis* mitochondria reveals dual targeting of antioxidant defenses in plants. J Biol Chem 278:46869–46877

Chiang CM, Chen LFO, Shih SW, Lin KH (2015a) Expression of eggplant ascorbate peroxidase increases the tolerance of transgenic rice plants to flooding stress. J Plant Biochem Biotechnol 24:257–267

Chiang CM, Chien HL, Chen LFO, Hsiung TC, Chiang MC, Chen SP, Lin KH (2015b) Overexpression of the genes coding ascorbate peroxidase from *Brassica campestris* enhances heat tolerance in transgenic *Arabidopsis thaliana*. Biol Plantarum 59:305–315

Clark D, Durner J, Navarre DA, Klessig DF (2000) Nitric oxide inhibition of tobacco catalase and ascorbate peroxidase. Mol Plant-Microbe Interact 13:1380–1384

Collen J, Rıo MJD, Garcıa-Reina G, Pedersen M (1995) Photosynthetic production of hydrogen peroxide by *Ulva rigida* C. Ag. (Chlorophyta). Planta 196:225–230

Corpas FJ (2015) What is the role of hydrogen peroxide in plant peroxisomes? Plant Biol (Stuttg) 17:1099–1103

Correa-Aragunde N, Foresi N, Delledonne M, Lamattina L (2013) Auxin induces redox regulation of ascorbate peroxidase 1 activity by S-nitrosylation/denitrosylation balance resulting in changes of root growth pattern in *Arabidopsis*. J Exp Bot 64:3339–3349

Cunha JR, Lima Neto MC, Carvalho FEL, Martins MO, Jardim-Messeder D, Margis-Pinheiro M, Silveira JAG (2016) Salinity and osmotic stress trigger different antioxidant responses related to cytosolic ascorbate peroxidase knockdown in rice roots. Environ Exp Bot 131:58–67

Danna CH, Bartoli CG, Sacco F, Ingala LR, Santa-Maria GE, Guiamet JJ, Ugalde RA (2003) Thylakoid-bound ascorbate peroxidase mutant exhibits impaired electron transport and photosynthetic activity. Plant Physiol 132:2116–2125

Das K, Roychoudhury A (2014) Reactive oxygen species (ROS) and response of antioxidants as ROS-scavengers during environmental stress in plants. Front Environ Sci 2:53

Dat J, Vandenabeele S, Vranová E, Van Montagu M, Inzé D, Van Breusegem F (2000) Dual action of the active oxygen species during plant stress responses. Cell Mol Life Sci 57:779–795

Davletova S, Rizhsky L, Liang HJ, Zhong SQ, Oliver DJ, Coutu J, Shulaev V, Schlauch K, Mittler R (2005) Cytosolic ascorbate peroxidase 1 is a central component of the reactive oxygen gene network of *Arabidopsis*. Plant Cell 17:268–281

De Leonardis S, Dipierro N, Dipierro S (2000) Purification and characterization of an APX from potato tuber mitochondria. Plant Physiol Biochem 38:773–779

De Pinto MC, Locato V, Sgobba A, Romero-Puertas MC, Gadaleta C, Delledonne M, De Gara L (2013) *S*-nitrosylation of ascorbate peroxidase is part of programmed cell death signaling in tobacco Bright Yellow-2 cells. Plant Physiol 163:1766–1775

del Río LA, Corpas FJ, Sandalio LM, Palma JM, Gómez M, Barroso JB (2002) Reactive oxygen species, antioxidant systems and nitric oxide in peroxisomes. J Exp Bot 53:1255–1272

Dong C, Zheng XF, Li GL, Pan C, Zhou MQ, Hu ZL (2011) Cloning and expression of one chloroplastic ascorbate peroxidase gene from *Nelumbo nucifera*. Biochem Genet 49:656–664

Duan M, Feng HL, Wang LY, Li D, Meng QW (2012) Overexpression of thylakoidal ascorbate peroxidase shows enhanced resistance to chilling stress in tomato. J Plant Physiol 169:867–877

Faize M, Burgos L, Faize L, Piqueras A, Nicolas E, Barba-Espin G, Clemente-Moreno MJ, Alcobendas R, Artlip T, Hernandez JA (2011) Involvement of cytosolic ascorbate peroxidase and Cu/Zn-superoxide dismutase for improved tolerance against drought stress. J Exp Bot 62:2599–2613

Faize M, Burgos L, Faize L, Petri C, Barba-Espin G, Díaz-Vivancos P, Clemente-Moreno MJ, Alburquerque N, Hernandez JA (2012) Modulation of tobacco bacterial disease resistance using cytosolic ascorbate peroxidase and Cu, Zn-superoxide dismutase. Plant Pathol 61:858–866

Fares A, Rossignol M, Peltier JB (2011) Proteomics investigation of endogenous *S*-nitrosylation in *Arabidopsis*. Biochem Biophys Res Commun 416:331–336

Foyer CH, Noctor G (2003) Redox sensing and signaling associated with reactive oxygen in chloroplasts, peroxisomes and mitochondria. Physiol Plant 119:355–364

Fryer MJ, Ball L, Oxborough K, Karpinski S, Mullineaux PM, Baker NR (2003) Control of Ascorbate Peroxidase 2 expression by hydrogen peroxide and leaf water status during excess light stress reveals a functional organization of *Arabidopsis* leaves. Plant J 33:691–705

Gadjev I, Vanderauwera S, Gechev TS, Laloi C, Minkov IN, Shulaev V, Apel K, Inzé D, Mittler R, Breusegem FV (2006) Transcriptomic footprints disclose specificity of reactive oxygen species signaling in *Arabidopsis*. Plant Physiol 141:436–445

Gelhaye E, Rouhier N, Navrot N, Jacquot JP (2005) The plant thioredoxin system. Cell Mol Life Sci 62:24–35

Gelhaye E, Navrot N, Macdonald IK, Rouhier N, Raven EL, Jacquot JP (2006) Ascorbate peroxidase-thioredoxin interaction. Photosynth Res 89:193–200

Gill SS, Tuteja N (2010) Reactive oxygen species and antioxidant machinery in abiotic stress tolerance in crop plants. Plant Physiol Biochem 48:909–930

Gou JY, Li K, Wu K, Wang X, Lin H, Cantu D, Uauy C, Dobon-Alonso A, Midorikawa T, Inoue K, Sánchez J, Fu D, Blechl A, Wallington E, Fahima T, Meeta M, Epstein L, Dubcovsky J (2015) Wheat stripe rust resistance protein WKS1 reduces the ability of the thylakoid associated ascorbate peroxidase to detoxify reactive oxygen species. Plant Cell 27:1755–1770

Guan Q, Xia D, Liu S (2010) *OsAPX4* gene response to several environmental stresses in rice (*Oryza sativa* L.). Afr J Biotechnol 9:5908–5913

Guan Q, Wang Z, Wang X, Takano T, Liu S (2015) A peroxisomal APX from *Puccinellia tenuiflora* improves the abiotic stress tolerance of transgenic Arabidopsis thaliana through decreasing of H_2O_2 accumulation. J Plant Physiol 175:183–191

Groden D, Beck E (1979) H_2O_2 destruction by ascorbate-dependent systems from chloroplasts. Biochim Biophys Acta 546:426–435

Hernández JA, Jiménez A, Mullineaux P, Sevilla F (2000) Tolerance of pea (*Pisum sativum* L.) to long-term salt stress is associated with induction of antioxidant defenses. Plant, Cell Environ 23:853–862

Hirooka S, Misumi O, Yoshida M, Mori T, Nishida K, Yagisawa F, Yoshida Y, Fujiwara T, Kuroiwa H, Kuroiwa T (2009) Expression of the *Cyanidioschyzon merolae* stromal ascorbate peroxidase in *Arabidopsis thaliana* enhances thermotolerance. Plant Cell Rep 28:1881–1893

Huang AHC, Trelease RN, Moore TS (1983) Plant peroxisomes. Academic Press, New York

Ishikawa T, Sakai K, Yoshimura K, Takeda T, Shigeoka S (1996) cDNAs encoding spinach stromal thylakoid bound ascorbate peroxidase, differing in the presence or absence of their 39-coding regions. FEBS Lett 384:289–293

Ishikawa T, Yoshimura K, Tamoi M, Takeda T, Shigeoka S (1997) Alternative mRNA splicing of 39-terminal exons generates ascorbate peroxidase isoenzymes in spinach (*Spinacia oleracea*) chloroplast. Biochem J 328:795–800

Ishikawa T, Shigeoka S (2008) Recent advances in ascorbate biosynthesis and the physiological significance of ascorbate peroxidase in photosynthesizing organisms. Biosci Biotechnol Biochem 72:1143–1154

Jardim-Messeder D, Caverzan A, Rauber R, de Souza Ferreira E, Margis-Pinheiro M, Galina A (2015) Succinate dehydrogenase (mitochondrial complex II) is a source of reactive oxygen species in plants and regulates development and stress responses. New Phytol 208:776–789

Jespersen HM, KjaersgardIvh OL, Welinder KG (1997) From sequence analysis of three novel APXs from *Arabidopsis thaliana* to structure, function and evolution of seven types of APX. Biochem J 326:305–310

Jiang G, Yin D, Zhao J, Chen H, Guo L, Zhu L, Zhai W (2016a) The rice thylakoid membrane-bound ascorbate peroxidase *OsAPX8* functions in tolerance to bacterial blight. Sci Rep 6:26104

Jiang M, Jiang JJ, He CM, Guan M (2016b) Broccoli plants over-expressing a cytosolic ascorbate peroxidase gene increase resistance to downy mildew and heat stress. J Plant Pathol 98:413–420

Jimenez A, Hernandez JA, Del Rio LA, Sevilla F (1997) Evidence for the presence of the ascorbate–glutathione cycle in mitochondria and peroxisomes of pea leaves. Plant Physiol 114:275–284

Kangasjärvi S, Lepistö A, Hännikäinen K, Piippo M, Luomala EM, Aro EM, Rintamäki E (2008) Diverse roles for chloroplast stromal and thylakoid-bound ascorbate peroxidases in plant stress responses. Biochem J 412:275–285

Kapoor D, Sharma R, Handa N, Kaur H, Rattan A, Yadav P, Gautam V, Kaur R, Bhardwaj R (2015) Redox homeostasis in plants under abiotic stress: role of electron carriers, energy metabolism mediators and proteinaceous thiols. Front Environ Sci 3:13

Karyotou K, Donaldson RP (2005) Ascorbate peroxidase, a scavenger of hydrogen peroxide in glyoxysomal membranes. Arch Biochem Biophys 434:248–257

Kavitha K, Venkatamaranm G, Parida A (2008) An oxidative and salinity stress induced peroxisomal ascorbate peroxidase from *Avicennia marina*: molecular and functional characterization. Plant Physiol Biochem 46:794–804

Kelly GJ, Latzko E (1979) Soluble ascorbate peroxidase, detection in plants and use in vitamin C estimation. Naturwissenschaften 66:617–619

Koussevitzky S, Suzuki N, Huntington S, Armijo L, Sha W, Cortes D, Shulaev V, Mittler R (2008) Ascorbate peroxidase 1 plays a key role in the response of *Arabidopsis thaliana* to stress combination. J Biol Chem 283:34197–34203

Kuźniak E, Skłodowska M (1999) The effect of *Botrytis cinerea* infection on ascorbate–glutathione cycle in tomato leaves. Plant Sci 148:69–76

Kuźniak E, Skłodowska M (2001) Ascorbate, glutathione and related enzymes in chloroplasts of tomato leaves infected by *Botrytis cinerea*. Plant Sci 160:723–731

Kuźniak E, Skłodowska M (2004) The effect of *Botrytis cinerea* infection on the antioxidant profile of mitochondria from tomato leaves. J Exp Bot 55:605–612

Lázaro JL, Jiménez A, Camejo D, Iglesias-Baena I, Martí MC, Lázaro-Payo A, Barranco-Medina S, Sevilla F (2013) Dissecting the integrative antioxidant and redox systems in plant mitochondria. Effect of stress and *S*-nitrosylation. Front Plant Sci 4:460

Lazzarotto F, Turchetto-Zolet AC, Margis-Pinheiro M (2015) Revisiting the non-animal peroxidase superfamily. Trends Plant Sci 20:1–7

Lee SH, Ahsan N, Lee KW, Kim DH, Lee DG, Kwak SS, Kwon SY, Kim TH, Lee BH (2007) Simultaneous overexpression of both CuZn superoxide dismutase and ascorbate peroxidase in transgenic tall fescue plants confers increased tolerance to a wide range of abiotic stresses. J Plant Physiol 164:1626–1638

Li YL, Hai RL, Du XH, Jiang XN, Lu H (2009) Over-expression of a *Populus* peroxisomal ascorbate peroxidase (PpAPX) gene in tobacco plants enhances stress tolerance. Plant Breed 128:404–410

Li Z, Su D, Lei B, Wang F, Geng W, Pan G, Cheng F (2015) Transcriptional profile of genes involved in ascorbate–glutathione cycle in senescing leaves for an early senescence leaf (esl) rice mutant. J Plant Physiol 176:1–15

Lisenbee CS, Heinze M, Trelease RN (2003) Peroxisomal APX resides within a subdomain of rough endoplasmic reticulum in wild-type *Arabidopsis* cells. Plant Physiol 132:870–882

Liu KL, Shen L, Wang JQ, Sheng JP (2008) Rapid inactivation of chloroplastic ascorbate peroxidase is responsible for oxidative modification to Rubisco in tomato (*Lycopersicon esculentum*) under cadmium stress. J Integr Plant Biol 50:415–426

Liu Z, Bao H, Cai J, Han J, Zhou L (2014) A novel thylakoid ascorbate peroxidase from *Jatropha curcas* enhances salt tolerance in transgenic tobacco. Int J Mol Sci 15:171–185

Locato V, de Pinto MC, De Gara L (2009) Different involvement of the mitochondrial, plastidial and cytosolic ascorbate-glutathione redox enzymes in heat shock responses. Physiol Plant 135:296–306

López-Huertas E, Corpas FJ, Sandalio LM, del Río LA (1999) Characterization of membrane polypeptides from pea leaf peroxisomes involved in superoxide radical generation. Biochem J 337:531–536

López-Huertas E, del Río LA (2014) Characterization of antioxidant enzymes and peroxisomes of olive (*Olea europaea* L.) fruits. J Plant Physiol 171:1463–1471

Lozano-Juste J, Colom-Moreno R, Leon J (2011) In vivo protein tyrosine nitration in *Arabidopsis thaliana*. J Exp Bot 62:3501–3517

Lu Z, Liu D, Liu S (2007) Two rice cytosolic ascorbate peroxidases differentially improve salt tolerance in transgenic *Arabidopsis*. Plant Cell Rep 26:1909–1917

Luo Q, Peng M, Zhang X, Lei P, Ji X, Chow W, Meng F, Sun G (2017) Comparative mitochondrial proteomic, physiological, biochemical and ultrastructural profiling reveal factors underpinning salt tolerance in tetraploid black locust (*Robinia pseudoacacia* L.). BMC Genomics 18:648

Madhusudhan R, Ishikawa T, Sawa Y, Shigeoka S, Shibata H (2003) Characterization of an ascorbate peroxidase in plastids of tobacco BY–2 cells. Physiol Plant 117:550–557

Mandelman D, Jamal J, Poulos TL (1998a) Identification of two electron-transfer sites in ascorbate peroxidase using chemical modification, enzyme kinetics crystallography. Biochemistry 37:17610–17617

Mandelman D, Schwarz FP, Li H, Poulos TL (1998b) The role of quaternary interactions on the stability activity of ascorbate peroxidase. Protein Sci 7:2089–2098

Mano S, Yamaguchi K, Hayashi M, Nishimura M (1997) Stromal thylakoid-bound ascorbate peroxidases are produced by alternative splicing in pumpkin. FEBS Lett 413:21–26

Mano J, Ohno C, Domae Y, Asada K (2001) Chloroplastic ascorbate peroxidase is the primary target of methylviologen-induced photooxidative stress in spinach leaves: its relevance to monodehydroascorbate radical detected with in vivo ESR. Biochim Biophys Acta 1504:275–287

Marchand C, Le MP, Meyer Y, Miginiac-Maslow M, Issakidis-Bourguet E, Decottignies P (2004) New targets of Arabidopsis thioredoxins revealed by proteomic analysis. Proteomics 4:2696–2706
Maruta T, Tanouchi A, Tamoi M, Yabuta Y, Yoshimura K, Ishikawa T, Shigeoka S (2010) *Arabidopsis* chloroplastic ascorbate peroxidase isoenzymes play a dual role in photoprotection and gene regulation under photooxidative stress. Plant Cell Physiol 51:190–200
Maruta T, Noshi M, Tanouchi A, Tamoi M, Yabuta Y, Yoshimura K, Ishikawa T, Shigeoka S (2012) H_2O_2-triggered retrograde signaling from chloroplasts to nucleus plays specific role in response to stress. J Biol Chem 287:11717–11729
Menezes-Benavente L, Teixeira FK, Kamei CLA, Margis-Pinheiro M (2004) Salt stress induces altered expression of genes encoding antioxidant enzymes in seedlings of a Brazilian indica rice (*Oryza sativa* L.) Plant Sci 166:323–331
Miller G, Suzuki N, Rizhsky L, Hegie A, Koussevitzky S, Mittler R (2007) Double mutants deficient in cytosolic and thylakoid ascorbate peroxidase reveal a complex mode of interaction between reactive oxygen species, plant development, and response to abiotic stresses. Plant Physiol 144:1777–17785
Mittler R, Zilinskas BA (1991) Purification and characterization of pea cytosolic ascorbate peroxidase. Plant Physiol 97:962–968
Mittler R (2002) Oxidative stress, antioxidants and stress tolerance. Trends Plant Sci 7:405–410
Mittler R, Vanderauwera S, Gollery M, Van Breusegem F (2004) Reactive oxygen gene network of plants. Trends Plant Sci 9:490–498
Mittler R, Vanderauwera S, Suzuki N, Miller G, Tognetti VB, Vandepoele K, Gollery M, Shulaev V, Breusegem VF (2011) ROS signaling: the new wave? Trends Plant Sci 16:300–309
Mittler R (2017) ROS are good? Trends Plant Sci 22:11–19
Mittova V, Tal M, Volokita M, Guy M (2003) Up-regulation of the leaf mitochondrial and peroxisomal antioxidative systems in response to salt-induced oxidative stress in the wild salt-tolerant tomato species *Lycopersicon pennellii*. Plant, Cell Environ 26:845–856
Mittova V, Guy M, Tal M, Volokita M (2004a) Salinity upregulates the antioxidative system in root mitochondria and peroxisomes of the wild salt-tolerant tomato species *Lycopersicon pennellii*. J Exp Bot 55:1105–1113
Mittova V, Theodoulou FL, Kiddle G, Volokita M, Tal M, Foyer CH, Guy M (2004b) Comparison of mitochondrial APX in the cultivated tomato, *Lycopersicon esculentum*, its wild, salt-tolerant relative, *L. pennellii*—role for matrix isoforms in protection against oxidative damage. Plant, Cell Environ 27:237–250
Miyagawa Y, Tamoi M, Shigeoka S (2000) Evaluation of the defense system in chloroplasts to photooxidative stress caused by paraquat using transgenic tobacco plants expressing catalase from *Escherichia coli*. Plant Cell Physiol 41:311–320
Miyake C, Michihata F, Asada K (1991) Scavenging of hydrogen peroxide in prokaryotic and eukaryotic algae: acquisition of ascorbate peroxidase during the evolution of cyanobacteria. Plant Cell Physiol 32:33–43
Miyake C, Asada K (1992) Thylakoid-bound ascorbate peroxidase in spinach chloroplasts and photoreduction of its primary product monodehydroascorbate radicals in thylakoids. Plant Cell Physiol 33:541–553
Miyake C, Cao WH, Asada K (1993) Purification and molecular properties of thylakoid-bound ascorbate peroxidase from spinach chloroplasts. Plant Cell Physiol 343:881–889
Miyake C, Asada K (1996) Inactivation mechanism of ascorbate peroxidase at low concentrations of ascorbate: hydrogen peroxide decomposes compound I of ascorbate peroxidase. Plant Cell Physiol 37:423–430
Miyake C, Shinzaki Y, Nishioka M, Horiguchi S, Tomizawa K (2006) Photoinactivation of ascorbate peroxidase in isolated tobacco chloroplasts: *Galdieria partita* APX maintains the electron flux through the water–water cycle in transplastomic tobacco plants. Plant Cell Physiol 47:200–210
Møller IM (2001) Plant mitochondria and oxidative stress: electron transport, NADPH turnover, and metabolism of reactive oxygen species. Annu Rev Plant Physiol Plant Mol Biol 52:561–591

Morita S, Nakatani S, Koshiba T, Masumura T, Ogihara Y, Tanaka K (2011) Differential expression of two cytosolic ascorbate peroxidases and two superoxide dismutase genes in response to abiotic stress in rice. Rice Sci 18:157–166

Mullen RT, Trelease RN (2000) The sorting signals for peroxisomal membrane-bound ascorbate peroxidase are within its C-terminal tail. J Biol Chem 275:16337–16344

Murgia I, Tarantino D, Vannini C, Bracale M, Carravieri S, Soave C (2004) *Arabidopsis thaliana* plants overexpressing thylakoidal ascorbate peroxidase show increased resistance to Paraquat-induced photooxidative stress and to nitric oxide-induced cell death. Plant J 8:940–953

Nakano Y, Asada K (1981) Hydrogen peroxide is scavenged by ascorbate-specific peroxidase in spinach chloroplasts. Plant Cell Physiol 22:867–880

Narendra S, Venkataramani S, Shen G, Wang J, Pasapula V, Lin Y, Kornyeyev D, Holaday AS, Zhang H (2006) The *Arabidopsis* ascorbate peroxidase 3 is a peroxisomal membrane-bound antioxidant enzyme and is dispensable for *Arabidopsis* growth and development. J Exp Bot 57:3033–3042

Nedelcu AM, Miles IH, Fagir AM, Karol K (2008) Adaptive eukaryote-to-eukaryote lateral gene transfer: stress-related genes of algal origin in the closest unicellular relatives of animals. J Evol Biol 21:1852–1860

Noctor G, Foyer CH (1998) Ascorbate and glutathione—keeping active oxygen under control. Annu Rev Plant Physiol Plant Mol Biol 49:249–279

Nulton-Persson AC, Szweda LI (2001) Modulation of mitochondrial function by hydrogen peroxide. J Biol Chem 276:23357–23361

Ogawa K, Kanematsu S, Takabe K, Asada K (1995) Attachment of CuZn-superoxide dismutase to thylakoid membranes at the site of superoxide generation (PSI) in spinach chloroplasts: detection by immunogold labeling after rapid freezing and substitution method. Plant Cell Physiol 36:565–573

Orvar L, Ellis BE (1997) Transgenic tobacco expressing antisense RNA for cytosolic ascorbate peroxidase show increased susceptibility to ozone injury. Plant J 11:1297–1305

Panchuk I, Volkov R, Schöffl F (2002) Heat stress and heat shock transcription factor-dependent expression and activity of APX in *Arabidopsis*. Plant Physiol 129:838–853

Panchuk II, Zentgraf U, Volkov RA (2005) Expression of the Apx gene family during leaf senescence of *Arabidopsis thaliana*. Planta 222:926–932

Pandey S, Fartyal D, Agarwal A, Shukla T, James D, Kaul T, Negi YK, Arora S, Reddy MK (2017) Abiotic stress tolerance in plants: myriad roles of ascorbate peroxidase. Front Plant Sci 8:581

Passaia G, Margis-Pinheiro M (2015) Glutathione peroxidases as redox sensor proteins in plant cells. Plant Sci 234:22–26

Passardi F, Bakalovic N, Teixeira FK, Margis-Pinheiro M, Penel C, Dunand C (2007) Prokaryotic origins of the non-animal peroxidase superfamily and organelle-mediated transmission to eukaryotes. Genomics 89:567–579

Per TS, Khan NA, Reddy PS, Masood A, Hasanuzzaman M, Khan MIR, Anjum NA (2017) Approaches in modulating proline metabolism in plants for salt and drought stress tolerance: phytohormones, mineral nutrients and transgenics. Plant Physiol Biochem 115:126–140

Pereira CS, Da Costa DS, Teixeira J, Pereira S (2005) Organ-specific distribution and subcellular localization of APX isoenzymes in potato *Solanum tuberosum* L. plants. Protoplasma 226:223–230

Pitcher LH, Repetti P, Zilinskas BA (1994) Overexpression of ascorbate peroxidase protects plants from oxidative stress. Plant physiol (suppl.) 105:116

Pnueli L, Liang H, Rozenberg M, Mittler R (2003) Growth suppression, altered stomatal responses, and augmented induction of heat shock proteins in cytosolic ascorbate peroxidase (Apx1)-deficient *Arabidopsis* plants. Plant J 34:185–201

Prasad TK, Anderson MD, Martin BA, Stewart CR (1994) Evidence for chilling-induced oxidative stress in maize seedlings and a regulatory role for hydrogen peroxide. Plant Cell 6:65–74

Ribeiro CW, Carvalho FEL, Rosa SB, Alves-Ferreira M, Andrade CMB, Ribeiro-Alves M, Silveira JAG, Margis R, Margis-Pinheiro M (2012) Modulation of genes related to specific metabolic

pathways in response to cytosolic ascorbate peroxidase knockdown in rice plants. Plant Biol 14:944–955
Ribeiro CW, Korbes AP, Garighan JA, Jardim-Messeder D, Carvalho FEL, Sousa RHV, Caverzan A, Teixeira FK, Silveira JAG, Margis-Pinheiro M (2017) Rice peroxisomal ascorbate peroxidase knockdown affects ROS signaling and triggers early leaf ssenescence. Plant Sci 263:55–65
Romero-Puertas MC, McCarthy I, Sandalio LM, Palma JM, Corpas FJ, Gómez M, del Río LA (1999) Cadmium toxicity and oxidative metabolism of pea leaf peroxisomes. Free Radic Res 31:S25–S31
Rosa SB, Caverzan A, Teixeira FK, Lazzarotto F, Silveira JA, Ferreira-Silva SL, Abreu-Neto J, Margis R, Margis-Pinheiro M (2010) Cytosolic APx knockdown indicates an ambiguous redox responses in rice. Phytochemistry 71:548–558
Rossel JB, Walter PB, Hendrickson L, Chow WS, Poole A, Mullineaux PM, Pogson BJ (2006) A mutation affecting ascorbate peroxidase 2 gene expression reveals a link between responses to high light and drought tolerance. Plant, Cell Environ 29:269–281
Sairam RK, Srivastava GC (2002) Changes in antioxidant activity in sub-cellular fractions of tolerant and susceptible wheat genotypes in response to long-term salt stress. Plant Sci 162:897–904
Sato Y, Murakami T, Funatsuki H, Matsuba S, Saruyama H, Tanida M (2001) Heat shock-mediated APX gene expression and protection against chilling injury in rice seedlings. J Exp Bot 52:45–151
Sato Y, Masuta Y, Saito K, Murayama S, Ozawa K (2011) Enhanced chilling tolerance at the booting stage in rice by transgenic overexpression of the ascorbate peroxidase gene *OsAPXa*. Plant Cell Rep 30:399–406
Saxena SC, Joshi P, Grimm B, Arora S (2011) Alleviation of ultraviolet-C induced oxidative through overexpression of cytosolic ascorbate peroxidase. Biologia 66:1052–1059
Scandalios JG (2005) Oxidative stress: molecular perception and transduction of signals triggering antioxidant gene defenses. Brazilian J Medic Biologic Res 38:995–1014
Sharma P, Dubey RS (2004) APX from rice seedlings, properties of enzyme isoforms, effects of stresses and protective roles of osmolytes. Plant Sci 167:541–550
Sharma P, Jha AB, Dubey RS, Pessarakli M (2012) Reactive oxygen species, oxidative damage, and antioxidative defense mechanism in plants under stressful conditions. J Botany 2012:1–26
Shi WM, Muramoto Y, Ueda A, Takabe T (2001) Cloning of peroxisomal ascorbate peroxidase gene from barley and enhanced thermotolerance by overexpressing in *Arabidopsis thaliana*. Gene 273:23–27
Shigeoka S, Ishikawa T, Tamoi M, Miyagawa Y, Takeda T, Yabuta Y, Yoshimura K (2002) Regulation and function of ascorbate peroxidase isoenzymes. J Exp Bot 53:1305–1319
Shigeoka S, Maruta T (2014) Cellular redox regulation, signaling, and stress response in plants. Biosci Biotechnol Biochem 78:1457–1470
Shikanai T, Takeda T, Yamauchi Y, Sano S, Tomizawa K, Yokota A, Shigeoka S (1998) Inhibition of ascorbate peroxidase under oxidative stress in tobacco having bacterial catalase in chloroplasts. FEBS Lett 428:47–51
Shrivastava DC, Kisku AV, Saxena M, Deswal R, Sarin NB (2015) Stress inducible cytosolic ascorbate peroxidase (Ahcapx) from *Arachis hypogaea* cell lines confers salinity and drought stress tolerance in transgenic tobacco. The Nucleus 58:3–13
Singh N, Mishra A, Jha B (2014) Over-expression of the Peroxisomal Ascorbate Peroxidase (SbpAPX) Gene Cloned from halophyte *Salicornia brachiata* confers salt and drought stress tolerance in transgenic tobacco. Mar Biotechnol 16:321–332
Song XS, Wang YJ, Mao WH, Shi K, Zhou YH, Nogués S, Yu JQ (2009) Effects of cucumber mosaic virus infection on electron transport and antioxidant system in chloroplasts and mitochondria of cucumber and tomato leaves. Physiol Plant 135:246–257
Sukweenadhi J, Kim YJ, Rahimi S, Silva J, Myagmarjav D, Kwon WS, Yang DC (2017) Overexpression of a cytosolic ascorbate peroxidase from *Panax ginseng* enhanced salt tolerance in *Arabidopsis thaliana*. Plant Cell Tiss Organ Cult 129:337–350
Taiz L, Zeiger E (2013) Fisiologia vegetal. Artmed, Porto Alegre

Tanou G, Filippou P, Belghazi M, Job D, Diamantidis G, Fotopoulos V, Molassiotis A (2012) Oxidative and nitrosative-based signaling and associated post-translational modifications orchestrate the acclimation of citrus plants to salinity stress. The Plant J 72:585–599

Teixeira FK, Menezes-Benavente L, Margis R, Margis-Pinheiro M (2004) Analysis of the molecular evolutionary history of the ascorbate peroxidase gene family, inferences from the rice genome. J Mol Evol 59:761–770

Teixeira FK, Menezes-Benavente L, Galvão VC, Margis R, Margis-Pinheiro M (2006) Rice ascorbate peroxidase gene family encodes functionally diverse isoforms localized in different subcellular compartments. Planta 224:300–314

Tretter L, Adam-Vizi V (2000) Inhibition of Krebs cycle enzymes by hydrogen peroxide: a key role of [alpha]-ketoglutarate dehydrogenase in limiting NADH production under oxidative stress. J Neurosci 20:8972–8979

Vaahtera L, Brosche M, Wrzaczek M, Kangasjarvi J (2014) Specificity in ROS signaling and transcript signatures. Antioxid Redox Signal 21:1422–1441

Verniquet F, Gaillard J, Neuburger M, Douce R (1991) Rapid inactivation of plant aconitase by hydrogen peroxide. Biochem J 276:643–648

Vighi IL, Benitez LC, Amaral MN, Moraes GP, Auler PA, Rodrigues GS, Deuner S, Maia LC, Braga EJB (2017) Functional characterization of the antioxidant enzymes in rice plants exposed to salinity stress. Biol Plantarum 61:540–550

Wada N, Kinoshita S, Matsuo M, Amako K, Miyake C, Asada K (1998) Purification and molecular properties of ascorbate peroxidase from bovine eye. Biochem Biophys Res Commun 242:256–261

Wang Y, Wisniewski M, Meilan R, Cui M, Webb R, Fuchigami L (2005) Overexpression of cytosolic ascorbate peroxidase in tomato confers tolerance to chilling and salt stress. JASHS 130:167–173

Wang Y, Wu J, Choi YW, Jun TH, Kwon SW, Choi IS, Kim YC, Gupta R, Kim ST (2015) Expression analysis of *Oryza sativa* ascorbate peroxidase 1 (OsAPx1) in response to different phytohormones and pathogens. J Life Sci 25:1091–1097

Wang L, Zhao R, Zheng Y, Chen L, Li R, Ma J, Hong X, Ma P, Sheng J, Shen L (2017) SlMAPK1/2/3 and Antioxidant enzymes are associated with H_2O_2-induced chilling tolerance in tomato plants. J Agri Food Chem 65:6812–6820

Webb RP, Allen RD (1996) Overexpression of pea cytosolic ascorbate peroxidase confers protection against oxidative stress in transgenic *Nicotiana tabacum*. Plant Physiol 99:912–918

Welinder KG (1992) Superfamily of plant, fungal and bacterial peroxidases. Curr Opin Struct Biol 2:388–393

Wood ZA, Poole LB, Karplus PA (2003) Peroxiredoxin evolution and the regulation of hydrogen peroxide signaling. Science 300:650–653

Wu G, Wang G, Ji J, Gao H, Guan W, Wu J, Guan C, Wang Y (2014) Cloning of a cytosolic ascorbate peroxidase gene from *Lycium chinense* Mill. and enhanced salt tolerance by overexpressing in tobacco. Gene 543:85–92

Xu WF, Shi WM, Liu F, Ueda A, Takabe T (2008) Enhanced zinc and cadmium tolerance and accumulation in transgenic Arabidopsis plants constitutively overexpressing a barley gene (HvAPX1) that encodes a peroxisomal ascorbate peroxidase. Botany 86:567–575

Xu L, Carrie C, Law SR, Murcha MW, Whelan J (2013) Acquisition, conservation, and loss of dual-targeted proteins in land plants. Plant Physiol 161:644–662

Yabuta Y, Motoki T, Yoshimura K, Takeda T, Ishikawa T, Shigeoka S (2002) Thylakoid membrane-bound ascorbate peroxidase is a limiting factor of antioxidative systems under photo-oxidative stress. Plant J 32:915–925

Yadav P, Yadav T, Kumar S, Rani B, Kumar S, Jain V, Malhotra SP (2014) Partial purification and characterization of ascorbate peroxidase from ripening ber (*Ziziphus mauritiana* L.) fruits. Afr J Biotechnol 13:3323–3331

Yamaguchi K, Mori H, Nishimura M (1995a) A novel isoenzyme of ascorbate peroxidase localized on glyoxysomal leaf peroxisomal membranes in pumpkin. Plant Cell Physiol 36:1157–1162

Yamaguchi K, Takeuchi Y, Mori H, Nishimura M (1995b) Development of microbody membrane proteins during the transformation of glyoxysomes to leaf peroxisomes in pumpkin cotyledons. Plant Cell Physiol 36:455–464

Yamazaki D, Motohashi K, Kasama T, Hara Y, Hisabori T (2004) Target proteins of the cytosolic thioredoxins in *Arabidopsis thaliana*. Plant Cell Physiol 45:18–27

Yan J, Wang J, Tissue D, Holaday AS, Allen R, Zhang H (2003) Photosynthesis and seed production under water-deficit conditions in transgenic tobacco plants that overexpress an *Arabidopsis* ascorbate peroxidase gene. Crop Sci 43:1477–1483

Yoshimura K, Ishikawa T, Nakamura Y, TamoiM Takeda T, Tada T, Nishimura K, Shigeoka S (1998) Comparative study on recombinant chloroplastic and cytosolic ascorbate peroxidase isozymes of spinach. Arch Biochem Biophys 353:55–63

Yoshimura K, Yabuta Y, Ishikawa T, Shigeoka S (2000) Expression of spinach ascorbate peroxidase isoenzymes in response to oxidative stresses. Plant Physiol 123:223–233

Zaharieva T, Abadía J (2003) Iron deficiency enhances the levels of ascorbate, glutathione related enzymes in sugar beet roots. Protoplasma 221:269–275

Zamocky M, Gasselhuber B, Furtmüller PG, Obinger C (2014) Turning points in the evolution of peroxidase–catalase superfamily: molecular phylogeny of hybrid heme peroxidases. Cell Mol Life Sci 71:4681–4696

Zhang H, Wang J, Nickel U, Allen RD, Goodman HM (1997) Cloning and expression of an *Arabidopsis* gene encoding a putative peroxisomal ascorbate peroxidase. Plant Mol Biol 34:967–971

Zhang Z, Zhang Q, Wu J, Zheng X, Zheng S, Sun X, Qiu Q, Lu T (2013) Gene knockout study reveals that cytosolic ascorbate peroxidase 2 (*OsAPX2*) plays a critical role in growth and reproduction in rice under drought, salt and cold stresses. PLoS ONE 8:e57472

Zhang Q, Ma C, Xue X, Xu M, Li J, Wu JX (2014) Overexpression of a cytosolic ascorbate peroxidase gene, *OsAPX2*, increases salt tolerance in transgenic alfalfa. J Integr Agric 13:2500–2507

Role of Reactive Oxygen Species Homeostasis in Root Development and Rhizotoxicity in Plants

Ayan Sadhukhan and Hiroyuki Koyama

Abstract Root growth is a vital process for plant survival under different environmental conditions, and roots are the primary sites of both nutrient uptake and encounter with toxic substances in the soil. Cell redox homeostasis plays a critical role in root growth both under control and rhizotoxic stress conditions. These include high salinity, several metals, and metalloid stresses as well as shortage or excess of essential nutrients like nitrogen, phosphate, potassium, and iron. Reactive oxygen species (ROS) formation primarily by plasma membrane NADPH oxidase/respiratory burst oxidase homolog (Rboh)s is necessary for maintaining essential developmental programs related to root growth mostly by influencing cell wall-related processes and molecular signaling. Specific species of reactive oxygen affects separate processes like cell elongation and differentiation at different root tissues, whereas excess ROS due to disrupted ROS balance and failure of antioxidant machinery cause extensive oxidative damage leading to root growth arrest. Critical molecules of the cell redox homeostasis network provide promise for engineering plant tolerance to rhizotoxic stressors.

1 Introduction

Reactive oxygen species (ROS) are produced by biological reactions as byproducts of metabolism as well as chemical reactions in the cell and participate in many important biological processes as signaling molecules. Superoxide ($\cdot O_2^-$), hydrogen peroxide (H_2O_2), singlet oxygen ($_1O^2$), and hydroxyl radicals ($\cdot OH$) are among different species of reactive oxygen. Production of these species is further enhanced in plant cells under every environmental stress encountered and cause oxidative damage to important biomolecules affecting plant metabolism and growth (Miller et al. 2010). The root is the most important organ exposed to several toxic substances in the soil including high salt (NaCl), aluminum (Al), protons (H^+), heavy metal(oids) including cadmium (Cd), mercury (Hg), arsenic (As), and other con-

A. Sadhukhan · H. Koyama (✉)
Faculty of Applied Biological Sciences, Gifu University, 1-1 Yanagido, Gifu 501-1132, Japan
e-mail: koyama@gifu-u.ac.jp

S. K. Panda and Y. Y. Yamamoto (eds.), *Redox Homeostasis in Plants*,
Signaling and Communication in Plants, https://doi.org/10.1007/978-3-319-95315-1_6

ditions like poor soils deficient in essential nutrients like potassium, sulfate, and phosphate. The redox balance in root tissues dedicated for growth and differentiation is severely affected by these rhizotoxic ions leading to overall growth defect of the entire plant. ROS like $\cdot O_2^-$ are primarily generated at the apoplast by plasma membrane NADPH oxidase/respiratory burst oxidase homologs (Rbohs) which gets dismuted to H_2O_2 by superoxide dismutase. $\cdot O_2^-$ and H_2O_2 interact with metals like Fe or Cu to generate the extremely reactive ·OH radical through the Fenton chemistry (Briat and Lebrun 1999). Other sources of ROS include chloroplast, mitochondria, and peroxisomes. Several antioxidant machineries including several enzymatic and non-enzymatic molecules in these cellular compartments ameliorate excess ROS. For example, the potential oxidant H_2O_2 is reduced by ascorbate peroxidase (APX) and catalase (CAT) enzymes to H_2O. The ascorbate (AsA)–glutathione (GSH) cycle maintains the ROS scavenging system by donating electrons (Nakano and Asada 1987; Gill et al. 2013). But under rhizotoxic stresses, these antioxidant machineries are sometimes inhibited leading to ROS overproduction and subsequent cellular damage.

2 Role of ROS Homeostasis in Root Development Under Non-stressful Conditions

Historically, it was known that knockout mutants of *Arabidopsis thaliana* respiratory burst oxidase homologs like *AtrbohF* and *AtrbohD* produced shorter root (Torres et al. 2002; Kwak et al. 2003). In the mutants of *root hair defective* (*RHD2*)/Atrboh*C*, root and root hair elongation are reduced, and calcium uptake and cell elongation are compromised due to lower ROS production. Treatment with extracellular ROS stimulated calcium acquisition channels suppressing the phenotype (Foreman et al. 2003). These results suggested that NADPH oxidase-produced ROS regulate cell expansion by activating calcium channels. Additionally, ROP GTPases are required for ROS production by NADPH oxidases for normal root hair growth (Jones et al. 2007). A signaling kinase (oxidative stress-inducible 1 (OXI1) receives signals from ROS and relays it via several mitogen-activated protein kinases (MAPKs) to control root hair development (Rentel et al. 2004) (Fig. 1). Different ROS like superoxide ($\cdot O_2^-$) and hydrogen peroxide (H_2O_2) are spatially distributed in different areas of the root. $\cdot O_2^-$ mainly localizes in the apoplast of cell elongation zone, and H_2O_2 in the differentiation zone and the cell wall of root hairs (Dunand et al. 2007). In the same study, a decrease in $\cdot O_2^-$ concentration by several treatments reduced root elongation and root hair formation, and scavenging of H_2O_2 improved root elongation but reduced root hair formation (Dunand et al. 2007). In a different study, it was found that $\cdot O_2^-$ accumulates in the meristematic region for maintaining cell proliferation, whereas H_2O_2 accumulates in the elongation zone for differentiation (Tsukagoshi et al. 2010) (Fig. 1). A transcription factor UPBEAT1 (UPB1) negatively regulates peroxidases which are responsible for ROS balance ($\cdot O_2^- \leftrightarrow H_2O_2$)

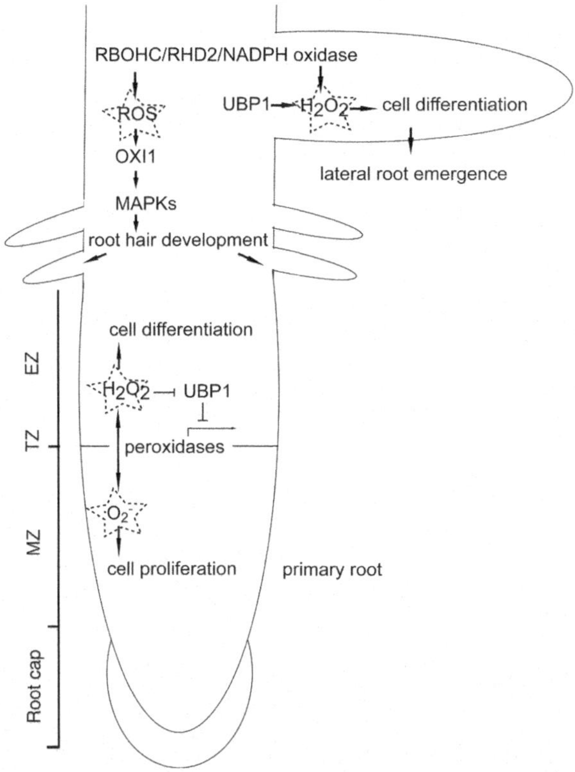

Fig. 1 Role of reactive oxygen species (ROS) homeostasis in root development under non-stressful conditions. Even without any stress, the cell membrane-bound enzyme NADPH oxidase/respiratory burst oxidase homolog C (RbohC)/root hair defective 2 (RHD2) produces ROS which participates in signal transduction via oxidative stress-inducible 1 (OXI1) and mitogen-activated protein kinases (MAPKs) to control root hair development. Superoxide radicals ($\cdot O_2^-$) accumulate in the meristematic zone (MZ) of the root to promote cell proliferation, while hydrogen peroxide (H_2O_2) accumulates in the elongation zone (EZ) to promote cell elongation. The transcription factor UBP1 controls expression levels of peroxidases in the transition zone (TZ) which in turn maintain a balance between $\cdot O_2^-$ and H_2O_2 leading to elongation of the primary root. UBP1 also control lateral root emergence by the same mechanism of promoting cell differentiation by controlling H_2O_2 production

at the transition between root elongation zone and meristem where differentiation begins (Tsukagoshi et al. 2010). *UPB1* expression is again negatively regulated by H_2O_2 in a feedback loop. UPB1 also regulates lignin synthesis suggesting its role in modification of cell walls for cell expansion. Loss of UBP1 disrupts ROS balance leading to increased $\cdot O_2^-$ and reduced H_2O_2 in the elongation zone, causing delayed differentiation and longer root with increased meristem size and longer cell length without any effect of phytohormone signaling. Conversely, overexpression of UBP1 leads to increased H_2O_2 in the elongation zone leading to shorter roots with shorter meristem and cell size. Recently, in monocots too (e.g., rice), ROS produced by several inducible NADPH oxidase homologs were demonstrated to be essential for root elongation during early stages of seedling establishment (Li et al. 2017).

Apart from the primary root, the lateral roots also form an essential part of the root system architecture and are under the control of ROS signaling. Several enzymes related to ROS homeostasis affect lateral root formation. Overexpression of peroxidases has been found to reduce the formation of lateral roots by oxidizing the auxins (Lagrimini et al. 1997). Glutathione peroxidases also repress lateral root formation (Passaia et al. 2014). Several other ROS production enzymes like cytochromes P450, AtrbohC NADPH oxidase, lipoxygenases, and electron carrier proteins were found to control lateral root emergence (Manzano et al. 2014). The transcription factor UBP1 controlling peroxidases in the model proposed by Tsukagoshi et al. (2010) was also found to be expressed in the lateral root primordia and control lateral root emergence by promoting the transition from proliferation to differentiation (Manzano et al. 2014).

3 Role of ROS Homeostasis in Root Development Under High Salt Stress

ROS metabolism (production as well as scavenging) in the roots has been observed to promote plant tolerance to high soil salinity (NaCl) by various mechanisms (Fig. 2). Under salinity stress, actin filaments are disrupted which disrupts the regulation of ROS levels produced by AtrbohC leading to extensive damage to the plant root (Liu et al. 2012). Jiang et al. (2016) showed that salt stress controls root meristem through changes in redox status and distribution of auxin transporters. Under control conditions, a sharp redox potential gradient exists between the proximal and distal ends of the meristem. The redox status of the root cap initials, quiescent center, and the distal part of the meristem are reduced while the redox state becomes more and more oxidized as we move toward the proximal meristem and further up basally to the transition and elongation zones. Under salt stress, the potential gradient ceases to exist, because the root cap initials and quiescent center also become oxidized and the localization of the auxin transporters also diffuses out. A ROS-dependent signaling pathway also controls water uptake by *A. thaliana* roots under NaCl stress. This is due to internalization of water conducting plasma membrane intrinsic proteins (PIPs) in small vesicles/vacuoles during salt stress, a process counteracted by application of exogenous ROS scavengers (Boursiac et al. 2008). Transgenic rice coexpressing ROS metabolism enzymes glutathione S-transferase and catalase showed lower root-to-shoot ratio but a higher number of adventitious roots under salt stress which was attributed to the altered redox states of components of the ascorbate–glutathione cycle (Zhao and Zhang 2006).

But ROS also acts as signaling molecules which actually promotes salt tolerance. For example, AtrbohF (soil salinity sensitive1-1)-induced ROS production in root vascular tissue leads to salinity tolerance and the loss of function mutant of this gene

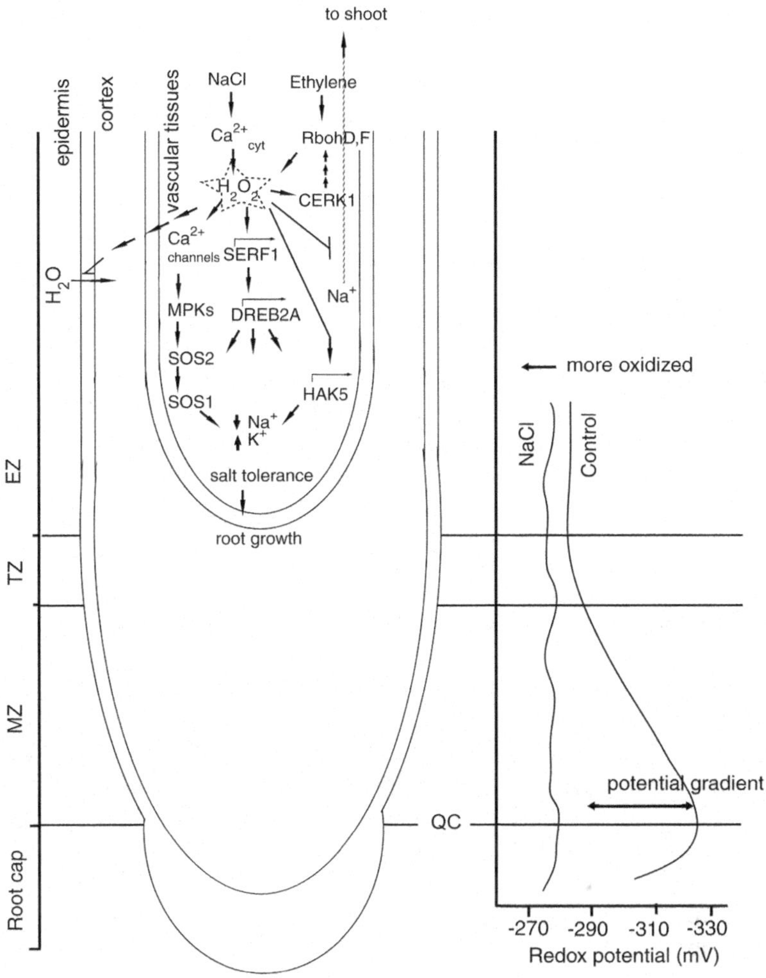

Fig. 2 Role of reactive oxygen species (ROS) homeostasis in root development under high salt stress. Under high salt stress, cell redox potential becomes more and more oxidized in different regions of the root. The potential gradient between proximal and distal ends of the meristematic zone under non-stressful condition becomes abolished under high salt (see graph at right, adapted from Jiang et al. 2016) leading to a flatter and more oxidized profile. High salt induces a sharp increase in cytosolic calcium ion concentration ($Ca^{2+}{}_{cyt}$) which induces production of ROS like hydrogen peroxide (H_2O_2). Under high salt, plant hormone ethylene also induces ROS production via respiratory burst oxidase homolog D and F (RbohD, F). ROS in moderate levels leads to signal transduction via kinases like CERK1 to control their own production in feedback loops. ROS induce transcription of several transcription factors such as salt-responsive ERF1 (SERF1) and dehydration responsive element-binding protein (DREB) leading to expression of salt tolerance genes. ROS signaling also regulates Na^+/K^+ homeostasis via control of the K^+ transporter HAK5, the salt overly sensitive (SOS) pathway, and calcium channels. All these lead to maintenance of root growth under high salt stress. ROS also control the root-to-shoot movement of Na^+ and water conduction by roots under salt stress through signal transduction. See main text for details. EZ elongation zone MZ meristematic zone QC quiescent center

is hypersensitive to salt (Jiang et al. 2012). Under salt stress, NaCl induces a sharp increase in cytosolic free Ca^{2+}. This lead to increased ROS (H_2O_2) production which in turn induces transcription factor salt-responsive ERF1 *SERF1* which regulates several salt response genes including dehydration responsive element-binding protein (DREB2A) (Schmidt et al. 2013). Several genes of the ROS response network are also induced in the root during salinity stress (Miller et al. 2010). These include signal transducers and transcription factors (TFs) promoting salt tolerance, e.g., CERK1, a receptor-like kinase controlling *RbohD* and *RbohF*, and TF WRKY70 which is involved in making APX1 mutants more salt tolerant (Ciftci-Yilmaz et al. 2007; Miller et al. 2007).

It was observed that a rise in ethylene production leads to salt tolerance. This is due to triggering of AtrbohF-dependent ROS production in root vasculature by ethylene signaling (Jiang et al. 2013; Ai et al. 2014). On another hand, ethylene signaling promotes salt tolerance by reducing ROS production through TFs JERF3 (Wu et al. 2008) and ethylene insensitive 3 (EIN3) (Peng et al. 2014). Na^+/K^+ homeostasis is important to maintain growth under high salt stress. While the beneficial K^+ has to enter and retained in the cell, the toxic Na^+ should be prevented from entering the cell or sequestered in a cellular compartment. ROS produced by AtrbohD and AtrbohF are essential in Na^+/K^+ homeostasis (reducing Na^+ and increasing K^+) as evidenced from the salt hypersensitivity of their mutants (Ma et al. 2012; Jiang et al. 2012). Particularly, under salt stress, ROS produced by AtrbohF in root vascular tissue reduces Na^+ in the xylem sap and further transport of Na^+ from root to the shoot, leading to salinity tolerance (Jiang et al. 2012). It was proposed that ROS may regulate Na^+/K^+ homeostasis by regulating the activity of cell membrane Ca^{2+} channels which propagate the signal via mitogen-activated protein kinases (MPKs), salt overly sensitive 2 (SOS2), and calcineurin B-like (CBL) (Mori and Schroeder 2004; Pottosin et al. 2014). The cell membrane Na^+/H^+ antiporter SOS1 and H^+-ATPase mRNA stability and/or activity are also controlled by ROS (Chung et al. 2008; Zhang et al. 2007). AtrbohF-dependent ROS production in the vascular tissue is again regulated by ethylene signaling which inhibits root-to-shoot Na^+ delivery (Jiang et al. 2013). Ethylene also controls AtrbohC-dependent ROS production leading to K^+ retention under salt stress through transcriptional regulation of K^+ transporter AtHAK5 (Shin and Schachtman 2004; Jung et al. 2009; Li et al. 2015).

Several other non-enzymatic methods of ROS scavenging in plant roots also promote salinity tolerance. Polyamine oxidase involved in polyamine catabolism in the root also regulates ROS balance (generation and scavenging) which in turn regulates expression of several drought-responsive genes (Kamada-Nobusada et al. 2008). A helicase PDH45 was found to minimize ROS production in roots and thereby confer salinity and drought tolerance in *Arabidopsis* and rice (Luo et al. 2009; Nath et al. 2016).

4 Role of ROS Homeostasis in Root Development Under Aluminum Stress

Aluminum stress alters the cellular ROS levels in different areas of the root which participate in controlling root growth (Fig. 3). In the root elongation zone, the ROS level decreases, demonstrated in *Arabidopsis* by a live-cell probe: hyper (Hernández-Barrera et al. 2015). Al increases cell wall components like polysaccharides, hemicelluloses, extensins, callose, and lignin (Jones et al. 2006) which produce a negative impact on ROS (superoxide radicals later dismuted to H_2O_2)-producing enzymes, such as apoplastic diamine oxidase or NADPH oxidases. Al also binds to the plasma membrane and affects receptor-like kinases (Blancaflor et al. 1998) that are important modulators of NADPH oxidases (Wolf and Hofte 2014; Boisson-Dernier et al. 2013; Duan et al. 2010). The decrease in intracellular ROS in this area leads to stunting of root growth under Al stress.

In a different part of the root, viz. the root distal transition zone (DTZ), 1–3 mm behind the root tip, the area of transition from cell division to cell elongation, Al reportedly induces higher accumulation of ROS in sorghum (Sivaguru et al. 2013). Coincidentally, the epidermal and outer cortical cells in the DTZ of the root accumulated higher transcript and protein levels of the citrate transporter, SbMATE. Hence, the purpose of the increased accumulation of ROS is to carry the signals from Al perception to the effector gene (SbMATE) expression leading to stress alleviation (Sivaguru et al. 2013). ROS production is induced in Al stress by several mechanisms including induction of ROS-producing enzymes like cytochrome P450 and cytochrome c oxidase (Richards et al. 1998; Milla et al. 2002; Goodwin and Sutter 2009; Huang et al. 2014). But ROS is mainly generated in Al stress by the plasma membrane NADPH oxidase which is further activated by Al-induced increase in cytoplasmic Ca^{2+} concentration (Bhuja et al. 2004). NADPH oxidase produces superoxide (Sagi and Fluhr 2001) which is subsequently dismuted to H_2O_2. Al together with other metals like Fe and Cu activates the Fenton reaction leading to highly reactive ·OH radical formation (Mujika et al. 2011; Ruipérez et al. 2012). ROS production may also result from disturbed redox homeostasis by acidification of the cytoplasm by Al (Moseyko and Feldman 2001).

In the elongation zone, for root growth maintenance, the cell wall is remodeled/loosened by the action of ROS like hydroxyl radicals (·OH) which break glycosidic bonds (Houde and Diallo 2008). Ascorbate (ASC) donates an electron when it is transformed to monodehydroascorbate in the apoplast. This electron reduces Cu^{2+} to Cu^+ (catalyzed by Al-inducible blue copper-binding proteins) which takes part in a Fenton reaction producing ·OH from H_2O_2. A balance of ASC levels is maintained in both apoplast and symplast. Al blocks the electron transfer from the cytoplasm to ASC or from ASC to Cu^{2+} (Maltais and Houde 2002); the diverted electron generates different ROS molecules for signaling or oxidative damage. Without fast enough regeneration, ASC is degraded to oxalate, which could lead to the production of H_2O_2 by Al-responsive oxalate oxidase (Hamel et al. 1998) in the apoplast, which goes on feeding the Fenton chemistry.

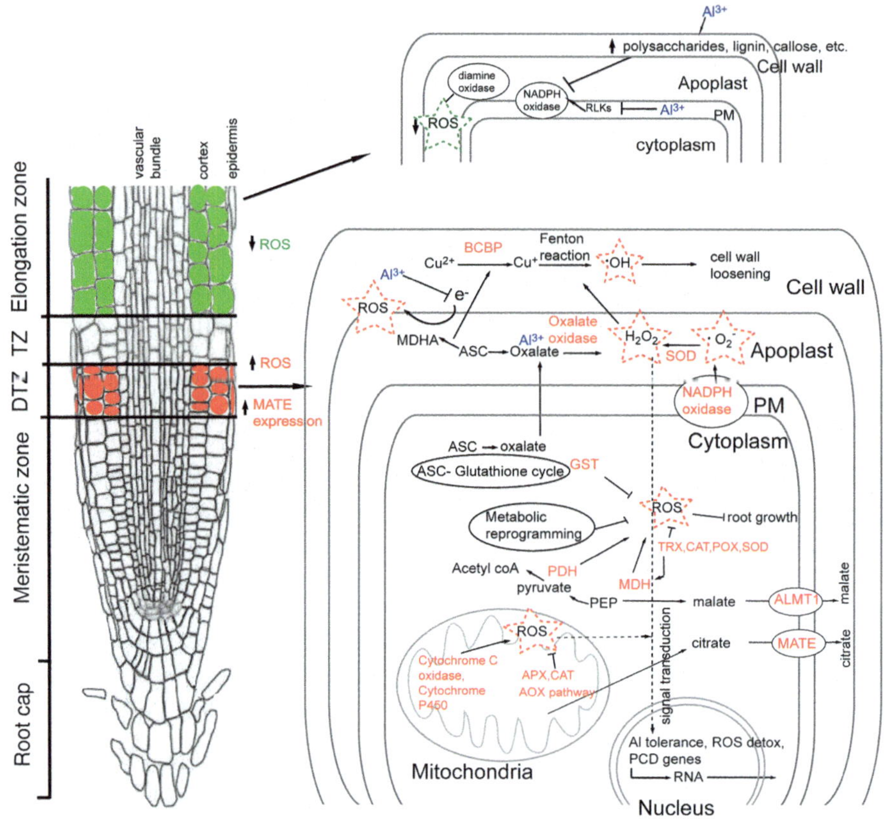
Al3+
polysaccharides, lignin, callose, etc.
Cell wall
diamine oxidase
NADPH oxidase
Apoplast
RLKs
Al3+
PM
ROS
cytoplasm
vascular bundle
cortex
epidermis
Elongation zone
TZ
DTZ
Meristematic zone
Root cap
ROS
ROS
MATE expression
Fenton reaction
BCBP
Cu2+
Cu+
OH
cell wall loosening
Cell wall
Al3+
e-
ROS
Oxalate oxidase
MDHA
Al3+
ASC
Oxalate
H2O2
O2-
SOD
Apoplast
NADPH oxidase
PM
Cytoplasm
ASC oxalate
ASC- Glutathione cycle
GST
ROS
root growth
Metabolic reprogramming
TRX,CAT,POX,SOD
Acetyl coA
PDH
pyruvate
MDH
PEP
malate
ALMT1
malate
ROS
citrate
MATE
citrate
Cytochrome C oxidase, Cytochrome P450
APX,CAT
AOX pathway
signal transduction
Al tolerance, ROS detox, PCD genes
RNA
Mitochondria
Nucleus

◀**Fig. 3 Role of reactive oxygen species (ROS) homeostasis in root development under aluminum stress.** Figure shows changes in cellular reactive oxygen species (ROS) levels and related metabolic processes in different regions of the growing root in response to aluminum (Al^{3+}) stress. In the elongation region of the root, ROS levels decrease (shown in green) due to an increase of cell wall components like polysaccharides and lignin, which inhibit ROS-generating apoplastic NADPH oxidase and diamine oxidases. Plasma membrane (PM)-bound Al also inhibits receptor-like kinases (RLKs) that regulate these ROS-producing enzymes. The epidermal and cortical cells of the distal transition (DTZ) zone of the root have an increase in ROS levels (shown in red) which coincides with higher expression and localization of malate releasing MATE transporter. ROS are generated in the mitochondria by cytochrome c oxidase and cytochrome P450 in the mitochondria or NADPH oxidase in the PM. Superoxide radicals ($\cdot O_2^-$) generated by NADPH oxidase are dismuted to hydrogen peroxide (H_2O_2) by superoxide dismutase (SOD). Divalent copper ions (Cu^{2+}) are reduced to monovalent copper (Cu^+) by blue copper-binding protein (BCBP) which along with H_2O_2 generates hydroxyl radicals ($\cdot OH$) by the Fenton reaction. The $\cdot OH$ loosen the cell wall by breaking glycosidic bonds. Electrons (e^-) for the reduction of Cu^{2+} are donated by ascorbate (ASC) (electron transfer being blocked and diverted by Al toward ROS production) when it is transformed to monodehydroascorbate (MDHA). ASC is degraded to oxalate which produces H_2O_2 by oxalate oxidase. ASC levels are maintained in both apoplast and cytoplasm. A key enzyme associated with ASC-GSH cycle, viz. glutathione S-transferase (GST), also takes part in ROS detoxification. Metabolic enzymes including pyruvate dehydrogenase (PDH) and malate dehydrogenase (MDH) also play role in ROS production/homeostasis. MDH is controlled by thioredoxin (TRX). Malate is released by PM transporter ALMT1 and citrate (produced in mitochondria) by MATE transporter for Al^{3+} chelation. Cytoplasmic and mitochondrial isoforms of catalase (CAT), peroxidase (POX), and ascorbate peroxidase (APX) help to detoxify ROS. Metabolic reprogramming and the alternative oxidase (AOX) pathways detoxify ROS. ROS including H_2O_2 also takes part in signal transduction to the nucleus (dotted arrow) and at high levels causes cellular damage and targets programmed cell death (PCD)

Since excessive accumulation of ROS in Al stress leads to oxidation of cellular molecules and subsequent cell death (Yamamoto et al. 2002; Panda et al. 2008; Huang et al. 2014) and sometimes stunted growth and metabolic shifts (Chowra et al. 2017), several ROS detoxifying systems manage the excess ROS by detoxification. Metabolic reprogramming was found to detoxify ROS (Baena-González et al. 2007). Apart from this, large gene families encoding enzymes for detoxification of ROS components are induced to alleviate such oxidative damage in prolonged Al stress (Richards et al. 1998; Chowra et al. 2017). In fact, several ROS scavenging enzymes (3 glutathione s-transferase (GST)s and 2 peroxidases (POX)s) were found to be commonly induced by many rhizotoxic ions like Na, Al, Cd, and Cu (Zhao et al. 2008). But, interestingly, some ROS scavenging enzymes like GST and catalase (CAT) were originally identified by screening Al-inducible genes in *Arabidopsis* (Ezaki et al. 2000). Members of the AOX family reduce mitochondrial ROS production (Clifton et al. 2006) leading to Al tolerance (Panda et al. 2008, 2013). Al-responsive enzyme malate dehydrogenase (MDH) that produces organic acid malate, exudated from the cell for aluminum chelation (Gaume et al. 2001; Ermolayev et al. 2003; Houde and Diallo 2008), is also a known enzyme of redox homeostasis. ROS including H_2O_2 on the other hand transcriptionally control organic acid transporters MATE and ALMT1 essential for aluminum tolerance (Kobayashi et al. 2013). One more enzyme of the ROS network, Al-inducible thioredoxin (TRX; Kumari et al. 2008), directly regulates cytosolic MDH (Hara et al. 2006; Hisabori et al. 2005). Heme-containing peroxidases of type III either generate or detoxify ROS, thereby modulating cell wall modification (Passardi et al. 2005) and controlling growth and cell elongation (Van et al. 1994; Kumari et al. 2008; Goodwin and Sutter 2009). Other ROS detoxifying enzymes were differentially regulated by Al, e.g., glutathione S-transferase (GST) (Ezaki et al. 2000; Kumari et al. 2008), POX (Ezaki et al. 1996), ascorbate peroxidase, glutathione reductase, and superoxide dismutase (SOD) (Kumari et al. 2008; Wu et al. 2013; Sunkar et al. 2006). Overexpression of several of the ROS detoxifying enzymes has been used to enhance aluminum tolerance with the improvement of root growth as tolerance index, establishing the firm link between oxidative stress and Al toxicity. Overexpression of NtPox (tobacco peroxidase) and parB (tobacco glutathione S-transferase) in *Arabidopsis* (Ezaki et al. 2000) and WMnSOD1 in wheat and mustard (Basu et al. 2001), and chloroplastic MDH in alfalfa (Tesfaye et al. 2001) enhanced Al resistance in transgenic plants. Manipulating non-enzymatic antioxidant defense molecules also provided Al tolerance. Overexpression of a dehydroascorbate reductase to increase the ascorbate levels in tobacco (Yin et al. 2010) conferred tolerance to Al, and reduction in polyamine synthesis by a mutation rendered Al sensitivity (Nezames et al. 2012), as well as exogenous polyamine application improved Al tolerance of saffron (Chen et al. 2008) because polyamines have protective roles against ROS (Alcázar et al. 2010).

5 Role of ROS Homeostasis in Root Development Under Cadmium and Metalloid Stress

Cadmium is a heavy metal that is highly toxic to the plant roots which is the primary site of nutrient mineral uptake from the soil (Lux et al. 2011; Huang et al. 2017). Cadmium enters roots through transporters for micronutrients like manganese and iron (Lux et al. 2011; Cappa and Pilon-Smits 2014). The growth and development of the root apex are affected due to ROS formed during cadmium stress (Xiong et al. 2009) causing extensive nuclear disintegration (Prasad 1995) (Fig. 4). Activities of several ROS scavenging enzymes like superoxide dismutase (SOD), ascorbate peroxidase (APX), peroxidase (POD), and catalase (CAT) are also inhibited by cadmium (Fidalgo et al. 2011; Sebastian and Prasad 2014) (Fig. 4). The activities of these enzymes recovered with iron or manganese supplement leading to lower ROS and less subsequent nuclear damage and higher growth of the roots (Sebastian and Prasad 2015). These supplements also enhanced the activity of the enzymes of the ascorbate (AsA)–glutiathione (GSH) cycle which manages ROS balance. Historically, a cadmium sensitive mutant of *Arabidopsis*, cadmium sensitive 2/root meristemless 1(cad2/rml1), unraveled a key developmental phenotype in the root under cadmium stress, viz. post-embryonic root growth arrest (Howden et al. 1995) due to lack of G1-S phase transition in the cell cycle. This locus encoded the first enzyme in GSH biosynthesis: gamma-glutamylcysteine synthetase, which is essential for maintain-

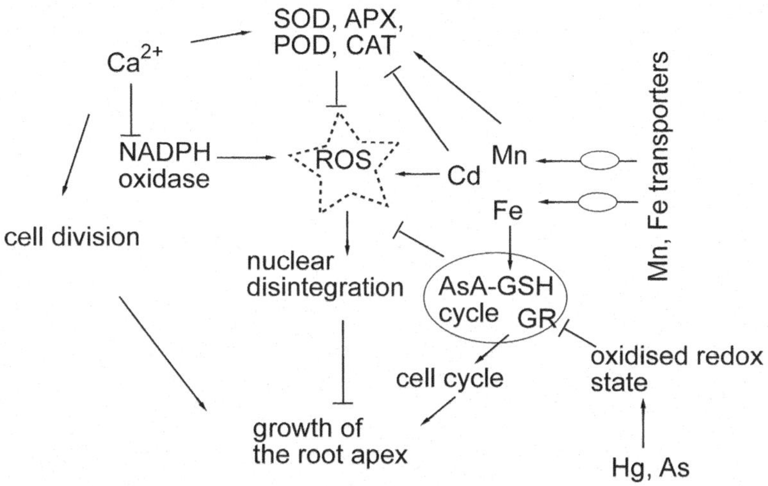

Fig. 4 Role of reactive oxygen species (ROS) homeostasis in root development under cadmium and metalloid stress. Cadmium (Cd) enters the cell via transporters for manganese (Mn) and iron (Fe) and induces ROS formation leading to nuclear disintegration and subsequent root growth arrest. Supplementation with Mn and Fe leads to improvement of root growth through ROS balance, due to enhanced activity of ROS scavenging enzymes such as superoxide dismutase (SOD), ascorbate peroxidase (APX), peroxidase (POD), and catalase (CAT) and also enzymes of the ascorbate (AsA)–glutathione (GSH) cycle. Calcium promotes root growth under high cadmium stress by inhibiting ROS production by NADPH oxidase and promoting cell division. Metalloids like mercury (Hg) and arsenic (As) oxidize the redox state of the cell inhibiting the activity of the enzyme glutathione reductase (GR) leading alteration of the GSH pool and subsequently to root growth arrest. See the main text for details

ing cellular GSH levels necessary for normal cell division and progression through the cell cycle in the roots (Vernoux et al. 2000).

Cadmium-induced root growth inhibition was also found to be rescued by the application of calcium, which improves cell division, decreases chromosomal aberrations (Mohamed 2012), and improves auxin transport in the root (Li et al. (2015). Under high cadmium stress, calcium also decreases ROS formation in the roots by regulating plasma membrane NADPH oxidases (Heyno et al. 2008), antioxidants, and ROS scavenging enzymes (Talukdar 2012; Farzadfar et al. 2013; Ahmad et al. 2015).

Metalloids include mercury and arsenic. Under metalloid stress, the redox status of the cell becomes oxidized. Several amino acid residues in ROS scavenging enzymes like glutathione reductase (GR) are sensitive to the cell redox status, and hence, the activities of these enzymes are altered in metalloid stress. Mercury can also bind to this enzyme due to its sulfur hydride affinity (Sharma and Dietz 2009; Sobrino-Plata et al. 2009, 2013). The substrate of GR, glutathione (GSH), is itself an antioxidant (Jozefczak et al. 2012) whose levels decrease under metalloid stress as GSH is oxidized to glutathione disulfide (GSSG). But under moderate metalloid stress, the GSH/GSSG ratio is not depleted as GSH is replenished by the sulfur assimilation pathways (Saito 2004) which are activated under metalloid stress (Nocito et al. 2006). Hence, maintenance of GSH is crucial for tolerance to metalloid (Foyer and Noctor 2011).

6 Role of ROS Homeostasis in Root Development Under Nutrient Deficiency

ROS as signal transduction molecules forms a common feature of all kinds of nutrient deprivation stresses, e.g., N, P, and K (Fig. 5). Induction of several genes including those encoding peroxidases was reported under N, P, and K deprivation (Shin and Schachtman 2004; Shin et al. 2005) which were abolished in the *rhd2* mutant (allelic to Atrboh*C*) indicating H_2O_2 required for signal transduction under nutrient deprivation. However, apart from common elements, each nutrient deficiency triggers specific components of the ROS pathway. Several root hair mutants were used keeping in mind the utility of the hairs in increasing the absorptive surface area for nutrient uptake particularly under conditions of deficiency (Shin et al. 2005). Under K deprivation, ROS production was reduced in the *rhd2* mutant, suggesting a role of NADPH oxidase AtrbohC in ROS production. The phytohormone ethylene also promotes tolerance to low K by inducing ROS production (Jung et al. 2009; Schachtman 2015). Under low N, the *rhd2* mutant maintains higher ROS levels, suggesting a role of other NADPH oxidase homolog under this condition. The localization pattern of ROS in the roots also differs between different nutrient deprivation conditions (Shin et al. 2005). Under N and K deprivation, ROS localized in the epidermis of the root and under low P localization occurs in the cortex (Fig. 5). These observations suggest a role of ROS in root hairs under N and K deprivation, while the role of ROS formed in the cortex for signaling to trigger lateral root growth from neighboring pericycle cells under low P (Shin et al. 2005). Longer root hairs and higher density of lateral roots form a hallmark of low P stress, since plants utilize these mechanisms to explore the deficient nutrient in new regions of the soil (Fig. 5). Lower NADPH oxidase and peroxidase activity reduce root growth under such condition pointing to the utility of ROS for root growth under low P. Under low P conditions, excess induction of ROS also causes root growth inhibition by affecting stem cells in the root apical meristem. The transcription factor sensitive to proton rhizotoxicity (STOP1), controls the expression of aluminum-activated malate transporter 1 (ALMT1), a plasma membrane-localized malate transporter, releasing malate into the apoplast. This induces ROS formation by an iron redox cycling pathway composed of low phosphate root 1 (LPR1) and ascorbate (Mora-Macías et al. 2017; Balzergue et al. 2017). The excess ROS activates callose formation which is deposited in the root stem cell niche blocking cell–cell communication through the plasmodesmata. This inhibits stem cell function and arrests root growth (Müller et al. 2015).

Normally, an essential nutrient, Fe, if present in excess in the soil, produces ROS including hydroxyl radicals (OH·) in the roots and causes primary root growth arrest

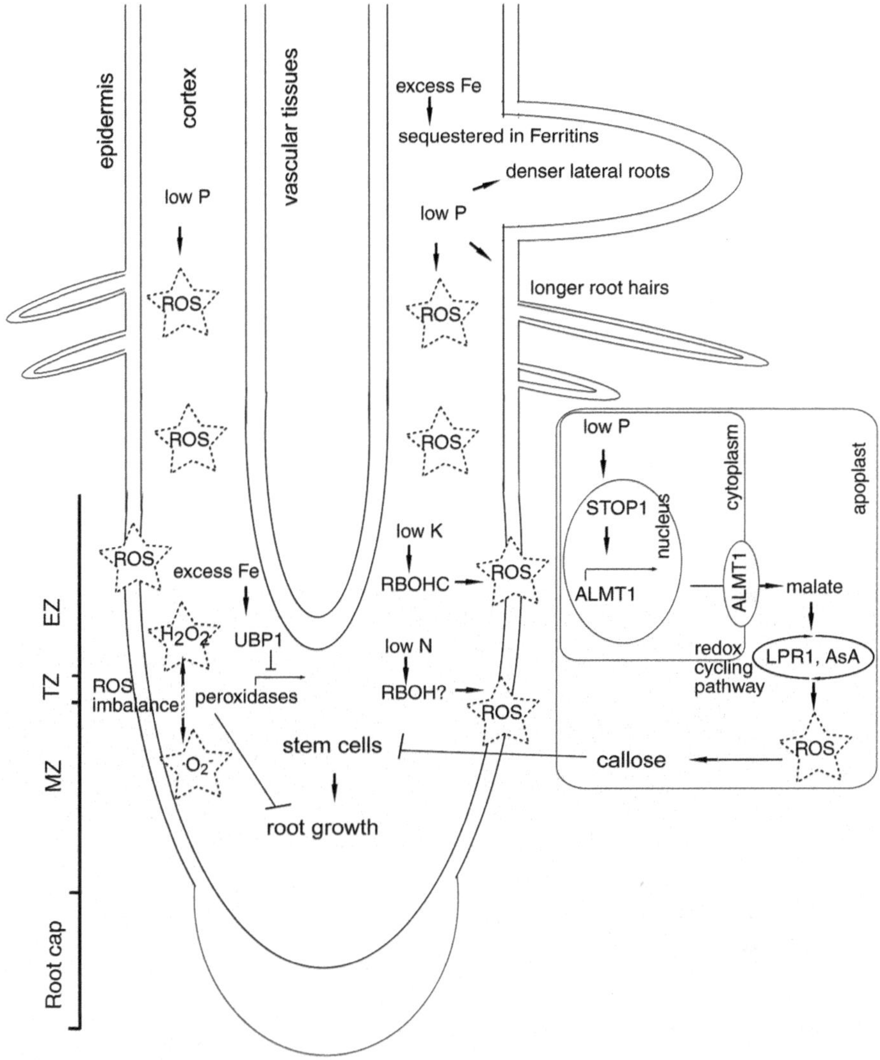

Fig. 5 Role of reactive oxygen species (ROS) homeostasis in root development under nutrient deficiency. Under conditions of phosphate (P) deficiency, a higher amount of ROS is generated in the root cortex, which leads to signaling events causing denser lateral roots and longer root hairs. Inside the cell, the transcription factor STOP1 controls the expression of the plasma membrane-localized malate transporter ALMT1, leading to malate exudation in the apoplast. Malate produces ROS by an iron redox recycling pathway, involving LPR1 and ascorbate (AsA), which leads to callose formation blocking cell-to-cell communication via plasmodesmata in the stem cell niche. This stops root growth. Low nitrogen (N) and low potassium (K) lead to ROS production in the epidermis by respiratory burst oxidase homolog (Rboh)s. An excess of iron (Fe) leads to ROS (superoxide $\cdot O_2^-$ $\leftrightarrow$ hydrogen peroxide H_2O_2) imbalance in the transition zone (TZ) between the elongation zone (EZ) and meristematic zone (MZ) via control of transcription factor UBP1 and peroxidases (see Fig. 1 and main text), leading to primary root growth arrest. In the lateral root, excess Fe is sequestered by ferritin proteins and hence cannot affect lateral root growth

(Graf et al. 1984). This is due to a decrease in primary root meristem size caused by an imbalance of ROS, viz. a decrease in $\cdot O_2^-$ in the root proliferation zone and increase in H_2O_2 in the transition zone, regulated by peroxidases under control of the UBP1 transcription factor (Reyt et al. 2015) similar to the model proposed by Tsukagoshi et al. (2010). However, iron excess cannot affect lateral root growth, as in those organs excess Fe is sequestered inside ferritin proteins which protect the cells from ROS-induced damage (Reyt et al. 2015). This is why ferritin-deficient mutants have more decreased lateral root length and density than wild-type under excess Fe. Apart from UBP1-controlled ROS balance in the control of lateral root meristem size, ROS-activated (and ferritin-repressed) cell cycle kinase inhibitors SMR5/SMR7 arrest the cell cycle to stop lateral root growth under excess Fe in ferritin-deficient plants (Reyt et al. 2015).

7 Concluding Remarks and Perspectives

The cellular ROS at different tissues of the growing root plays a critical role in plant survival during several rhizotoxic environmental stresses. At the same time, higher ROS production as a consequence of these stresses causes cellular damage and growth arrest. Therefore, a fine balance between production and scavenging of the different species of reactive oxygen is necessary, and it determines the tolerance indices of the plants to environmental stress. This directly implies the possibility of genetic improvement of plants through manipulation of the molecular machinery for ROS balance through breeding and/or biotechnology.

References

Ahmad P, Sarwat M, Bhat NA, Wani MR, Kazi AG, Tran LS (2015) Alleviation of cadmium toxicity in *Brassica juncea* L. (Czern. & Coss.) by calcium application involves various physiological and biochemical strategies. PLoS One 10:e0114571

Ai LF, Li CH, Wang G, Zhao JL, Zhang LQ, Han YF, Sun DY, Zhang SW, Sun Y (2014) The receptor-like kinase SIT1 mediates salt sensitivity by activating MAPK3/6 and regulating ethylene homeostasis in rice. Plant Cell 26:2538–2553

Alcázar R, Altabella T, Marco F et al (2010) Polyamines: molecules with regulatory functions in plant abiotic stress tolerance. Planta 231:1237–1249

Baena-González E, Rolland F, Thevelein JM, Sheen J (2007) A central integrator of transcription networks in plant stress and energy signalling. Nature 448:938–942

Balzergue C, Dartevelle T, Godon C, Laugier E, Meisrimler C, Teulon J-M, Creff A, Bissler M, Brouchoud C, Hagège A, Müller J, Chiarenza S, Javot H, Becuwe-Linka N, David P, Péret B, Delannoy E, Thibaud M-C, Armengaud J, Abel S, Pellequer J-L, Nussaume L, Desnos T (2017) Low phosphate activates STOP1-ALMT1 to rapidly inhibit root cell elongation. Nat Commun 8:15300

Basu U, Good AG, Taylor GJ (2001) Transgenic *Brassica napus* plants overexpressing aluminum-induced mitochondrial manganese superoxide dismutase cDNA are resistant to aluminum. Plant Cell Environ 24:1269–1278

Blancaflor EB, Jones DL, Gilroy S (1998) Alterations in the cytoskeleton accompany aluminum-induced growth inhibition and morphological changes in primary roots of maize. Plant Physiol 118:159–172

Boisson-Dernier A, Lituiev DS, Nestorova A et al (2013) ANXUR receptor-like kinases coordinate cell wall integrity with growth at the pollen tube tip via NADPH oxidases. PLoS Biol 11:1–15

Boursiac Y, Boudet J, Postaire O, Luu D-T, Tournaire-Roux C, Maurel C (2008) Stimulus-induced downregulation of root water transport involves reactive oxygen species-activated cell signalling and plasma membrane intrinsic protein internalization. Plant J. 56:207–218

Bhuja P, McLachlan K, Stephens J, Taylor GJ (2004) Accumulation of 1,3-ß-D-glucans, in response to aluminum and cytosolic calcium in Triticum aestivum. Plant Cell Physiol 45:543–549

Briat J, Lebrun M (1999) Plant responses to metal toxicity. Comptes Rendus de l'Académie des Sciences-Series III-Sciences de la Vie 322:43–54

Cappa JJ, Pilon-Smits EAH (2014) Evolutionary aspects of elemental hyperaccumulation. Planta 239:267–275

Chen W, Xu C, Zhao B, Wang X, Wang Y (2008) Improved Al tolerance of saffron (*Crocus sativus* L.) by exogenous polyamines. Acta Physiol Plant 30:121–127

Chowra U, Yanase E, Koyama H, Panda SK (2017) Aluminium-induced excessive ROS causes cellular damage and metabolic shifts in black gram Vigna mungo (L.) Hepper. Protoplasma 254:293–302

Chung JS, Zhu JK, Bressan RA, Hasegawa PM, Shi HZ (2008) Reactive oxygen species mediate Na^+-induced SOS1 mRNA stability in *Arabidopsis*. Plant J 53:554–565

Ciftci-Yilmaz S, Morsy MR, Song L, Coutu A, Krizek BA, Lewis MW, Warren D, Cushman J, Connolly EL, Mittler R (2007) The EARmotif of the Cys2/His2-type zinc finger protein Zat7 plays a key role in the defense response of *Arabidopsisto* salinity stress. J Biol Chem 282:9260–9268

Clifton R, Millar AH, Whelan J (2006) Alternative oxidases in *Arabidopsis*: a comparative analysis of differential expression in the gene family provides new insights into function of non-phosphorylating bypasses. Biochim Biophys Acta 1757:730–741

Duan Q, Kita D, Li C, Cheung AY, Wu HM (2010) FERONIA receptor-like kinase regulates RHO GTPase signaling of root hair development. Proc Natl Acad Sci U S A 107:17821–17826

Dunand C, Crèvecoeur M, Penel C (2007) Distribution of superoxide and hydrogen peroxide in *Arabidopsis* root and their influence on root development: possible interaction with peroxidases. New Phytol 174:332–341

Ermolayev V, Weschke W, Manteuffel R (2003) Comparison of Al induced gene expression in sensitive and tolerant soybean cultivars. J Exp Bot 54:2745–2756

Ezaki B, Gardner RC, Ezaki Y, Matsumoto H (2000) Expression of aluminum-induced genes in transgenic *Arabidopsis* plants can ameliorate aluminum stress and/or oxidative stress. Plant Physiol 122:657–665

Ezaki B, Tsugita S, Matsumoto H (1996) Expression of a moderately anionic peroxidase is induced by aluminum treatment in tobacco cells: possible involvement of peroxidase isozymes in aluminum ion stress. Physiol Plant 96:21–28

Farzadfar S, Zarinkamar F, Modarres-Sanavy S, Hojati M (2013) Exogenously applied calcium alleviates cadmium toxicity in *Matricaria chamomilla* L. plants. Environ Sci Pollut Res 20(3):1413–1422

Fidalgo F, Freitas R, FerreiraR Pessoa AM, Teixeira J (2011) *Solanum nigrum* L. antioxidant defence system isozymes are regulated transcriptionally and posttranslationally in Cd-induced stress. Environ Exp Bot 72:312–319

Foreman J, Demidchik V, Bothwell JH et al (2003) Reactive oxygen species produced by NADPH oxidase regulate plant cell growth. Nature 422:442–446

Foyer CH, Noctor G (2011) Ascorbate and glutathione: the heart of the redox hub. Plant Physiol 155:2–18

Gaume A, Mächler F, Frossard E (2001) Aluminum resistance in two cultivars of Zea mays L.: root exudation of organic acids and influence of phosphorus nutrition. Plant Soil 234:73–81

Gill SS, Anjum NA, Hasanuzzaman M, Gill R, Trivedi DK, Ahmad I, Pereira E, Tuteja N (2013) Glutathione and glutathione reductase: a boon in disguise for plant abiotic stress defense operations. Plant Physiol Biochem 70:204–212

Goodwin SB, Sutter TR (2009) Microarray analysis of *Arabidopsis* genome response to aluminum stress. Biol Plant 53:85–99

Graf E, Mahoney JR, Bryant RG, Eaton JW (1984) Iron-catalyzed hydroxyl radical formation: stringent requirement for free iron coordination site. J Biol Chem 259:3620–3624

Hamel F, Breton C, Houde M (1998) Isolation and characterization of wheat aluminum-regulated genes: possible involvement of aluminum as a pathogenesis response elicitor. Planta 205:531–538

Hara S, Motohashi K, Arisaka F, Romano PGN, Hosoya-Matsuda N, Kikuchi N, Fusada N, Hisabori T (2006) Thioredoxin-h1 reduces and reactivates the oxidized cytosolic malate dehydrogenase dimer in higher plants. J Biol Chem 281:32065–32071

Hernández-Barrera A, Velarde-Buendía A, Zepeda I et al (2015) Hyper, a hydrogen peroxide sensor, indicates the sensitivity of the *Arabidopsis* root elongation zone to aluminum treatment. Sensors (Basel) 15:855–867

Heyno E, Klose C, Krieger-Liszkay A (2008) Origin of cadmium induced reactive oxygen species production: mitochondrial electron transfer versus plasma membrane NADPH oxidase. New Phytol 179(3):687–699

Hisabori T, Hara S, Fujii T, Yamazaki D, Hosoya-Matsuda N, Motohashi K (2005) Thioredoxin affinity chromatography: a useful method for further understanding the thioredoxin network. J Exp Bot 56:1463–1468

Huang D, Gong X, Liu Y et al (2017) Effects of calcium at toxic concentrations of cadmium in plants. Planta 245:863–873

Huang W, Yang X, Yao S, LwinOo T et al (2014) Reactive oxygen species burst induced by aluminum stress triggers mitochondria-dependent programmed cell death in peanut root tip cells. Plant Physiol Biochem 82:76–84

Houde M, Diallo AO (2008) Identification of genes and pathways associated with aluminum stress and tolerance using transcriptome profiling of wheat near-isogenic lines. BMC Genom 9:400

Howden R, Andersen CR, Goldsbrough PB, Cobbett C (1995) A cadmium-sensitive, glutathione-deficient mutant of *Arabidopsis thaliana*. Plant Physiol 107:1067–1073

Jiang C, Belfield EJ, Cao Y, Smith JAC, Harberd NP (2013) An *Arabidopsis* soil salinity-tolerance mutation confers ethylene mediated enhancement of sodium/potassium homeostasis. Plant Cell 25:3535–3552

Jiang C, Belfield EJ, Mithani A, Visscher A, Ragoussis J, Mott R, Smith JAC, Harberd NP (2012) ROS-mediated vascular homeostatic control of root-to-shoot soil Na delivery in *Arabidopsis*. EMBO J 31:4359–4370

Jiang K, Moe-Lange J, Hennet L, Feldman LJ (2016) Salt stress affects the redox status of *Arabidopsis* root meristems. Front Plant Sci 7:81

Jones DL, Blancaflor EB, Kochian LV, Gilroy S (2006) Spatial coordination of aluminium uptake, production of reactive oxygen species, callose production and wall rigidification in maize roots. Plant, Cell Environ 29:1309–1318

Jones MA, Raymond MJ, Yang Z, Smirnoff N (2007) NADPH oxidase-dependent reactive oxygen species formation required for root hair growth depends on ROP GTPase. J Exp Bot 8:1261–1270

Jozefczak M, Remans T, Vangronsveld J, Cuypers A (2012) Glutathione is a key player in metal-induced oxidative stress defenses. Int J Mol Sci 13:3145–3175

Jung JY, Shin R, Schachtman DP (2009) Ethylene mediates response and tolerance to potassium deprivation in *Arabidopsis*. Plant Cell 21:607–621

Kobayashi Y, Kobayashi Y, Sugimoto M, Lakshmanan V, Iuchi S, Kobayashi M, Bais HP, Koyama H (2013) Characterization of the complex regulation of AtALMT1 expression in response to phytohormones and other inducers. Plant Physiol 162:732–740

Kwak JM, Mori IC, Pei J-M, Leonhardt N, Torres MA, Dangl JL, Bloom RE, Bodde S, Jones JDG, Schroeder JI (2003) NADPH oxidase AtrbohD and AtrbohF genes function in ROS-dependent ABA signaling in *Arabidopsis*. EMBO J 22:2623–2633

Lagrimini LM, Joly RJ, Dunlap JR, Liu TT (1997) The consequence of peroxidase overexpression in transgenic plants on root growth and development. Plant Mol Biol 33:887–895

Li W-Y, Chen B-X, Chen Z-J, Gao Y-T, Chen Z, Liu J (2017) Reactive oxygen species generated by NADPH oxidases promote radicle protrusion and root elongation during rice seed germination. Int J Mol Sci 18:110

Li G, Xu W, Kronzucker HJ, Shi W (2015) Ethylene is critical to the maintenance of primary root growth and Fe homeostasis under Fe stress in *Arabidopsis*. J Exp Bot 66:2041–2054

Liu SG, Zhu DZ, Chen GH, Gao X-Q, Zhang XS (2012) Disrupted actin dynamics trigger an increment in the reactive oxygen species levels in the *Arabidopsis* root under salt stress. Plant Cell Rep 31:1219–1226

Luo Y, Liu YB, Dong YX, Gao XQ, Zhang XS (2009) Expression of a putative alfalfa helicase increases tolerance to abiotic stress in *Arabidopsis* by enhancing the capacities for ROS scavenging and osmotic adjustment. J Plant Physiol 166:385–394

Kamada-Nobusada T, Hayashi M, FukazawaM Sakakibara H, Nishimura M (2008) Aputative peroxisomal polyamine oxidase, AtPAO4, is involved in polyamine catabolism in *Arabidopsis thaliana*. Plant Cell Physiol 49:1272–1282

Kumari M, Taylor GJ, Deyholos MK (2008) Transcriptomic responses to aluminum stress in roots of *Arabidopsis thaliana*. Mol Genet Genomics 279:339–357

Lux A, Martinka M, Vaculik M, White PJ (2011) Root responses to cadmium in the rhizosphere: a review. J Exp Bot 62:21–37

Ma L, Zhang H, Sun L, Jiao Y, Zhang G, Miao C, Hao F (2012) NADPH oxidase AtrbohD and AtrbohF function in ROS dependent regulation of Na +/K + homeostasis in *Arabidopsis* under salt stress. J Exp Bot 63:305–317

Maltais K, Houde M (2002) A new biochemical marker for aluminium tolerance in plants. Physiol Plant 115:81–86

Manzano C, Pallero-Baena M, Casimiro I et al (2014) The emerging role of reactive oxygen species signaling during lateral root development. Plant Physiol 165:1105–1119

Miller G, Suzuki N, Rizhsky L, Hegie A, Koussevitzky S, Mittler R (2007) Double mutants deficient in cytosolic and thylakoid ascorbate peroxidase reveal a complex mode of interaction between reactive oxygen species, plant development, and response to abiotic stresses. Plant Physiol 144:1777–1785

Miller GN, Suzuki N, Ciftci-Yilmaz S, Mittler R (2010) Reactive oxygen species homeostasis and signaling during drought and salinity stresses. Plant Cell Environ 33:453–467

Milla MA, Butler E, Huete AR et al (2002) Expressed sequence tag-based gene expression analysis under aluminum stress in rye. Plant Physiol 130:1706–1716

Mohamed FI (2012) Protective effects of some antioxidant metals against chromosomal damage induced by cadmium in Vicia faba plants. Int J Agric Res 7:376–387

Mora-Macías J, Ojeda-Rivera JO, Gutiérrez-Alanís D, Yong-Villalobos L, Oropeza-Aburto A, Raya-González J, Jiménez- Domínguez G, Chávez-Calvillo G, Rellán-Álvarez R, Herrera-Estrella L (2017) Malate-dependent Fe accumulation is a critical checkpoint in the root developmental response to low phosphate. Proc Nat Acad Sciences 114:E3563–E3572

Mori IC, Schroeder JI (2004) Reactive oxygen species activation of plant Ca^{2+} channels. A signaling mechanism in polar growth, hormone transduction, stress signaling, and hypothetically mechanotransduction. Plant Physiol 135:702–708

Moseyko N, Feldman LJ (2001) Expression of pH-sensitive green fluorescent protein in *Arabidopsis thaliana*. Plant, Cell Environ 24:557–563

Mujika J, Ruipérez F, Infante I et al (2011) Pro-oxidant activity of aluminum: stabilization of the aluminum superoxide radical ion. J Phys Chem A 115:6717–6723

Müller J, Toev T, Heisters M, Teller J, Moore KL, Hause G, Dinesh DC, Bürstenbinder K, Abel S (2015) Iron-dependent callose deposition adjusts root meristem maintenance to phosphate availability. Dev Cell 33:216–230

Nakano Y, Asada K (1987) Purification of ascorbate peroxidase in spinach chloroplasts; its inactivation in ascorbate-depleted medium and reactivation by monodehydroascorbate radical. Plant Cell Physiol 28:131–140

Nath M, Yadava S, Sahoo RK, Passricha N, Tuteja R, Tuteja N (2016) PDH45 transgenic rice maintain cell viability through lower accumulation of Na^+, ROS and calcium homeostasis in roots under salinity stress. J Plant Physiol 191:1–11

Nezames CD, Ochoa V, Larsen PB (2012) Mutational loss of *Arabidopsis* SLOW WALKER2 results in reduced endogenous spermine concomitant with increased aluminum sensitivity. Funct Plant Biol 40:67–78

Nocito FF, Lancilli C, Crema B, Fourcroy P, Davidian J, Sacchi GA (2006) Heavy metal stress and sulfate uptake in maize roots. Plant Physiol 141:1138–1148

Panda SK, Sahoo L, Katsuhara M, Matsumoto H (2013) Overexpression of alternative oxidase gene confers aluminum tolerance by altering the respiratory capacity and the response to oxidative stress in tobacco cells. Mol Biotechnol 54:551–563

Panda SK, Yamamoto Y, Kondo H, Matsumoto H (2008) Mitochondrial alterations related to programmed cell death in tobacco cells under aluminium stress. C R Biol 331:597–610

Passaia G, Queval G, Bai J, Margis-Pinheiro M, Foyer CH (2014) The effects of redox controls mediated by glutathione peroxidases on root architecture in *Arabidopsis thaliana*. J Exp Bot 65:1403–1413

Passardi F, Cosio C, Penel C, Dunand C (2005) Peroxidases have more functions than a Swiss army knife. Plant Cell Rep 24:255–265

Peng J, Li Z, Wen X, Li W, Shi H, Yang L, Zhu H, Guo H (2014) Salt-induced stabilization of EIN3/EIL1 confers salinity tolerance by deterring ROS accumulation in *Arabidopsis*. PLoS Genet 10:e1004664

Pottosin I, Buendía AMV, Bose J, Jazo IZ, Shabala S, Dobrovinskaya O (2014) Cross-talk between reactive oxygen species and polyamines in regulation of ion transport across the plasma membrane: implications for plant adaptive responses. J Exp Bot 65:1271–1283

Prasad MNV (1995) Cadmium toxicity and tolerance in vascular plants. Environ Exp Bot 35:525–545

Rentel MC, Lecourieux D, Ouaked F et al (2004) OXI1 kinase is necessary for oxidative burst-mediated signalling in *Arabidopsis*. Nature 427:858–861

Reyt G, Boudouf S, Boucherez J, Gaymard F, Briat J-F (2015) Iron- and Ferritin-dependent reactive oxygen species distribution: impact on *Arabidopsis* root system architecture. Mol Plant 8:439–453

Richards KD, Schott EJ, Sharma YK et al (1998) Aluminum induces oxidative stress genes in *Arabidopsis thaliana*. Plant Physiol 116:409–418

Ruipérez F, Mujika J, Ugalde J, Exley C, Lopez X (2012) Pro-oxidant activity of aluminum: promoting the Fenton reaction by reducing Fe (III) to Fe (II). J Inorg Biochem 117:118–123

Saito K (2004) Sulfur assimilatory metabolism. The long and smelling road. Plant Physiol 136:2443–2450

Sagi M, Fluhr R (2001) Superoxide production by plant homologues of the gp91phox NADPH oxidase. Modulation of activity by calcium and by tobacco mosaic virus infection. Plant Physiol 126:1281–1290

Schachtman DP (2015) The role of ethylene in plant responses to K+ deficiency. Front Plant Sci 6:1153

Schmidt R, Mieulet D, Hubberten H-M et al (2013) SALT-RESPONSIVE ERF1 regulates reactive oxygen species-dependent signaling during the initial response to salt stress in rice. Plant Cell 25:2115–2131

Sebastian A, Prasad MNV (2014) Red and blue lights induced oxidative stress tolerance promote cadmium rhizocomplexationin *Oryza sativa*. J Photochem Photobiol B Biol 137:135–143

Sebastian A, Prasad MNV (2015) Iron- and manganese-assisted cadmium tolerance in *Oryza sativa* L.: lowering of rhizotoxicity next to functional photosynthesis. Planta 241:1519–1528

Sharma SS, Dietz K (2009) The relationship between metal toxicity and cellular redox imbalance. Trends Plant Sci 14:43–50

Shin R, Berg RH, Schachtman DP (2005) Reactive oxygen species and root hairs in *Arabidopsis* root response to nitrogen, phosphorus and potassium deficiency. Plant Cell Physiol 46:1350–1357
Shin R, Schachtman DP (2004) Hydrogen peroxide mediates plant root cell response to nutrient deprivation. Proc Natl Acad Sci U S A 101:8827–8832
Sivaguru M, Liu J, Kochian LV (2013) Targeted expression of SbMATE in the root distal transition zone is responsible for sorghum aluminum resistance. Plant J 76:297–307
Sobrino-Plata J, Herrero J, Carrasco-Gil S, Pérez-Sanz A, Lobo C, Escobar C, Millán R, Hernández LE (2013) Specific stress responses to cadmium, arsenic and mercury appear in the metallophyte Silene vulgaris when grown hydroponically. RSC Advances 3:4736–4744
Sobrino-Plata J, Ortega-Villasante C, Laura Flores-Cáceres M, Escobar C, Del Campo FF, Hernández LE (2009) Differential alterations of antioxidant defenses as bioindicators of mercury and cadmium toxicity in alfalfa. Chemosphere 77:946–954
Sunkar R, Kapoor A, Zhu JK (2006) Posttranscriptional induction of two Cu/Zn superoxide dismutase genes in *Arabidopsis* is mediated by downregulation of miR398 and important for oxidative stress tolerance. Plant Cell 18:2051–2065
Talukdar D (2012) Exogenous calcium alleviates the impact of cadmium-induced oxidative stress in Lens culinaris Medic. Seedlings through modulation of antioxidant enzyme activities. J Crop Sci Biotechnol 15(4):325–334
Tesfaye M, Temple SJ, Allan DL, Vance CP, Samac DA (2001) Overexpression of malate dehydrogenase in transgenic alfalfa enhances organic acid synthesis and confers tolerance to aluminum. Plant Physiol 127:1836–1844
Torres MA, Dangl JL, Jones JDG (2002) *Arabidopsis* gp91(phox) homologues AtrbohD and AtrbohF are required for accumulation of reactive oxygen intermediates in the plant defense response. Proceedings of the National Academy of Sciences, USA 99:517–522
Tsukagoshi H, Busch W, Benfey PN (2010) Transcriptional regulation of ROS controls transition from proliferation to differentiation in the root. Cell 143:606–616
Van HL, Kuraishi S, Sakurai N (1994) Aluminum-induced rapid root inhibition and changes in cell-wall components of squash seedlings. Plant Physiol 106:971–976
Vernoux T, Wilson RC, Seeley KA et al (2000) The ROOT MERISTEMLESS1/CADMIUM SENSITIVE2 Gene defines a glutathione-dependent pathway involved in initiation and maintenance of cell division during postembryonic root development. Plant Cell 12:97–109
Wolf S, Hofte H (2014) Growth control: a saga of cell walls, ROS, and Peptide receptors. Plant Cell 26:1848–1856
Wu K, Xiao S, Chen Q et al (2013) Changes in the activity and transcription of antioxidant enzymes in response to Al stress in black soybeans. Plant Mol. Biol. Rep. 31:141–150
Wu L, Zhang Z, Zhang H, Wang XC, Huang R (2008) Transcriptional modulation of ethylene response factor protein JERF3 in the oxidative stress response enhances tolerance of tobacco seedlings to salt, drought, and freezing. Plant Physiol 148:1953–1963
Xiong J, Lu H, Lu K, Duan Y, An L, Zhu C (2009) Cadmium decreases crown root number by decreasing endogenous nitric oxide which is indispensable for crown root primordia initiation in rice seedlings. Planta 230:599–610
Yin LN, Wang SW, Eltayeb AE et al (2010) Overexpression of dehydroascorbate reductase, but not monodehydroascorbate reductase, confers tolerance to aluminum stress in transgenic tobacco. Planta 231:609–621
Yamamoto Y, Kobayashi Y, Devi SR, Rikiishi S, Matsumoto H (2002) Aluminum toxicity is associated with mitochondrial dysfunction and the production of reactive oxygen species in plant cells. Plant Physiol 128:63–72
Zhang F, Wang Y, Yang Y, Wu H, Wang D, Liu J (2007) Involvement of hydrogen peroxide and nitric oxide in salt resistance in the calluses from *Populus euphratica*. Plant Cell Environ 30:775–785

Zhao C-R, Ikka T, Sawaki Y et al (2008) Comparative transcriptomic characterization of aluminum, sodium chloride, cadmium and copper rhizotoxicities in *Arabidopsis thaliana*. BMC Plant Biol 9:32

Zhao FY, Zhang H (2006) Salt and paraquat stress tolerance results from coexpression of the *Suaeda salsa* glutathione-S-transferase and catalase in transgenic rice. Plant Cell Tissue Organ Cult 86:349–358

Advances in Chlorophyll Fluorescence Theories: Close Investigation into Oxidative Stress and Potential Use for Plant Breeding

Etsuko Watanabe, Rym Fekih and Ichiro Kasajima

Abstract Solar irradiation is the source of lives: animals including us humans live depending on the chemical energy produced by photosynthesis of plants. Then, what if solar irradiation is always toxic to plants, especially to chloroplasts in plant leaf cells? As well as the production of chemical energy in the forms of high-energy phosphate bond and reducing power, chlorophyll excitation energy excites oxygen molecules to produce reactive oxygen species. To alleviate the toxic effect of chlorophyll excitation energy, plants are equipped with detoxification mechanisms. Chloroplast avoidance movement reduces the absorption of light energy by chlorophyll. Once the light energy is absorbed by chlorophyll, excessive excitation energy causes acidification of chloroplast lumen through cyclic electron transfer. This is the trigger of the *qE* component of non-photochemical quenching (NPQ), which releases chlorophyll excitation energy in the form of heat. NPQ components are measured by the changes in the yield of chlorophyll fluorescence. Chlorophyll fluorescence parameters are also influenced by chloroplast movement and oxidative stress; therefore, such errors in estimated parameter values are corrected based on fluorescence theories. Finally, the molecular breeding study of plants is recently revealing the possible effects of oxidative stress tolerance in boosting crop production.

Sections 2, 5, and 6 were written by EW and RF. The other sections were written by IK. Texts and figures were edited by all authors.

E. Watanabe · R. Fekih · I. Kasajima (✉)
Faculty of Agriculture, Iwate University, Morioka, Japan
e-mail: kasajima@iwate-u.ac.jp; kasajima2008@live.jp

I. Kasajima
Laboratory of Plant Pathology, Faculty of Agriculture, Iwate University, Ueda 3-18-8, Morioka, Iwate 020-8550, Japan

S. K. Panda and Y. Y. Yamamoto (eds.), *Redox Homeostasis in Plants*,
Signaling and Communication in Plants, https://doi.org/10.1007/978-3-319-95315-1_7

1 Introduction: Generation of Reactive Oxygen Species by Excessive Light Energy

The toxicity of light energy is caused by the energy itself. Excitation of chlorophyll pigments is the initial step of photosynthesis, and this chlorophyll excitation energy is transformed to chemical energy in the photosynthetic electron transfer system. Here, reactive oxygen species are produced as inevitable by-products of electron transfer. In photosynthetic electron transfer system, electrons are excited by chlorophyll excitation energy twice: once in photosystem II and once in photosystem I (Heldt 2005). As well as excitation of electrons, chlorophyll excitation energy gives rise to singlet oxygen $\left({}^1O_2\right)$ in photosystem II and gives rise to superoxide radical anion $\left(O_2^{\cdot -}\right)$ in photosystem I (Apel and Hirt 2004). Superoxide radical anion reduces metal ions, which react with hydrogen peroxide (H_2O_2) to form highly toxic hydroxyl radical ($^{\cdot}OH$; Heldt 2005). Then superoxide is catabolized to hydrogen peroxide by superoxide dismutase to avoid accumulation of hydroxyl radical. Hydrogen peroxide is also catabolized to water and/or oxygen by ascorbate peroxidase, glutathione peroxidase, and catalase (Apel and Hirt 2004). Enforcement of these enzymatic systems to degrade reactive oxygen species is a good candidate for the enhancement of crop production, but its expositions are left to other chapters and literature. The generation of reactive oxygen species is not solely caused by high light. They are also produced by other stresses such as cold, heat, salt, drought, and aluminum (Lin et al. 2016; Pandey et al. 2016; Rachoski et al. 2015; Xu et al. 2015; Printó-Marijuan and Munné-Bosch 2014; Demidchik 2015). Thus, oxidative stress tolerance benefits stable crop production under various conditions.

In this chapter, we will focus on how plants reduce the amount of excessive light energy absorbed by chlorophylls, and how plants safely release excessive chlorophyll excitation energy in photosystem II. These mechanisms are expected to reduce the formation of reactive oxygen species. Chlorophyll fluorescence theory is the key to understanding energy release and the degree of oxidative stress. Genetic analysis and genetic modification by researchers have even started to target such anti-oxidative processes for plant breeding.

In the meantime, how 'excessive' is chlorophyll excitation energy? This is well demonstrated by the chlorophyll fluorescence analysis of photosystem II (Oxborough 2004; Baker 2008). Its principles of measurements will be explained in the following sections, but an example of the measurement of electron transport rate (ETR) and quantum yield of photosystem II (Φ_{II}) in rice (*Oryza sativa*) leaves will be introduced here. ETR indicates how many electrons are excited at photosystem II, and Φ represents the fraction of chlorophyll excitation energy in photosystem II which is utilized for excitation of electrons (Fig. 1). As shown in Fig. 1a, photosynthesis rate is saturated at around the light intensity (PPFD, photosynthetic photon flux density) of 1000 μmol m^{-2} s^{-1}. The rate of photosynthesis is also much lower than the total chlorophyll excitation energy generated in photosystem II. In Fig. 1b, the fraction of chlorophyll excitation energy which is used for photosynthesis is far below 100%. Its value is only 32% at the PPFD of 1000 μmol m^{-2} s^{-1}. Much more energy may be

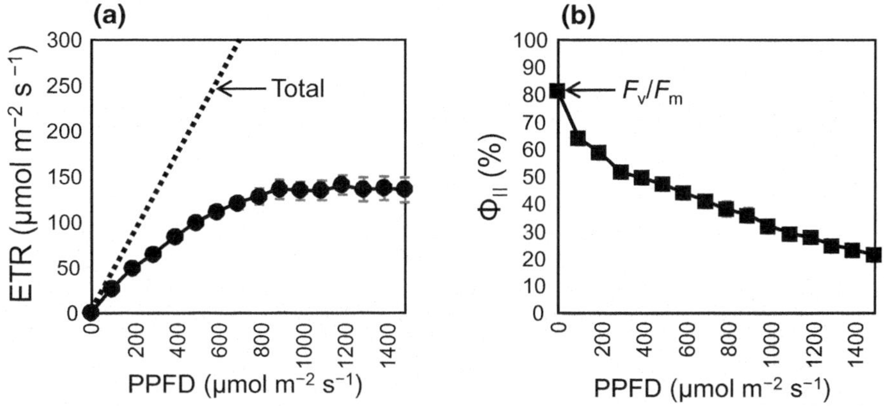

Fig. 1 ETR and Φ_{II} values of rice cultivar 'Koshihikari'. **a** ETR values (continuous line). Dotted line shows the estimated total amount of chlorophyll excitation energy in photosystem II. **b** Φ_{II} values. 'F_v/F_m' indicates the maximum quantum yield of photosystem II. Leaf disks were excised from the uppermost-expanded leaves of rice plants, dark-adapted, and measured by two-dimensional pulse amplitude modulation (PAM) fluorometer FluorCam (Photon Systems Instruments, Drasov, Czech Republic). PPFD, photosynthetic photon flux density. Data represent means and standard deviations ($n = 8$). 'Koshihikari' is the most popular rice cultivar in Japan

wasted in paddy field, because the intensity of sunlight exceeds 2000 μmol m^{-2} s^{-1} around the noon on sunny days in summer. These data will demonstrate how chlorophyll excitation energy becomes highly excessive within plant chloroplast, depending on the conditions such as seasons and climates. The fluorescence parameter value 'F_v/F_m' shown in Fig. 1b represents the maximum quantum yield of photosystem II. F_v/F_m value is usually 80% in rice plants (e.g., Takahara et al. 2010; Kasajima 2017), but it is around 85% in thale cress (*Arabidopsis thaliana*; Kasajima et al. 2009). We would tend to describe as if 100% of light energy is used for photosynthesis in plants, but actually, the rate of light energy usage never reaches 100% under any weak illumination.

2 Chloroplast Movement

The primary step to attenuate photodamage will be chloroplast movement. Indeed, chloroplasts which usually accumulate at the periclinal side of the cell surface, under optimum light intensity, relocate to the anticlinal side of cell surface under strong light conditions (Fig. 2a; Kagawa and Wada 2002). In this way, some plants can reduce the excess amount of absorbed incident light through a strategy called chloroplast avoidance movement (Kagawa and Wada 2002; Kong and Wada 2014; Suetsugu et al. 2017; Suetsugu and Wada 2007; Wada 2013; Wada et al. 2003).

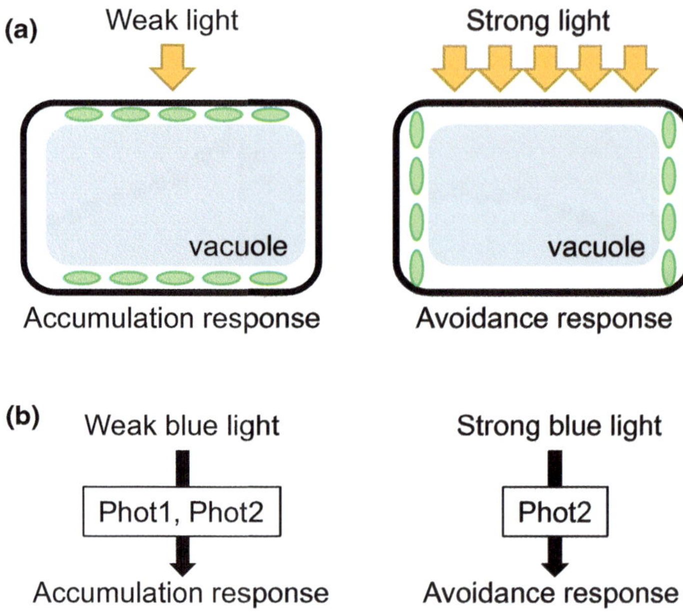

Fig. 2 Schematic representation of chloroplast movement. **a** Schematic models of accumulation response under weak light (left) and strong light (right). Chloroplasts are illustrated as green ellipses. **b** Phot1 and Phot2 mediate the accumulation response of chloroplast under weak light condition (left). Phot2 mediates the light-dependent avoidance under strong light condition (right)

In higher plants, chloroplast movement is known to be mediated by phototropic photoreceptor family of proteins (Phots; Fig. 2b; Kagawa and Wada 2002): Blue light is recognized by Phots, and the blue-light receptors are associated with plasma membrane (Suetsugu and Wada 2007). In *Arabidopsis*, there are two Phots, Phot1, and Phot2. Phot2 localizes at plasma membrane, Golgi apparatus, and cytosol (Kong et al. 2006), depending on conditions. Phot1 is localized not only to the plasma membrane but also to the chloroplast envelope during the chloroplast movement (Kong and Wada 2014). Chloroplast relocation movement is under the control of *Chloroplast unusual positioning1* (*Chup1*), a gene encoding for a CHUP1 protein (Oikawa et al. 2008). CHUP1 is a multi-domain protein including an actin-binding motif and a proline-rich region. CHUP1 is attached to the chloroplast outer envelope through the *N*-terminal hydrophobic region (Suetsugu et al. 2010, 2016).

The above reactions are for harvesting light efficiently and for protection of photosynthetic apparatus from strong light. Phot also enhances photosynthesis and plant growth at low temperature. When *Arabidopsis* wild-type (non-mutated) plants, *phot1*, and *phot2* single mutants, or *phot1 phot2* double mutant are grown under red light, there is no significant difference in growth between the plants. However, if a very low intensity of blue light is mixed with red light, large amount of plant size increases is observed in those plants with functional Phot1. This result suggests that the mixture of blue and red lights contributes to the growth of plants through chloroplast accumulation response (Takemiya et al. 2005).

Beside *Arabidopsis*, chloroplast movement was also investigated in several other plant species. Unlike the other plants, the leaves of climbing plants grown under strong sunlight show very low or no chloroplast photo-relocation responses to blue light. When dark-grown leaves are illuminated with light, rapid leaf transmittance changes are observed, indicating the occurrence of chloroplast movement. The question of whether chloroplast avoidance response is functional under direct sunlight in perennial climbing plants needs further investigations (Wada and Kong 2011). In case of Japanese cayratia herb (*Cayratia japonica*) leaves, the number of chloroplasts in the upper side of periclinal walls becomes reduced after blue-light irradiation. Most of the chloroplasts remain at the side of the anticlinal wall, the state of chloroplast avoidance response (Higa and Wada 2016). After all, the essential role of chloroplast movement in tolerating strong illumination at least in some plant species is clearly demonstrated by fatal photodamage in the mutant plant (Kasahara et al. 2002).

3 Metabolism of Chlorophyll Excitation Energy in Photosystem II

Light energy is captured by chlorophyll molecules bound to light-harvesting complexes (LHCs). These excitation energies are carried to the reaction center chlorophyll for excitation of electrons in photosynthetic electron transfer system. By application of chlorophyll fluorescence analysis, metabolism of chlorophyll excitation energy is studied in photosystem II, because chlorophyll fluorescence is emitted from photosystem II. Then what are the pathways (reactions/processes) through which chlorophyll excitation energy is used (de-excited)?

The use of chlorophyll excitation energy in photosynthetic electron transfer system (i.e., photosynthesis) is called 'photochemistry'. The de-excitation processes are physicochemical reactions, so rate constants are adopted to describe and calculate their reaction rates. The rate constant for photochemistry is 'k_p'. Photochemistry is an intermolecular reaction (EET—excitation energy transfer) for excitation of electrons.

Reactions other than photochemistry are dissipations. There are basal dissipations that are intramolecular reactions taking place anytime. Basal dissipation is collectively described as 'k_{fid}' for ease of understanding, and its components are chlorophyll fluorescence (k_f), intersystem crossing (k_{isc}), and basal non-radiative decay (k_d). We observe intensities (yields) of chlorophyll fluorescence generated by k_f reaction. The k_{isc} reaction represents the conversion of first singlet state of chlorophyll molecule to the first triplet state. This reaction is the source of singlet oxygen, because the first triplet chlorophyll generates singlet oxygen by EET, together with de-excitation to the ground state by phosphorescence or intersystem crossing. The k_d reaction is internal conversion, where chlorophyll excitation energy is released as heat.

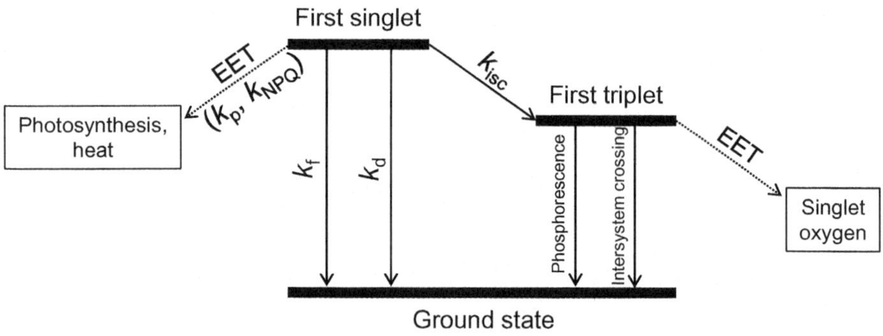

Fig. 3 Simplified Jablonski diagram of chlorophyll molecule in photosystem II. Chlorophyll molecules are excited to first singlet state by light energy. Arrows represent de-excitation processes. k_f, chlorophyll fluorescence. k_d, basal non-radiative decay. k_{isc}, intersystem crossing from first singlet state to first triplet state. k_p, photochemistry. k_{NPQ}, NPQ. EET, intermolecular excitation energy transfer

In addition to basal dissipation, there are also induced intermolecular reactions to release the chlorophyll excitation energy as heat. These reactions are called 'non-photochemical quenching' (NPQ, with the rate constant of k_{NPQ}). 'Quenching' is an intermolecular de-excitation process. Photochemistry is photochemical quenching, so the other EETs are non-photochemical quenching. Components of NPQ will be described in other section of this chapter. The de-excitation processes discussed above are summarized in the Jablonski diagram of chlorophyll energy states (Kasajima et al. 2009; Fig. 3).

Chlorophyll excitation energies are de-excited in each reaction mentioned above, in proportion to their rate constants. When it comes to reducing the rate of the generation of singlet oxygen in photosystem II under high light conditions, and in adverse environments under which photosynthesis is inhibited, it is important to increase the sum of all rate constants (k_s): If the generation of singlet oxygen takes place depending on the k_{isc} reaction, it is proportional to the value of k_{isc}/k_s. NPQ reactions are induced under such stressful conditions for photosystem II to reduce the value of k_{isc}/k_s, to release excessive chlorophyll excitation energy as heat, and to reduce the production of singlet oxygen.

4 Measurement of the Yield of Chlorophyll Fluorescence

Relative values of rate constants for chlorophyll de-excitation processes, such as photochemistry, basal dissipation, and NPQ, are calculated based on the measurement of the intensities (yields) of chlorophyll fluorescence. The yields of chlorophyll fluorescence are variable depending on the physiological states of chloroplasts, because the reaction of chlorophyll fluorescence emission competes with the other de-excitation processes as described in the previous section. Equations relating between chlorophyll fluorescence intensities and rate constants of de-excitation processes are used to derive chlorophyll fluorescence parameters, which measure ratios between rate constants (Kramer et al. 2004; Kasajima et al. 2009). There are also other kinds of parameters, but these quantitative fluorescence parameters (e.g., F_v/F_m, NPQ, Φ_{II}, and qL) would be recommended rather than the other non-quantitative fluorescence parameters. The variety of chlorophyll fluorescence parameters have been generated by many researchers (Kitajima and Butler 1975; van Kooten and Snel 1990; Bilger and Björkman 1990; Laisk et al. 1997; Oxborough and Baker 1997; Sonoike 1999; Maxwell and Johnson 2000; Kramer et al. 2004; Hendrickson et al. 2004, 2005; Kornyeyev and Holaday 2008; Stefanov and Terashima 2008; Miyake et al. 2009; Kasajima et al. 2009, 2011a). This section explains how some representative quantitative parameters can be calculated from chlorophyll fluorescence yields.

Pulse amplitude modulation (PAM) is the technique to measure chlorophyll fluorescence yields. The advantage of PAM is that chlorophyll fluorescence yields can be measured even under a strong light illumination. Further details are written in a literature (Kasajima et al. 2011a), but 'measuring pulse light' is supplemented to plant leaves during analysis of fluorescence. The pulse causes relatively small increase in the intensity of chlorophyll fluorescence, and the value of the increased fluorescence intensity corresponds to fluorescence yield of plant leaves at that moment. Fluorescence yield is written as 'F' in general. Then, the value of F is mathematically explained by rate constants as follows:

$$F = S \cdot k_f / \left(k_{\text{fid}} + k_{\text{NPQ}} + k_p\right) \qquad (1)$$

Please check the previous section for the meaning of rate constants. 'S' is the sensitivity factor, and a non-measurable constant. Then, absolute values of rate constants cannot be calculated from fluorescence yields, although their relative values are calculated by combinations of fluorescence yields. In one of our research papers (Kasajima et al. 2009), we utilized inverse fluorescence intensity to compare rate constants by simple calculations. This way of calculation is based on the following equation, derived from Eq. 1:

$$k_{\text{fid}} + k_{\text{NPQ}} + k_p = \left(S \cdot k_f\right) \cdot F^{-1} \qquad (2)$$

Considering that rate constants of basal dissipations (k_{fid} and k_f) and the sensitivity factor (S) are stable, Eq. 2 is a simple linear function between '$k_{\text{NPQ}} + k_p$' and 'F^{-1}'.

During measurement of chlorophyll fluorescence, 'actinic light', a continuous light for photosynthesis is supplemented to plant leaves. NPQ is induced during illumination of actinic light. k_{NPQ} value is nothing (0) in the initial dark condition, and it increases in light condition. Photochemistry is partly saturated by illumination of actinic light, and thus, k_p value decreases by actinic light illumination from the initial k_{pi} value in the dark. 'Saturation pulse' is also illuminated at a high light

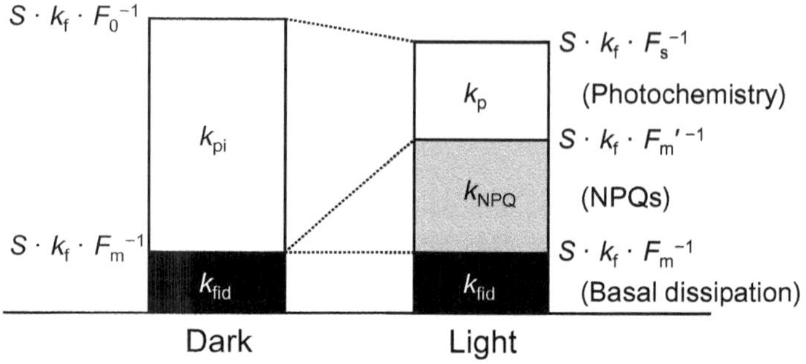

Fig. 4 Schematic representation of the relationship between fluorescence intensities and rate constants. Modified from Kasajima et al. (2009)

intensity to temporarily saturate photochemistry. The k_p value can be viewed to be nothing (0) under saturation pulse light. Taking these rules into account, the most basic fluorescence values and their correlated rate constants are as follows, in the form of Eq. 2:

$$k_{\text{fid}} + k_{\text{pi}} = S \cdot k_f \cdot F_0^{-1} \tag{3}$$

$$k_{\text{fid}} = S \cdot k_f \cdot F_m^{-1} \tag{4}$$

$$k_{\text{fid}} + k_{\text{NPQ}} = S \cdot k_f \cdot F_m^{\prime -1} \tag{5}$$

$$k_{\text{fid}} + k_{\text{NPQ}} + k_p = S \cdot k_f \cdot F_s^{-1} \tag{6}$$

Let us also check the relationship between these values in Fig. 4.

To compare the states of de-excitation processes in plant leaves, relative values of rate constants can be calculated as described above and as shown in Fig. 4. Alternatively, there are quantitative fluorescence parameters to describe the states/sizes of each de-excitation process. Following are these parameters and their formulae:

$$F_v/F_m = \left(F_0^{-1} - F_m^{-1}\right)/F_0^{-1} \left(= k_{\text{pi}}/\left(k_{\text{pi}} + k_{\text{fid}}\right)\right) \tag{7}$$

$$\text{NPQ} = \left(F_m^{\prime -1} - F_m^{-1}\right)/F_m^{-1} \left(= k_{\text{NPQ}}/k_{\text{fid}}\right) \tag{8}$$

$$\text{qL} = \left(F_s^{-1} - F_m^{\prime -1}\right)/\left(F_0^{-1} - F_m^{-1}\right) \left(= k_p/k_{\text{pi}}\right) \tag{9}$$

$$\Phi_{\text{II}} = \left(F_s^{-1} - F_m^{\prime -1}\right)/F_s^{-1} \left(= k_p/\left(k_p + k_{\text{NPQ}} + k_{\text{fid}}\right)\right) \tag{10}$$

$$\Phi_{\text{NPQ}} = \left(F_m^{\prime -1} - F_m^{-1}\right)/F_s^{-1} \left(= k_{\text{NPQ}}/\left(k_p + k_{\text{NPQ}} + k_{\text{fid}}\right)\right) \tag{11}$$

$$\Phi_{\text{NO}} = F_m^{-1}/F_s^{-1}\left(= k_{\text{fid}}/\left(k_p + k_{\text{NPQ}} + k_{\text{fid}}\right)\right) \tag{12}$$

$$\text{ETR} = \Phi_{\text{II}} \cdot \text{PAR} \cdot 0.84 \cdot 0.5\left(\mu\text{mol}\,\text{m}^{-2}\text{s}^{-1}\right) \tag{13}$$

The parameter NPQ represents relative size of the rate constant of NPQ in comparison to the size of basal dissipation, and the parameter qL represents the ratio of 'open' (reactive) photosystem II under illumination of actinic light. F_v/F_m is the quantum yield of photochemistry in the dark (or under 'dark-adapted' state). Φ_{II}, Φ_{NPQ}, and Φ_{NO} represent quantum yields of photochemistry, NPQ, and basal dissipation, respectively, in the light conditions (Kasajima et al. 2009). Electron transport rate (ETR) represents the rate of photosynthetic electron transfer. This value is calculated from Φ_{II} and PAR (photosynthetically active radiation), under the assumption that 84% of the illuminated photon is absorbed by chlorophyll (0.84), and that halves of light energies are divided between photosystem I and photosystem II (0.5; Sonoike 2003). Calculation of these seven parameters is the purposes of chlorophyll fluorescence analyses in many cases.

5 Molecular Mechanisms for Regulation of *qE* Quenching

Previous reports described more than one components of NPQ (Quick and Stitt, 1989). This idea derived from the observation that the timescales for NPQ relaxation in the dark, after NPQ induction under the light condition, consist of more than one kinetic curve. The idea will be also supported by the observation that a 'slow-relaxing' component of NPQ is still observed in the *Arabidopsis npq4* mutant impaired in the induction of 'fast-relaxing' NPQ, because of the mutation in the *AtPsbS* gene (Li et al. 2000). The identity of slow-relaxing NPQ(s) is not so much clear at the moment. The fast-relaxing NPQ is the main component of NPQ in higher plants and called *qE*. The impact of *qE* on reducing oxidative stress in plants would be well demonstrated by overaccumulation of superoxide and hydrogen peroxide in the *Arabidopsis npq4* mutant under high light condition (Jänkänpää et al. 2013).

PGR5 was identified as a thylakoid membrane protein that is involved in the transfer of electrons from ferredoxin to plastoquinone around photosystem I. PGR5 pathway is involved in the generation of the proton gradient (ΔpH) that induces thermal dissipation (*qE* quenching) when Calvin cycle activity is reduced, or light intensity is too strong. PGR5 pathway limits the overreduction of the acceptor side of photosystem I, preventing photosystem I photodamage. The PGR5 pathway around photosystem I is also called 'cyclic electron transport'. Acidification of chloroplast lumen as the result of cyclic electron transport is also the trigger of *qE* (Fig. 5; Munekage et al. 2002).

In both the model green alga *Chlamydomonas reinhardtii* and higher plants, *qE* requires the de-epoxidation of violaxanthin into zeaxanthin by violaxanthin de-epoxidase (VDE) and re-organization of the photosystem II protein complex. PsbS

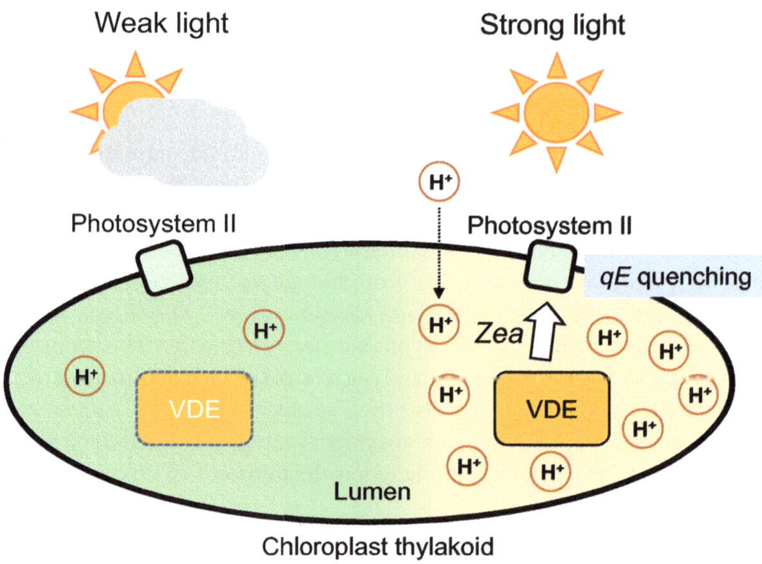

Fig. 5 Simplified illustration of the mechanism leading to the induction of *qE* quenching in higher plants. Light energy is absorbed by chlorophyll, under strong light conditions, and excessive energy causes acidification of chloroplast lumen through cyclic electron transfer which triggers the *qE* component of non-photochemical quenching (NPQ). Zea, zeaxanthin; VDE, violaxanthin de-epoxidase

protein is involved in this re-organization in higher plants, but in many green algae, PsbS is replaced by stress-related LHCSR3 (light-harvesting complex stress-related 3) proteins, the ancient members of the LHC (light-harvesting chlorophyll protein complex) protein. However, in contrast to higher plants, significant amounts of *qE* are induced only 4 h after application of strong light and are not considered constitutive in *Chlamydomonas*. Thus, the antenna-based approach for the fast regulation of the release of photosystem II excitation energy has been important for the survival response of photosynthesizers since the advent of photosystem II (Derks et al. 2015).

6 Boosting Crop Production Through Enhancement of *qE* Quenching

As already described, plants have a protective mechanism to adapt to chlorophyll overexcitation by using photosynthetic antenna, when exposed to strong light intensity. This is also true for crop plants. Further improvement of crop productivity through genetic modification of *qE* quenching might be important for agriculture (Niyogi and Truong 2013; Niyogi 2017).

There is actually a report on manipulating plant gene expression to improve crop productivity. Tobacco (*Nicotiana tabacum*) plants overexpressing *Arabidopsis ZEP*

(*Zeaxanthin epoxidase*), *VDE,* and *PsbS* genes were generated. Interestingly, the incubation at stable temperature does not cause any difference in growth, by contrast to temperature fluctuating (changing) conditions, which causes a difference in the growth of the overexpressing plants improved in the NPQ capacity (Kromdijk et al. 2016). There is even natural variation in NPQ (*qE*) capacity between rice cultivars. Kasajima et al. (2011b) reported that NPQ is stronger in cold-adapted *Japonica* rice cultivars, compared to non-adapted *Indica* cultivars. This phenomenon is caused by the difference in the expression levels of *OsPsbS1* gene, as a result of transposon insertion in the upstream region of this gene (Nuruzzaman et al. 2014). Selection and usage of such natural mutations regulating NPQ capacity will be also beneficial to breed high light and stress-tolerant crops.

Previous studies for understanding the enzymes involved in the regulation of NPQ have relied primarily on the time-consuming generation of stable transgenic lines and mutant approaches. To enable rapid functional testing of NPQ-related genes from diverse organisms, *Agrobacterium tumefaciens*-mediated transient expression assays were adopted in tobacco to test if NPQ kinetics in fully expanded leaves can be tested. By expressing *Arabidopsis* genes known to be involved in NPQ, it was confirmed that the viability of this method for studying dynamic photosynthetic processes (Leonelli et al. 2016). This newly developed method offers a powerful alternative to traditional gene characterization methods by providing a fast and easy platform for assessing gene function in planta. This technology, combined with variation in NPQ-related genes between different cultivars, will also contribute to improve NPQ capacity in crops.

'Truncated light-harvesting chlorophyll antenna size' (TLA), in another word reduced amount in LHCII (light-harvesting chlorophyll protein complex II), in all classes of photosynthetic organisms would help to alleviate excess absorption of sunlight and the ensuing wasteful non-photochemical dissipation of excitation energy. This TLA concept was previously shown in green microalgae and cyanobacteria, as well as in crop plants. As much as 25% of improvement in stem and leaf biomass accumulation was observed for the TLA tobacco canopies (Kirst et al. 2017). Thus, in addition to the improvement of the photosynthetic reaction itself, breeding approaches from the viewpoint of the protection against photooxidative stress will no doubt be essential to realize even better improvement in crop production.

7 Fluctuation of the Sensitivity Factor (*S*) and Basal Dissipation (K_{fid}) and Their Corrections

Calculations of basic chlorophyll fluorescence parameters were introduced in the above sections. During these experiments, it is hypothesized that the sensitivity factor (*S*) and basal dissipation (k_{fid}) are stable. But in recent years, it has been clarified that these values can change (fluctuate) according to the states of plant samples and the conditions of experiments. In this section, the reasons for these fluctuations are explained, together with the mathematical corrections of the fluctuated values based on the inverse chlorophyll fluorescence theory (Eq. 2), for further application of chlorophyll fluorescence analysis.

A clear example of the fluctuation of S value (S fluctuation) is related to chloroplast avoidance movement. As already described in this chapter, chloroplasts hide behind each other under strong illumination (Fig. 2). Under such high illuminations for relatively long period (10 min or longer), there is clear discrimination in NPQ values between wild-type plants and *phot2* mutant plants of *Arabidopsis* (Cazzaniga et al. 2013); the NPQ value of *phot2* is much lower than wild-type. This is caused by decrease in the ratio of light absorption in chloroplasts, which was illuminated on the leaf (Kasajima et al. 2015). If the light absorption rate is reduced, S value is reduced in proportion to the changes in the absorption rate, and so are the F_s and F'_m values. The decrease in F_s and F'_m values results in the increase in the apparent values of k_{fid}, k_{NPQ}, and k_p. In other words, reduction of chlorophyll fluorescence yields which is caused by chloroplast avoidance movement is erroneously attributed to the increase in rate constants of all de-excitation processes under illumination of actinic light, because S fluctuation is not considered in the basic, simple calculations of fluorescence parameters. If the fluctuated S value is written as $S_{(f)}$, the degree of S fluctuation (σ) is:

$$\sigma = S_{(f)}/S \tag{14}$$

Thus, S fluctuation reasonably explains at least part of the mechanisms for slow-relaxing NPQ: only superficial increase in k_{fid} and k_{NPQ} values caused by chloroplast avoidance movement. Such changes in apparent values of rate constants are, certainly, not favorable to know the actual photosynthetic status. Identification of the σ value enables correction of S fluctuation, but measurement of the σ value is quite difficult. This value can be calculated by comparing F'_m values of wild-type and *phot2* mutant (Kasajima et al. 2015), although this will not be applicable to plant species other than *Arabidopsis*. In addition, S fluctuation also takes place when chlorophyll is degraded under stressful conditions (Kasajima 2017). As a matter of fact, we can do no better than avoiding induction of S fluctuation. Then, we had better treat leaf samples with mild conditions for short period. In addition, S fluctuation by chloroplast movement does not happen when red light is used as the actinic light (Cazzaniga et al. 2013). Two-dimensional PAM system also reduces S fluctuation by chloroplast movement to nearly negligible levels, probably because actinic lights are not lit from directly above (i.e., from the sides of leaves to some extent; Kasajima et al. 2015).

Strong light and other stressful conditions to plant leaves cause not only S fluctuation during measurement of chlorophyll fluorescence, but also fluctuations in the k_{pi} and k_{fid} values even after long-term (hours of) adaptation in the dark. NPQ components are relaxed in the dark, but this relaxation is not complete when leaves are severely photodamaged. This remaining slow NPQ is counted as a part of basal dissipation (k_{fid}) by mistake, when F_m is measured in the dark at the beginning of fluorescence measurements. Similarly, stress-damaged and non-repaired photo-

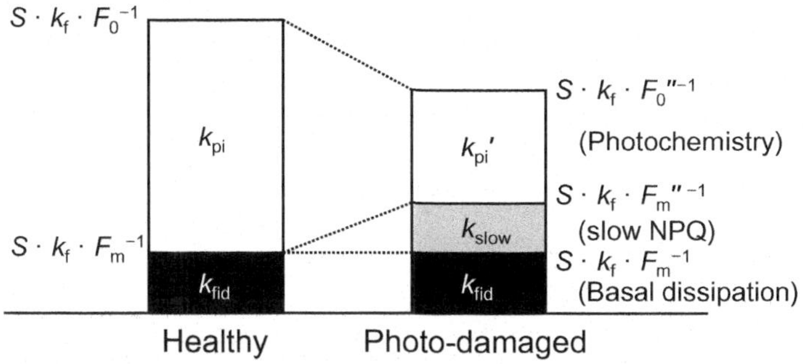

Fig. 6 Schematic representation of the relationship between fluorescence intensities and rate constants before and after photodamage. 'k_{slow}' represents the rate constant of slow-relaxing NPQ. Modified from Kasajima et al. (2015)

chemistry cause an illegal decrease in 'full' photochemistry (k_{pi}) in F_0 measurement. However, these changes by chloroplast damages can be detectable as the decrease in the F_v/F_m value.

F_v/F_m is the most popular parameter and measured in nearly all experiments of chlorophyll fluorescence analysis. When there are clear differences in F_v/F_m values between plant samples, or when the F_v/F_m values are lower than the standard values (e.g., 0.80 for rice and 0.85 for *Arabidopsis*), k_{fid} and k_{pi} values should be corrected. The following chlorophyll fluorescence parameters are used in the analysis of stress-related damages and correction of the rate constants (Kasajima et al. 2009, 2015; Kasajima 2017):

$$F_v/F_m(\text{after photodamage}) = \left(F_0''^{-1} - F_m''^{-1}\right)/F_0''^{-1} \tag{15}$$

$$qSlow = \left(F_m''^{-1} - F_m^{-1}\right)/F_m^{-1} \tag{16}$$

$$qPI = \left(F_0''^{-1} - F_m''^{-1}\right)/\left(F_0^{-1} - F_m^{-1}\right) \tag{17}$$

$$qSI = F_0''^{-1}/F_0^{-1} \tag{18}$$

Here, 'F_v/F_m (after photodamage)' represents the fluctuated (reduced) F_v/F_m value after photodamage (photoinactivation). *qSlow* represents the remaining, non-relaxed slow component of NPQ. *qPI* represents relative size of photochemistry after photodamage, and *qSI* represents relative size of total de-excitation processes after photodamage. F_0'' and F_m'' are fluorescence yields after photodamage, measured under the same conditions as F_0 and F_m. Relationship between rate constants and fluorescence values is also illustrated in Fig. 6.

8 Two-Dimensional *qPI-qSlow* Plotting of Fluorescence Parameters for Measurement of Oxidative Stress

Looking at Fig. 6, you will notice that the rate constants of both photochemical and non-photochemical de-excitation processes are greatly changed from their original values by severe photodamage. One of the problems with these changes is that the size of NPQ which is induced by the actinic light is underestimated when the parameter NPQ is calculated by Eq. (8), due to the overestimation of the rate constant of basal dissipation. Other parameters are also influenced by photodamage. This problem is solved if the original F_0 and F_m values are calculated from F_0'' and F_m'' values. These calculations are possible if *qSlow*, *qPI*, and *qSI* values are deduced from the F_v/F_m value after photodamage. And empirically, there is a curvilinear correlation between the changes in F_v/F_m values and *qSlow, qPI*, or *qSI* values during photodamage (Kasajima et al. 2015; Kasajima 2017). These regression curves can be used to calculate *qSlow, qPI*, and *qSI* values from photodamaged F_v/F_m values as follows (Kasajima 2017):

$$qPI = 3.34 \cdot (F_v/F_m)^4 - 1.36 \cdot (F_v/F_m)^3 - 2.80 \cdot (F_v/F_m)^2 + 3.16 \cdot (F_v/F_m) - 0.504 \quad (19)$$

$$qSI = 5.86 \cdot (F_v/F_m)^4 - 7.73 \cdot (F_v/F_m)^3 + 2.67 \cdot (F_v/F_m)^2 + 0.547 \cdot (F_v/F_m) + 0.346 \quad (20)$$

$$qSlow = 17.1 \cdot (F_v/F_m)^4 - 36.3 \cdot (F_v/F_m)^3 + 27.3 \cdot (F_v/F_m)^2 - 11.6 \cdot (F_v/F_m) + 3.45 \quad (21)$$

The values of F_0 and F_m are calculated as:

$$F_0 = \mathrm{qSI} \cdot F_0'' \quad (22)$$

$$F_m = (1 + qSlow) \cdot F_m'' \quad (23)$$

The parameters estimating photodamage (*qPI*, *qSI*, and *qSlow*) also benefits comparison of stress tolerance between different plant cultivars. Stress tolerance of plants has been almost always compared by the F_v/F_m values in chlorophyll fluorescence analysis. On the other hand, changes in F_v/F_m values integrate changes both in photochemistry and non-photochemical processes. F_v/F_m also does not reflect *S* fluctuation caused by chlorophyll degradation under higly stressful conditions. Two-dimensional plotting of *qPI* and *qSlow* values allows for comparison of the changes in photochemistry (*qPI*) and changes in non-photochemical pathways (*qSlow*) separately and also detects *S* fluctuation as simultaneous increase in both *qPI* and *qSlow* values (Kasajima 2017). Figure 7 shows the *qPI-qSlow* plot of 67 rice cultivars under oxidative stress. In this graph, the original coordinate before damage is (1, 0).

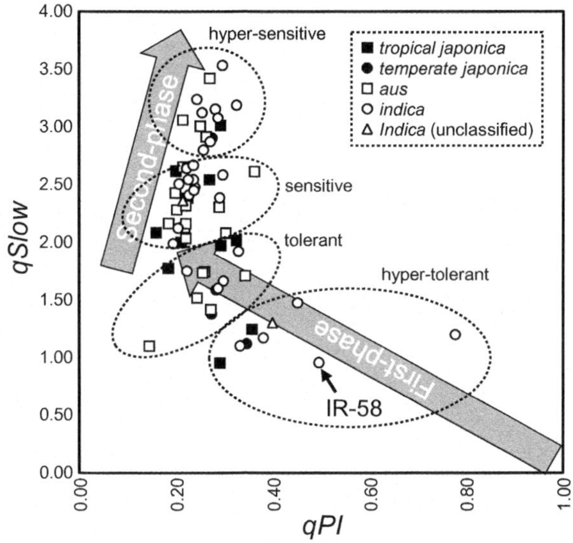

Fig. 7 *qPI-qSlow* plot during oxidative stress in 67 rice cultivars. The graph was slightly modified from Kasajima (2017). This data is used under the terms of Creative Commons license (CC BY 4.0; http://creativecommons.org/licenses/by/4.0/)

In the first phase of the changes in parameters, *qPI* decreases and *qSlow* increases at the same time. This is caused by damage to photochemistry and induction of slow NPQ. In the second phase, both *qPI* and *qSlow* increases where *qPI* and '1 + *qSlow*' are proportional. This change is caused by chlorophyll degradation. In the analysis in Fig. 7, it will be noteworthy that the only modern breeding line 'IR-58' among the tested cultivars was quite tolerant to oxidative stress. In another experiment, Japanese major modern cultivar 'Koshihikari' was also hyper-tolerant to oxidative stress. It could be deduced that oxidative stress tolerance attaches some favorable traits required for modern rice cultivars, and then we suggest that the analysis of oxidative stress tolerance should be introduced into rice breeding program.

9 Conclusion: Anti-oxidative Breeding to Feed the World

Chlorophyll fluorescence analysis, combined with fluorescence theories, provides immediate access to the analysis of photosynthetic states of plant leaves, photoprotection, and photodamages. It seems even possible to enhance crop production by stiffening the light protection system by genetic modifications. On the contrary, it is not yet understood at all whether such a strategy is also applicable in traditional breeding by crossing between different cultivars. Identification of stronger genes for genetic modifications and effective loci and mutations for crossing will give us new strategies for breeding even better crops. Pest and weed management are well recognized as integral treatments to save full crop production. For example, there are potentially as much as 35–55% losses of rice production in China due to diseases, insects/animals, and weeds, respectively. Similar levels of potential losses of rice production are also estimated for the other Asian countries (Oerke 1994). On the other hand, the degrees of crop losses caused by abiotic stresses are not well understood. The study of stress tolerance in crops will unravel this figure, and breeding for stress-tolerant crops will contribute to stable food production to feed the world population. Knowledge of photooxidative stress and photoprotection in plants has been accumulated by excellent studies and efforts of a large number of researchers. It is our great honor to have had the chance to introduce part of them, as one of the contributors to this field.

References

Apel K, Hirt H (2004) Reactive oxygen species: metabolism, oxidative stress, and signal transduction. Annu Rev Plant Biol 55:373–399

Baker NR (2008) Chlorophyll fluorescence: a probe of photosynthesis in vivo. Annu Rev Plant Biol 59:89–113

Bilger W, Björkman O (1990) Role of the xanthophyll cycle in photoprotection elucidated by measurements of light-induced absorbance changes, fluorescence and photosynthesis in leaves of *Hedera canariensis*. Photosynth Res 25:173–185

Cazzaniga S, Dall'Osto L, Kong SG et al (2013) Interaction between avoidance of photon absorption, excess energy dissipation and zeaxanthin synthesis against photooxidative stress in *Arabidopsis*. Plant J 76:568–579

Demidchik V (2015) Mechanisms of oxidative stress in plants: from classical chemistry to cell biology. Environ Exp Bot 109:212–228

Derks A, Schaven K, Bruce D (2015) Diverse mechanisms for photoprotection in photosynthesis. Dynamic regulation of photosystem II excitation in response to rapid environmental change. Biochim Biophys Acta 1847:468–485

Heldt HW (2005) Plant biochemistry. In: Chapter 3 'Photosynthesis is an electron transport process', 3rd edn. Elsevier Academic Press, Burlington, USA, pp 67–114

Hendrickson L, Furbank RT, Chow WS (2004) A simple alternative approach to assessing the fate of absorbed light energy using chlorophyll fluorescence. Photosynth Res 82:73–81

Hendrickson L, Förster B, Pogson BJ et al (2005) A simple chlorophyll fluorescence parameter that correlates with the rate coefficient of photoinactivation of Photosystem II. Photosynth Res 84:43–49

Higa T, Wada M (2016) Chloroplast avoidance movement is not functional in plants grown under strong sunlight. Plant, Cell Environ 39:871–882

Jänkänpää HJ, Frenkel M, Zulfugarov I et al (2013) Non-photochemical quenching capacity in *Arabidopsis thaliana* affects herbivore behaviour. PLoS ONE 8:e53232

Kagawa T, Wada M (2002) Blue light-induced chloroplast relocation. Plant Cell Physiol 43:367–371

Kasahara M, Kagawa T, Oikawa K et al (2002) Chloroplast avoidance movement reduces photodamage in plants. Nature 420:829–832

Kasajima I (2017) Difference in oxidative stress tolerance between rice cultivars estimated with chlorophyll fluorescence analysis. BMC Res Notes 10:168

Kasajima I, Takahara K, Kawai-Yamada M et al (2009) Estimation of the relative sizes of rate constants for chlorophyll de-excitation processes through comparison of inverse fluorescence intensities. Plant Cell Physiol 50:1600–1616

Kasajima I, Kawai-Yamada M, Uchimiya H (2011a) Theories in PAM chlorophyll fluorescence measurement of plant leaves. In: Maes KJ, Willems JM (eds) Photochemistry: UV/VIS spectroscopy, photochemical reactions and photosynthesis. Nova Science Publishers, Hauppauge, USA, pp 351–371

Kasajima I, Ebana K, Yamamoto T et al (2011b) Molecular distinction in genetic regulation of nonphotochemical quenching in rice. Proc Natl Acad Sci USA 108:13835–13840

Kasajima I, Suetsugu N, Wada M et al (2015) Collective calculation of actual values of non-photochemical quenching from their apparent values after chloroplast movement and photoinhibition. Am J Plant Sci 6:1792–1805

Kirst H, Gabilly ST, Niyogi KK et al (2017) Photosynthetic antenna engineering to improve crop yields. Planta 245:1009–1020

Kitajima M, Butler WL (1975) Quenching of chlorophyll fluorescence and primary photochemistry in chloroplasts by dibromothymoquinone. Biochim Biophys Acta 376:105–115

Kong SG, Suzuki T, Tamura K et al (2006) Blue light-induced association of phototropin 2 with the Golgi apparatus. Plant J 45:994–1005

Kong SG, Wada M (2014) Recent advances in understanding the molecular mechanism of chloroplast photorelocation movement. Biochim Biophys Acta 1837:522–530

Kornyeyev D, Holaday AS (2008) Corrections to current approaches used to calculate energy partitioning in photosystem 2. Photosynthetica 46:170–178

Kramer DM, Johnson G, Kiirats O et al (2004) New fluorescence parameters for the determination of Q_A redox state and excitation energy fluxes. Photosynth Res 79:209–218

Kromdijk J, Glowacka K, Leonelli L et al (2016) Improving photosynthesis and crop productivity by accelerating recovery from photoprotection. Science 354:857–861

Laisk A, Oja V, Rasulov B et al (1997) Quantum yields and rate constants of photochemical and nonphotochemical excitation quenching. Plant Physiol 115:803–815

Leonelli L, Erickson E, Lyska D et al (2016) Transient expression in *Nicotiana benthamiana* for rapid functional analysis of genes involved in non-photochemical quenching and carotenoid biosynthesis. Plant J 88:375–386

Li XP, Björkman O, Shih C et al (2000) A pigment-binding protein essential for regulation of photosynthetic light harvesting. Nature 403:391–395

Lin KC, Jwo WS, Chandrika NNP et al (2016) A rice mutant defective in antioxidant-defense system and sodium homeostasis possesses increased sensitivity to salt stress. Biol Plant 60:86–94

Maxwell K, Johnson GN (2000) Chlorophyll fluorescence—a practical guide. J Exp Bot 51:659–668

Miyake C, Amako K, Shiraishi N et al (2009) Acclimation of tobacco leaves to high light intensity drives the plastoquinone oxidation system—relationship among the fraction of open PSII centers, non-photochemical quenching of Chl fluorescence and the maximum quantum yield of PSII in the dark. Plant Cell Physiol 50:730–743

Munekage Y, Hojo M, Meurer J et al (2002) PGR5 is involved in cyclic electron flow around photosystem I and is essential for photoprotection in *Arabidopsis*. Cell 110:361–371

Niyogi KK, Truong TB (2013) Evolution of flexible non-photochemical quenching mechanisms that regulate light harvesting in oxygenic photosynthesis. Curr Opin Plant Biol 16:307–314

Niyogi KK (2017) Editorial overview: physiology and metabolism: light responses from photoreceptors to photosynthesis and photoprotection. Curr Opin Plant Biol 37:iv–vi

Nuruzzaman M, Kanno T, Amada R et al (2014) Does the upstream region possessing MULE-like sequence in rice upregulate *PsbS1* gene expression? PLoS ONE 9:e102742

Oerke EC (1994) Estimated crop losses in rice. In: Oerke EC, Dehne HW, Schönbeck F, Weber A. (ed) Crop Production and crop protection—estimated losses in major food and cash crops. Elsevier Science B.V., Amsterdam, The Netherlands, pp 89–179

Oikawa K, Yamasato A, Kong SG et al (2008) Chloroplast outer envelope protein CHUP1 is essential for chloroplast anchorage to the plasma membrane and chloroplast movement. Plant Physiol 148:829–842

Oxborough K, Baker N (1997) Resolving chlorophyll a fluorescence images of photosynthetic efficiency into photochemical and non-photochemical components—calculation of qP and F_v'/F_m' without measuring F_0'. Phosynth Res 54:135–142

Oxborough K (2004) Imaging of chlorophyll *a* fluorescence: theoretical and practical aspects of an emerging technique for the monitoring of photosynthetic performance. J Exp Bot 55:1195–1205
Pandey P, Srivastava RK, Rajpoot R et al (2016) Water deficit and aluminum interactive effects on generation of reactive oxygen species and responses of antioxidative enzymes in the seedlings of two rice cultivars differing in stress tolerance. Environ Sci Pollut Res 23:1516–1528
Printó-Marijuan M, Munné-Bosch S (2014) Photo-oxidative stress markers as a measure of abiotic stress-induced leaf senescence: advantages and limitations. J Exp Bot 65:3845–3857
Quick WP, Stitt M (1989) An examination of factors contributing to non-photochemical quenching of chlorophyll fluorescence in barley leaves. Biochim Biophys Acta 977:287–296
Rachoski M, Gazquez A, Calzadilla P et al (2015) Chlorophyll fluorescence and lipid peroxidation changes in rice somaclonal lines subjected to salt stress. Acta Physiol Plant 37:117
Sonoike K (1999) The different roles of chilling temperatures in the photoinhibition of photosystem I and photosystem II. J Photochem Photobiol B: Biol 48:136–141
Sonoike K (2003) Measurement of photosynthesis by using pulse amplitude modulation and chlorophyll fluorescence. http://www.photosynthesis.jp/fluo1.html
Stefanov D, Terashima I (2008) Non-photochemical loss in PSII in high- and low-light-grown leaves of *Vicia faba* quantified by several fluorescence parameters including L_{NP}, F_0/F_m', a novel parameter. Physiol Plant 133:327–338
Suetsugu N, Wada M (2007) Chloroplast photorelocation movement mediated by phototropin family proteins in green plants. Biol Chem 388:927–935
Suetsugu N, Yamada N, Kagawa T et al (2010) Two kinesin-like proteins mediate actin-based chloroplast movement in *Arabidopsis thaliana*. Proc Natl Acad Sci USA 107:8860–8865
Suetsugu N, Higa T, Gotoh E et al (2016) Light-Induced Movements of Chloroplasts and Nuclei Are Regulated in Both Cp-Actin-Filament-Dependent and Independent Manners in *Arabidopsis thaliana*. PLoS ONE 11:e0157429
Suetsugu N, Higa T, Wada M (2017) Ferns, mosses and liverworts as model systems for light-mediated chloroplast movements. Plant, Cell Environ 40:2447–2456
Takahara K, Kasajima I, Takahashi H et al (2010) Metabolome and photochemical analysis of rice plants overexpressing *Arabidopsis* NAD kinase gene. Plant Physiol 152:1863–1873
Takemiya A, Inoue S, Doi M et al (2005) Phototropins promote plant growth in response to blue light in low light environments. Plant Cell 17:1120–1127
van Kooten O, Snel JFH (1990) The use of chlorophyll fluorescence nomenclature in plant stress physiology. Photosynth Res 25:147–150
Wada M, Kagawa T, Sato Y (2003) Chloroplast movement. Annu Rev Plant Biol 54:455–468
Wada M, Kong SG (2011) Analysis of chloroplast movement and relocation in Arabidopsis. Methods Mol Biol 774:87–102
Wada M (2013) Chloroplast movement. Plant Sci 210:177–182
Xu K, Chun S, Li T et al (2015) *OsGRAS23*, a rice GRAS transcription factor gene, is involved in drought stress response through regulating expression of stress-responsive genes. BMC Plant Biol 15:141

Water Stress and Redox Regulation with Emphasis on Future Biotechnological Prospects

B. Loedolff and C. van der Vyver

Abstract Water deficit conditions can disturb the redox homeostasis in plants and result in increased production of reactive oxygen species (ROS), which could ultimately lead to plant death. Many regulatory mechanisms exist in plants to overcome the damaging effects of free radical accumulation. These include morphological changes and the production of enzymatic and non-enzymatic anti-oxidants. Anti-oxidants accumulate in an organized manner in different cellular compartments of which the chloroplast especially is vital in maintaining redox homeostasis during water deficit conditions. The signaling mechanisms in plants during water deficit are complex. This ranges from hormonal signaling to where sugars, such as sucrose, RFOs, trehalose and fructans, are suggested to contribute significantly to the redox homeostasis mechanisms through ROS scavenging. These sugars display characteristics of osmoprotectants and anti-oxidants. Worth investigating is the modulation of sugars and key metabolic enzymes during the development of drought-tolerant crops. The use of direct genome-editing techniques such as CRISPR/Cas9 could be very useful in developing new tolerant crop varieties. Also, a multitude of health beneficial phenolic and sugar compounds accumulate in plant vacuolar compartments as a part of the redox homeostasis mechanism. Future studies could seek to implement environmental stresses in a positive light to enhance food quality and establish the concept of producing functional foods.

1 The Oxygen Paradox and Mechanisms Required to Overcome Oxidative Damage

Global climate changes are accompanied by severe and prolonged periods of water deficit affecting processes associated with plant growth and development (Fayez and Bazaid 2014). These processes are mostly linked to photosynthesis, transpiration and other biochemical pathways (Shannon 1997). The obligate aerobic conditions

B. Loedolff · C. van der Vyver (✉)
Department of Genetics, Institute for Plant Biotechnology, Stellenbosch University, Stellenbosch, South Africa
e-mail: cvdv@sun.ac.za

S. K. Panda and Y. Y. Yamamoto (eds.), *Redox Homeostasis in Plants*,
Signaling and Communication in Plants, https://doi.org/10.1007/978-3-319-95315-1_8

on earth result in excess amounts of oxygen (O_2). O_2 can be extremely unstable in an unpaired electron state, resulting in severe structural and non-structural damage in both plants and animals. This phenomenon is referred to as the "oxygen paradox". Plants require a constant supply of O_2 to maintain energy production in subcellular compartments, specifically the mitochondria. This is achieved during the process of photosynthesis where plants continuously use carbon dioxide to produce O_2 in green tissues. The O_2 demand-and-supply results in continuous fluctuations of sub-cellular O_2 concentrations and can lead to severe hyperoxic (>700 μM) or hypoxic (<1 mM) conditions, especially during the periods of abiotic stress. Consequently, severe periods of abiotic stress result in an oxidative cellular environment which leads to an increased electron leakage toward O_2 during photosynthetic processes. This ultimately enhances the production of reactive oxygen species (ROS), which can cause oxidative damage to cellular and non-cellular structures leading to cell death (Mittler 2002; Asada 2006).

Plants have developed highly regulated anti-oxidant defense systems and during optimal growth conditions govern a fine balance between ROS generation and consumption (Hameed et al. 2011). One of these defense systems includes a multitude of anti-oxidant enzymes. Considerable resistance against oxidative damage caused by ROS has been implicated in plants high in anti-oxidant enzyme activities (Gapinska et al. 2008).

Adaptive evolution in plants resulted in a cascade of metabolic mechanisms to reduce oxidative damage resulting from water stress, relying on the production of anti-oxidants. These mechanisms are considered the orchestrators of redox homeostasis in plants and include increased synthesis of (i) vitamins, (ii) phenolic compounds, (iii) lipophilic anti-oxidants, (iv) hormone interactions, (v) osmoprotectants and (vi) transcriptional regulatory mechanisms.

A steady-state cellular ROS level prevails under normal conditions because of a balance between ROS generation and scavenging. Stresses including water deficit shift this balance with more ROS being produced than scavenged, creating an oxidative stress. Before this imbalance is reached, an early increase in ROS level can actually play an important role as signaling molecule for acclimation or defense responses (Lamb and Dixon 1997). How is the intracellular redox state governed during periods of stress? The topics in this chapter will be focused toward metabolic and molecular responses in a plant during periods of water deficit/water stress (which in turn leads to an over-production of damaging oxygen molecules) and mechanisms utilized to overcome the resulting oxidative stress.

2 A Myriad of Metabolic and Molecular Elements Responds to Oxidative Stress in a Synchronized Manner

2.1 Vitamins

The first line of oxidative damage defense is ascorbic acid (AA, vitamin C). Despite being an ubiquitous anti-oxidant in all plant tissues, AA is mostly transported from the inner mitochondrial membrane to the cytoplasm and apoplast where it is most abundant (Barnes et al. 2002). A requirement of the redox homeostasis is the ability of anti-oxidant molecules to stabilize free radicals (like ROS) by donating electrons. AA is a powerful anti-oxidant because it can donate electrons freely to a wide variety of enzymatic and non-enzymatic reactions.

The mitochondria are responsible for the production of the majority of AA in plant cells, catalyzed by L-galactono-γ-lactone dehydrogenase (the Smirnoff–Wheeler pathway), with the remaining being generated from D-galacturonic acid. When AA is oxidized, a by-product termed MDA is produced after reacting with ROS. MDA can be recycled to ascorbate with the assistance of the photosynthetic machinery which functions in another very important cellular compartment, the chloroplast, where it can serve as a ROS scavenger. At the water–membrane interface, AA will react with ROS, reducing it to water, and is further regenerated into tocopherols. Subsequently, membranes are protected from oxidative damage through the generation of these lipophilic anti-oxidants such as α-tocopherol (Igamberdiev et al. 2004; Shao et al. 2005). Plants challenged with water deficit conditions display an over-production of AA enabling the plant to overcome physiological barriers and maintain redox homeostasis.

2.2 Lipophilic Anti-oxidants

Tocopherols (considered to be methylated phenols with vitamin E activity) are also ubiquitous compounds in all plant cells and belong to a family of lipophilic anti-oxidants. It has the ability to protect membrane constituents (such as lipids), specifically of the chloroplasts to protect the photosynthetic machinery, both structurally and functionally. Tocopherols only occur in green tissues of plants because it is only synthesized in photosynthetic organisms. Among the tocopherols, α-tocopherol is the most efficient scavenger of ROS, rendering this one of the most essential components during oxidative protection of biological membranes (Holländer-Czytko et al. 2005; Das and Roychoudhury 2014).

Carotenoids, localized in the plastids of both photosynthetic and non-photosynthetic plant tissues, also belong to a family of lipophilic anti-oxidants. These molecules form part of a group of antennae molecules which absorbs light (450–570 nm) and transfers the energy to the chlorophyll molecule. Carotenoids also exhibit their anti-oxidative activity by protecting the photosynthetic machinery through scavenging a multitude of free radicals, including ROS.

2.3 Phenolic Compounds

Considerable resistance to oxidative damage caused by ROS has been demonstrated in plants with high anti-oxidant enzyme activities (Gapinska et al. 2008). However, these enzymatic anti-oxidant activities can become depleted very fast during periods of water deficit. Higher plants have also developed different adaptive mechanisms to reduce oxidative damage during periods of abiotic stress, through the biosynthesis of a cascade of non-enzymatic anti-oxidants. Numerous studies demonstrated that water deficit, specifically, can induce a wide range of anti-oxidants in a number of plant species (Keles and Öncel 2002; Munné-Bosch et al. 2001; Sofo et al. 2005). Among the non-enzymatic anti-oxidants are the phenolic compounds, known to accumulate during environmental stresses, which also play a critical role in free radical scavenging to protect plants against the damaging effects of increased ROS levels (Dixon and Paiva 1995; Grace and Logan 2000; Petridis et al. 2012). Phenolic compounds form part of the secondary metabolism of a plant, and although required for various functions such as plant reproduction, its biosynthesis via the phenylpropanoid pathway, is often an energy-depriving process. The first committed enzyme in the phenylpropanoid pathway is PAL (phenylalanine ammonia lyase), which yields a wide range of phenolic compounds (Diallinas and Kanellis 1994; Liu et al. 2006). It is noteworthy that some studies have indicated that water deficit conditions were associated with an increased activation of PAL, concomitant with an increase in phenolic compounds (Oh and Rajashekar 2009).

Flavonoids (including anthocyanins), a subgroup of the phenolics, are ubiquitous in the plant kingdom and occur commonly in leaves, flowers, fruits, seeds and pollen grains. They have diverse roles in plant fertility and germination of pollen and defense against plant pathogens. Flavonoids have been suggested to play a secondary ROS-scavenging role to protect the photosynthetic apparatus against free radical damage, due to the excess excitation energy (Dixon and Paiva 1995; Fortes et al. 2011; Fini et al. 2011; Garrido et al. 2016). It has also been implemented as having a role in scavenging superoxide to relieve the damages caused to the outer envelope of the chloroplastic membrane (Agati et al. 2012).

2.4 Phyto-Hormones

Phyto-hormones (particularly abscisic acid, ethylene and salicylic acid) are considered major regulators of plant responses to environmental stresses and have been implicated in controlling physiological and metabolic processes (Kwak et al. 2006). Water deficit triggers an increase in the production of abscisic acid (ABA), which is considered one of the most important phyto-hormones to play a signaling role to confer downstream processes (including transcriptional regulation). ABA induces stomatal closure during periods of water deficit. These responses are mediated by ROS, particularly H_2O_2, which is synthesized by plasma membrane-bound NADPH oxidase (Zhang et al. 2001; Kwak et al. 2003).

Ethylene is often associated with water deficit, where its synthesis is dependent on ROS due to increased accumulation during periods of osmotic stresses (Ke and Sun 2004). Ethylene has been implicated in stomatal closure, at least to some extent. Research done on ETR1, an ethylene receptor, indicated its involvement in sensing H_2O_2 which stimulates the subsequent closure of the stomata, providing suggestive evidence to the importance of ethylene during water deficit (Desikan et al. 2005).

Salicylic acid (SA), a hormonelike substance, fulfills an important function during periods of water deficit and has been demonstrated to participate in the regulation of (i) photosynthetic reactions, (ii) stomatal regulation and (iii) increasing anti-oxidative protection (Khan et al. 2003; Arfan et al. 2007). Exogenous application of SA has been shown to modulate anti-oxidant enzyme activities resulting in an increased plant tolerance to abiotic stresses (Fayez and Bazaid 2014). It appears that hormones may have a finely regulated cross talk signaling cascade during physiological and developmental processes. During periods of water deficit, ROS either induces hormonal cross talk or participates during signaling cascades to maintain redox homeostasis by controlling stress responses (Fig. 1).

2.5 Water-Soluble Carbohydrates and Other Compatible Solutes

Thus far, it is evident that plants, upon stress perception, are able to induce signaling cascades which leads to modifications in gene expression and the subsequent accumulation of ROS, phyto-hormones, transcription factors and compatible solutes (Schramm et al. 2008; Yamaguchi and Blumwald 2005; Yamaguchi-Shinozaki and Shinozaki 2006). Metabolic adaptation, which enabled plants to cope with water and osmotic stresses, includes an increased synthesis of compatible solutes, such as proline and soluble sugars, which not only contribute to osmoregulation but also protect the structural integrity of biomolecules and membranes and act as free radical scavengers that protect DNA from damaging effects of ROS (Ashraf and Foolad 2007). Compatible solutes (like some amino acids, sugars and sugar alcohols) have been proposed to function as one of the most important components in stress response mechanisms because it does not interfere with normal cellular metabolic processes even when they occur in substantial concentrations. Proline, widely distributed in plants, has been reported to counterbalance the deleterious effects of water deficit due to its increased accumulation during periods of water deficit (Fayez and Bazaid 2014). It may also be involved in the recovery mechanism from stress, where it serves as an organic nitrogen reserve (Sairam and Tyagi 2004). The

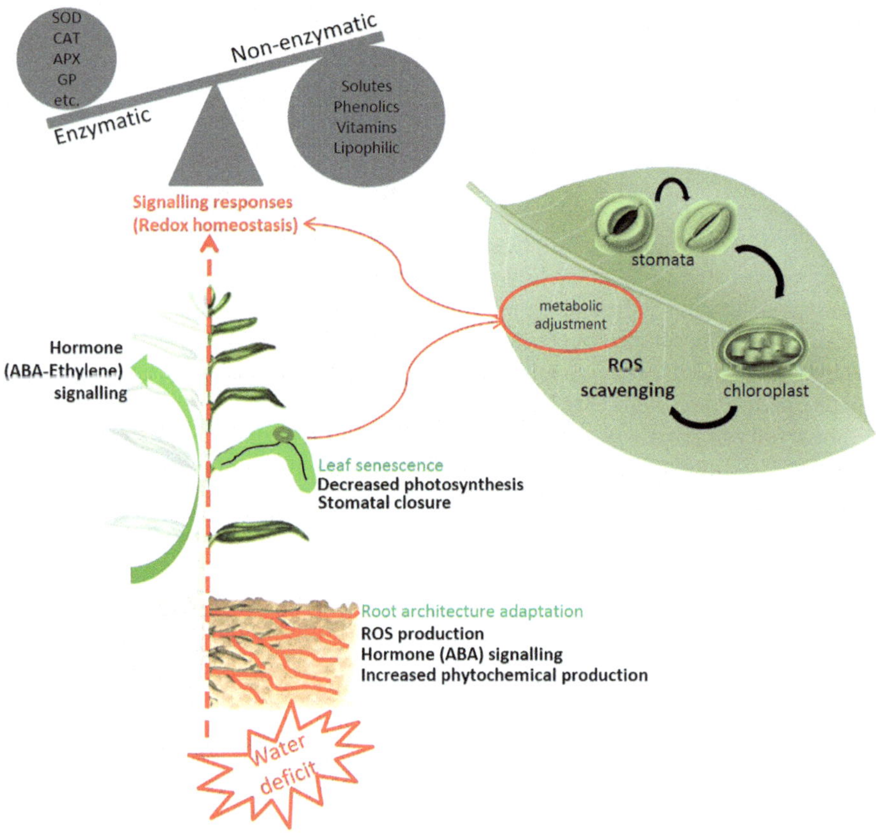

Fig. 1 Key components in a plant's response to drought stress in order to maintain redox homeostasis, including morphological, physiological and biochemical responses

accumulation of soluble sugars (e.g., trehalose, raffinose family of oligosaccharides and fructans) is thought to serve in osmotic adjustment and cellular protection (membrane stabilization and ROS scavenging) (Hoekstra et al. 2001; Valluru and Van den Ende 2011). Water deficit stress has been associated with increased concentrations of soluble sugars in many plants (Schubert et al. 1995; Kameli and Losel 1996; Pinheiro et al. 2001; Pinheiro and Chaves 2011). Soluble sugars are considered the most important signaling molecules during periods of environmental stress.

Trehalose, a non-reducing disaccharide of glucose, is synthesized by numerous organisms and may function as a compatible solute in the stabilization of biological structures during periods of water deficit (Elbein 2010). Despite earlier research indicating the absence of trehalose from higher plants, it has been reported to accumulate in high amounts in some drought- and desiccation-tolerant plant species, implicating a potentially important role during water deficit (Drennan et al. 1993; Sah et al. 2016). Rather than regulating water potential, trehalose may fulfill physiological roles including (i) efficient stabilization of dehydrated enzymes, proteins and lipid membranes, and (ii) protection of biological structures (Wingler 2002).

The raffinose family of oligosaccharides (RFOs) are considered key compatible solutes during a myriad of abiotic stress conditions (Koster and Leopold 1988; Bartels and Sunkar 2005) to stabilize cell proteins and membranes, and to support cell turgor (Hoekstra et al. 2001). During periods of water deficit, it has been demonstrated in resurrection plant species that RFOs accumulate as a part of a desiccation tolerance mechanism to possibly: (i) protect plants from membrane and subcellular damage during dehydration (Peters et al. 2007) and (ii) supply plant cells with the necessary energy levels to assist during rehydration and damage control (*Xerophyta viscosa*; Peters et al. 2007). The role of RFOs as ROS scavengers has recently been demonstrated (Nishizawa et al. 2008; Van den Ende and Valluru 2009; Valluru and Van den Ende 2011; Peshev et al. 2013) in transgenic Arabidopsis plants (Nishizawa et al. 2006, 2008), providing further evidence that RFOs may play a much larger role in protection and signaling mechanisms during periods of water deficit.

Furthermore, fructans are also considered to act as compatible solutes during conditions of abiotic stress (Valluru and Van den Ende 2008). The correlation of drought tolerance and fructan accumulation was primarily demonstrated in transgenic *Nicotiana tabacum* (tobacco; Pilon-Smits et al. 1995) and *Beta vulgaris* (sugar beet; Pilon-Smits et al. 1999). Recent exploration into environmentally induced fructan accumulation provided insight that fructans may also play a role in ROS scavenging (Peshev et al. 2013; Peukert et al. 2014). Similar to RFOs, fructan-based oligomers are able to stabilize membranes during periods of water deficit by interacting with lipid layers (Vereyken et al. 2001; Hincha et al. 2003). However, due to its restricted localization to the vacuole, it is proposed to only contribute to the stabilization of the tonoplast and possibly the plasma membrane (Valluru et al. 2008).

2.6 *Transcriptional Regulatory Elements*

Response to water deficit can be regulated at the transcriptional, posttranscriptional and translational levels. The majority of research has been done to explore the regulation of gene expression in response to water deficit (transcriptional regulation). Cis-acting DNA elements, of which there are more than three classes, control the transcription of specific water deficit-induced genes. Numerous transcriptomic studies have identified the regulatory water deficit-responsive genes and key transcription factors which orchestrate water deficit-responsive gene transcription. These transcription factors include, among others, MYB, MYC, DREB/CBF (drought-responsive cis-element binding protein/C-repeat binding factor), ABF/AREB, NAC and WRKY (Stockinger et al. 1997; Nakashima et al. 2009; Rushton et al. 2010). The corresponding cis-elements, DRE/CRT and ABRE (ABA-responsive cis-element), were subsequently discovered in the promoters of many stress-responsive genes (Yamaguchi-Shinozaki and Shinozaki 2006).

As mentioned previously, ABA is a key phyto-hormone during water deficit signaling and downstream responses. Genes regulated by ABA contain the ABA response element (ABRE), with the consensus RYACGTGGYR, bound by a bZIP transcription factor. An ABA-independent DNA element, the drought-responsive element (DRE), is also involved in transcriptional regulation during water deficit and has the consensus TACCGACAT, bound by a DREB transcription factor. Interestingly, the DREB transcription factor contains a DNA-binding domain involved in ethylene-regulated gene expression which could be extrapolated into a role during ethylene-mediated water deficit responses (Abebe et al. 2010; Dugas et al. 2011).

Novel types of transcription factors involved in stomatal regulation may play a critical role in future plant modification to withstand periods of water deficit. In rice, the drought and salt tolerance (DST) transcription factor was demonstrated to regulate gene expression involved in free radical homeostasis and mediates ROS-induced stomatal closure and abiotic stress tolerance (Huang et al. 2009). Furthermore, the over-expression of stress-responsive NAC1 (SNAC1) gene in rice guard cells enhanced ABA sensitivity, stomatal closure and upregulation of DST (Hu et al. 2006). In Arabidopsis, the NFYA5 (drought-inducible nuclear) transcription factor was demonstrated to control stomatal aperture and play a role in drought tolerance (Li et al. 2008). Other transcription factors in Arabidopsis that regulate stomatal aperture, MYB60 and MYB61 expressed mainly in guard cells, have been demonstrated to improve drought tolerance in plants (Cominelli et al. 2005). In this case, MYB60 negatively regulates stomatal closure which is important during redox homeostasis (Cominelli et al. 2005; Oh et al. 2011). Further studies to determine the molecular targets and signaling systems associated with these transcription factors in stomatal responses will assist in our understanding of the regulatory networks required for the control of plant water deficit responses and how it could potentially enable growth adjustment during these periods.

3 Major Intracellular Organelles Generate Enzymatic Anti-oxidant Species and are Regulatory Compartments During Redox Homeostasis

The advent of aerobic metabolism (with reference to the oxygen paradox) is the sole reason for production of ROS in plants. ROS is produced during normal and stressful environmental conditions, with the latter condition being much more aggravated and causing severe damage to plant structure and impinging on growth (Dat et al. 2000). Plants have thus developed both enzymatic and non-enzymatic (discussed previously) defense mechanisms to scavenge ROS by anti-oxidant systems. The enzymes responsible for anti-oxidant activities are localized in the different sub-cellular compartments and respond accordingly to water deficit through mechanisms of cross talk and signaling to maintain redox homeostasis. This anti-oxidant machinery includes the following major enzymes: superoxide dismutase (SOD), catalase (CAT), ascorbate peroxidase (APX), monodehydroascorbate reductase (MDHAR), dehydro-ascorbate reductase (DHAR), glutathione reductase (GR), glutathione peroxidase (GP) and guaiacol peroxidase (GPX). Here, we will merely mention some of the well-studied anti-oxidant enzymes involved in water deficit responses and how it may be involved directly, or indirectly, to maintain redox homeostasis in different cellular compartments.

The chloroplast, peroxisomes and mitochondria are considered the major contributors to the pool of ROS (superoxide and hydrogen peroxide) produced within the cells, caused by the reduction of oxygen during key metabolic reactions, such as photorespiration and electron transport processes (Foyer et al. 1994; Noctor et al. 2002; Foyer and Noctor 2003; del Río et al. 2006). The first line of defense against ROS is the anti-oxidant enzyme SOD. SOD is present in all subcellular locations and has the ability to dismutate O_2^- into more stable hydrogen peroxide (H_2O_2), which is subsequently detoxified by enzymes like CAT, APX and GP. CAT can scavenge H_2O_2 directly, while the latter enzymes require ascorbate and glutathione, respectively, as electron donors (Dat et al. 2000; Alscher et al. 2002). The SODs require metal cofactors which have classed them into three groups: Fe-SOD, Mn-SOD and Cu-Zn-SOD. The three groups are independently localized in different compartments where they fulfill their anti-oxidant functions. Fe-SODs are located in the chloroplasts, Mn-SODs in mitochondria and peroxisomes and Cu-Zn-SODs in the chloroplasts, the cytosol and the extracellular (apoplast) space (Alscher et al. 2002). Other anti-oxidant enzymes have been described to be compartment-specific due to their affinity for H_2O_2. CAT which has a low affinity for H_2O_2 is localized mostly in peroxisomes. APX which has a higher affinity for H_2O_2 is localized in chloroplasts, mitochondria, peroxisomes and the cytosol. Because APX is localized in every ROS-producing compartment, it has been suggested that this enzyme may function as the master regulator of intracellular redox homeostasis (Cruz de Carvalho 2008).

It is evident that the different cellular compartments maintain an appropriate ROS balance (redox homeostasis) within plant cells, mainly as a result of efficient coordination of ROS production and scavenging. This coordination in cellular compartments is governed by a complex network of pro-oxidant and anti-oxidant systems (Quan et al. 2008). During periods of water deficit, this balance is perturbed which could ultimately lead to plant death. Based on the ability of a plant to maintain or re-establish the redox homeostasis, it may be referred to as tolerant to water deficit. Tolerant plants have been the focal point of selection in crop species for future breeding strategies. Selection criteria often rely on the levels of compatible solutes and comparative changes in anti-oxidant enzyme activities in response to water stress (Sinhababu and Kar 2003; Nandi et al. 2004). Essential elements to keep in mind include: (i) cellular compartments and (ii) physiological state of a plant, as these elements determine the differential ROS signature of the stressed cell or plant. Preferential scavenging of particular ROS occurs in a very specific manner and often depends on (i) and (ii). Differential sensitivity and affinity, of all the previously mentioned anti-oxidant enzymes, to ROS are suggestive for their differential role in ROS scavenging—for example, CAT counteracts excessive production under severe water deficit and reduces the ROS level that is scavenged subsequently by APX and the ascorbate-glutathione cycle.

4 Plant Structures Affected by Internal Redox Programming During Water Deficit

4.1 Seeds Are Naturally in a State of Water Deficit

Seeds are divided into two categories based on their ability to survive moisture loss associated with maturation (Roberts 1973). These categories have been termed recalcitrant (desiccation-sensitive, can not survive water loss beyond 40% of water content) and orthodox (desiccation-tolerant, can survive water loss of up to 95% of water content; Farrant et al. 1989; Finch-Savage 1992). Mature seeds are quiescent organs and accumulate compounds such as sugars, lipids and anti-oxidants. Proteins, lipids and glutathione are the most abundant components contained within seeds (contributing approximately 40% each to seed dry weight) and accumulate during the mid-maturation phases. Conversely, sucrose (Suc) and RFOs are deposited during late maturation phases (contributing approximately 2–20% of seed dry weight; Baud et al. 2002; Fait et al. 2006).

The major anti-oxidants present within seeds include AA and GSH, which fulfill the role of ROS scavenging during the constant state of water deficit, and are usually transferred from maternal plant to seed during seed development (Tommasi et al. 1999). Unfortunately, AA is a very unstable anti-oxidant and is therefore not stored in orthodox seeds (De Tullio and Arrigoni 2003), but rather synthesized on a need basis during germination when it is required for cell cycle activity (Pallanca and Smirnoff 1999). Recalcitrant seeds on the other hand maintain high levels of AA from maturation to germination (Tommasi et al. 1999). In contrast to AA, GSH is stable in the desiccated state and accumulates to high concentrations in both orthodox and recalcitrant seeds (Kranner et al. 2010). Mature seeds (in the "dry" state) contain more glutathione than any other tissue in the plant, particularly relevant to orthodox seeds, and could potentially be ascribed to the ROS protection required during this period of water deficit (Klapheck 1988).

Seed orthodoxy has long been associated with RFO (raffinose and stachyose) accumulation. Recalcitrant seeds accumulate no RFOs (or very low amounts). It is thus thought that RFO accumulation facilitates a protective role during seed maturation (when tissue desiccation occurs), initiating a glassy state in orthodox seeds (Blackman et al. 1995; Bailly et al. 2001; Hoekstra et al. 2001). This glassy state is proposed to prevent cytoplasmic crystallization and the consequent drastic pH change which is detrimental to seed viability (Caffrey et al. 1988).

RFOs have also been implicated for their role as anti-oxidants and consequently impact longevity of orthodox seeds (Horbowicz and Obendorf 1994; Lin and Huang

1994). During oxidative stress (an active process during seed imbibition and early germination), a major burst of reactive oxygen species (ROS) destabilizes macromolecules such as proteins, lipids and DNA. RFOs are abundant in most seeds and rapidly decline as germination proceeds, suggesting a complementary role to AA and GSH as ROS scavengers protecting plant cells as anti-oxidant agents to maintain redox homeostasis (Mittler 2002; Nishizawa et al. 2008). ROS, including H_2O_2, have been demonstrated to play an important role in radicle emergence during seed germination (Garnczarska and Wojtyla 2008). This concluded the fine balance required to maintain redox homeostasis because, apart from its destructive role during various environmental stresses, a positive role of ROS during seed maturation and germination is critical to subsequent plant growth and development.

4.2 *Immature Plants are Better Equipped to React to Oxidative Stress due to Increased Content of Phenolics*

Typically, a carbon limitation during seed germination results in germinating seeds and young, immature seedlings to shift to secondary metabolism. Subsequently, the accumulation of secondary metabolites, especially anti-oxidants, has been proposed to aid young seedlings against biotic and abiotic threats (Bryant 1992; Clarke and Davison 2004). The complementary role between RFOs and ROS during seed germination also provides seedlings with immediate protection against these environmental stresses. However, this is strictly applicable to younger seedlings of about two to four weeks old. Recent studies have demonstrated that young seedlings had an increased phenolic content (and thus anti-oxidant capacity), four times higher when compared to older plants, upon water deficit exposure (Oh et al. 2010). The accumulation of phenolic compounds during water deficit is equally reversed to decreased amounts in response to re-watering. This clearly indicates that young seedlings are more responsive to water stress than older plants and suggests that phenolic compounds may play a signaling role, as well as a protective anti-oxidant role, in younger seedlings during periods of water stress.

4.3 *Mature Plants Require a Modulated Signaling Cascade Between Molecular and Morphological Adaptations*

At whole plant level, water deficit can elicit a chain of events (from root to shoot), which generally includes molecular and/or metabolite signals and signaling cascades, as well as morphological changes. These terms are often used in research during the analyses of its interaction (between signal, signaling cascade and morphological responses) and whether the observations are due to a direct or indirect stimulus of water deficit conditions. Water deficit is known to affect plant leaf area, biomass accumulation, organ number and nutrient partitioning (Chaves et al. 2002; Cattivelli et al. 2008). How do these signaling events lead to changes in metabolites and how does this affect the photosynthetic machinery of a plant during water deficit? What role do morphological responses play? Many questions such as these remain elusive.

Classically, plant responses during water deficit occur in one of four ways, some of which have been discussed previously in this chapter, including: (i) stomatal adjustments to decrease water loss (Tanaka et al. 2005); (ii) reducing the size, shape and/or amount of leaves (Reymond et al. 2003); (iii) increase of water uptake by modification of root structure (Turner et al. 2001; Chaves et al. 2002); (iv) osmotic adjustment (involving sugars) (Hummel et al. 2010); and (v) transcriptional and posttranscriptional regulation (Chaves and Oliveira 2004).

Evidently, water-soluble sugars (also referred to here as the compatible solutes) are critical carbon sources with a central role in plant metabolism and energy production (for growth and development). During conditions of water deficit, on a whole plant level, soluble sugars are in a constant flux of supply-and-demand and regulated in such a way to ensure a balanced distribution and production of carbon sources, throughout the plant. It is well recognized that soluble sugars function as signaling molecules during a myriad of environmental stresses (reviewed in Hanson and Smeekens 2009). Plants respond to very small changes in the carbon status, which initiates the supply of sugars in signaling cascades (including the activation and cross talk of hormone pathways), and ultimately play a big role in a plant adaptation and acclimation to its environment (Usadel et al. 2008; Ramel et al. 2009). This was demonstrated in a quantitative stress tolerance study where biochemical and physiological parameters indicated the importance of sugars, especially sucrose, as signaling molecules (Ramel et al. 2009).

The rate of photosynthesis is regulated by feedback control mechanisms exerted through intracellular sugar concentrations. The latter mechanisms lead to significant transcriptional and posttranscriptional changes which are regulated very differently in sink and source tissues (Koch et al. 1996). Low sugar concentrations upregulate source tissue photosynthetic rates, nutrient mobilization and nutrient export to support the rest of the plant (the opposite is true for increased amounts of sugars in source tissues) during water deficit conditions. In contrast, the expression of photosynthesis-associated genes is repressed by high concentrations of sugars (Smeekens 1998), and in sink tissue, high amounts of sugars upregulate growth and storage of abundant carbon reserves (Pego et al. 2000). During water deficit, this photosynthetic interplay between sink and source tissues favors osmoregulation in such a way that metabolic activity can be regained much faster in young leaves, allowing for plant survival (of course, this occurs in combination with older leaf shedding). The photosynthetic machinery (which controls carbon production) of a plant is vital in minimizing harmful effects during early stages of water deficit, which is often induced by high concentrations of sugars (Eberhard et al. 2008). This indicates that photosynthesis, during water deficit, constitutes an important role in the sugar-signaling cascade (sugar production vs. utilization), which modulates redox homeostasis in different plant organs to confer protective roles during oxidative stress (sugar fluctuations alter ROS production in all plant organs; Couée et al. 2006). Other metabolic and signaling pathways are also influenced by sugar modulation, including those involving hormones such as ABA and ethylene (Smeekens 1998).

Plants can also endure water deficit conditions through morphological adaptations. Water potential is regulated via the stomata (or by shedding of older source leaves) to avoid unnecessary water loss (Ehleringer and Cooper 1992). A key survival strategy of plants is the shedding of older leaves (also termed senescence), which allows for the re-allocation of nutrients to sink tissue or to the stem. The strategy of senescence (leading to a short supply of sugars) is very important during fruit abortion (allowing for the re-allocation of sugars) under water deficit conditions. This could benefit source-limited crops where stem sugar reserves are insufficient for fruit production during water deficit, making a delayed-senescence phenotype desirable. Despite the loss of leaves during senescence, this seems to modulate future behavior of plant responses to water deficit (when sink becomes source), where the sink leaves that survived these conditions seem to be better equipped in its photosynthetic machinery (Ludlow and Ng 1974). This is suggestive of an epigenetic regulatory mechanism where plants are faced with the "adapt-or-die" approach.

Furthermore, to increase water uptake plants have developed strategies of adjustment which involves investment into root architecture (Poorter and Nagel 2000; Gill and Jackson 2000). Apart from metabolic signals (sugars and phyto-hormones such as ABA), roots depend on signals from the leaves (regulated by photosynthesis) and nutrient assimilation from the leaves to support the development of elongated root structures. Enabling plants to thrive during water deficit could thus reside in adequate trade-off between root growth allowing water uptake and shoot growth allowing adequate photosynthesis. However, the exact mechanism of root/shoot interaction during water deficit is not completely understood and might be a valuable area of developing tolerant crops. Many of the morphological plant adaptation strategies during water deficit are constitutive and not necessarily stress induced (e.g., deep rooting and shoot reserves). However, these adaptive strategies are central to the plant and its ability to protect cells and organs during periods of oxidative stress. Interestingly, in roots non-enzymatic anti-oxidants do not play a role in ROS scavenging (Fan et al. 2014). This leads to increased ROS concentrations during water deficit, activating all the classical enzymatic anti-oxidant activities (as discussed previously in this chapter). Studies of root physiology and key genes that regulate redox homeostasis during water deficit, complemented with traditional studies of shoot physiology and stomatal control, are fundamental for understanding the whole plant response during water deficit and how it is able to modulate and maintain a redox balance in all organs.

5 Food Security and CRISPR/Cas9: A Potential Strategy for Future Crop Improvement During Times of Water Deficit

Current climate changes require the application of biotechnological approaches (from research to commercialization) to assist in the improvement of staple crop species (such as maize, rice, wheat and sugarcane) with environmentally tolerant characteristics. Specifically, drought tolerance (withstanding water deficit conditions) is a key area of research because it induces dramatic yield penalties and threatens food security. This is especially of concern in developing countries. A new era of precise genome editing has come of age in crop modification strategies (Griggs et al. 2013). The most recent genome-editing technique, clustered regularly interspaced short palindromic repeat-associated Cas9 nuclease (CRISPR/Cas9), utilizes the DNA repair system already existing inside each plant cell and is based on engineered sequence-specific nucleases (Jain 2015). Briefly, foreign DNA is flanked by host repeat sequences and transcribed as a CRISPR RNA which directs a sequence-specific nuclease Cas9 to create double-stranded nicks in the genome. These nicks will consequently be repaired by non-homologous end joining (NHEJ) resulting in the same species with new traits (Shi et al. 2017).

The use of CRISPR/Cas9 as an editing tool in plants only recently emerged (Shan et al. 2013). Since then it has been successfully used for genome editing of several plant species, such as rice (Li et al. 2017; Ren et al. 2017), maize (Svitashev et al. 2016; Shi et al. 2017; Zong et al. 2017) and even in the hexaploid wheat (Gil-Humanes et al. 2017). CRISPR/Cas9 has also been recently developed for targeted genome editing in sugarcane, but not yet experimentally demonstrated (Jung and Altpeter 2016; Mohan 2016; Ferreira et al. 2017). Sugarcane is considered an important crop for sugar and ethanol production, but is sensitive to water deficit conditions (Lakshmanan and Robinson 2014). Numerous attempts have been made to develop new sugarcane varieties which are water stress tolerant and commercially useful. These strategies have been either done via breeding or by introducing transgenes. As discussed throughout this chapter, a variety of complex mechanisms occur within sugarcane during periods of water deficit including: (i) morphological changes in leaf area and root structure, (ii) photosynthetic systems produce more anti-oxidant enzymes, (iii) accumulation of osmolytes such as sucrose and proline in the aerial tissues, and trehalose in the roots, (iv) activation of the ABA hormone pathway, (v) increased concentrations of ROS and consequently non-enzymatic anti-oxidants and (vi) transcriptional activation of stress response genes (Rodrigues et al. 2011; Delorge et al. 2014). All of these mechanisms in sugarcane have been used in attempts to create water stress tolerant varieties. In conclusion, there is a definitive correlation between water stress and anti-oxidant system responses in sugarcane. The activity of ROS-scavenging enzymes could be useful as a marker for water stress tolerance in sugarcane, and the genes involved in these pathways could potentially be used as targets for crop improvement using CRISPR/Cas9 techniques. Current data on

CRISPR/Cas9 in plants advocate that genome editing may even be feasible in a complex polyploidy (such as sugarcane).

A single successful study has recently been reported in a similar attempt in maize (Shi et al. 2017). This study used a negative regulator of ethylene response (*AGOS8*), a gene related to water deficit in maize, as the target (of transcriptional regulation). The strategy involved replacement of the native *AGOS8* promoter with a constitutive *GOS2* promoter (also from maize). Results demonstrated higher *AGOS8* transcript levels in several tissues concomitant with an increased grain yield during water deficit conditions and no yield penalty during optimal growth conditions. Further studies need to demonstrate the effect of this strategy on plant redox homeostasis. Nonetheless, this indicates a promising tool for the future of crop improvement.

6 Harnessing the Natural Potential of Plant Abiotic Stress Responses to Address Nutritional Food Security

In the recent health-conscious society, the importance of fruits and vegetables is becoming more prevalent due to their health-promoting effects. Probably, the biggest global concern currently is nutritional food security and the availability of foods that confer these health benefits to humans (beyond basic nutrition). Phytochemicals (especially phenolic compounds) have gained considerable attention during the last decade due to its ability to act as an anti-oxidant, not only in plants, but in humans and animals as well. Phytochemicals play an important role in human health and nutrition and could potentially be essential in the fight against non-communicable (oxidative stress-related) diseases, such as cancer (Manach et al. 2004).

Consider the use of immature plants, naturally equipped with a powerful myriad of ROS-scavenging compounds. Compared with mature plants, immature plants are known to accumulate high amounts of phytochemicals which are known for their health-promoting properties and strong protective functions against a multitude of degenerative diseases (Gill et al. 2004; Barillari et al. 2006; Oh and Rajashekar 2009). These immature plants (often termed micro-greens) can be bio-fortified with the use of environmental stresses (such as water deficit), thus enhancing the health beneficial phytochemical content. A recent study done in lettuce micro-greens indicated the use of mild water deficit conditions effectively enhanced the phytochemical content (Oh et al. 2010). In this context, micro-greens are better equipped (as stated previously) to respond to environmental stresses in order to maintain redox homeostasis, mainly for survival. This plant survival strategy can be translated into nutritional food security, whereby subsistence farmers can bio-fortify their small crops improving health-promoting attributes. These approaches suggest the potential use of regulated water deficits as a crop management strategy to enhance nutritive value of plants (immature, micro-green plants) for human health.

7 Concluding Remarks

Environmental stresses, especially water deficit conditions, can disturb the redox homeostasis in plants and result in an increased ROS production which could ultimately lead to plant death. Many regulatory mechanisms exist in plants to overcome the damaging effects of free radical accumulation. These include morphological changes (in root, stem and leaf structures) and the production of enzymatic and non-enzymatic anti-oxidants all of which participate in the complex redox homeostasis mechanism. All of the anti-oxidants accumulate in a much orchestrated manner, in different cellular compartments, where it either functions on site or is distributed throughout the plant. It is evident that, from all the cellular compartments, the chloroplast (hosting the photosynthetic machinery) plays a vital role in redox homeostasis during water deficit and seems to be the powerhouse of producing both enzymatic and non-enzymatic anti-oxidants upon stress perception.

The signaling mechanisms in plants during water deficit are quite complex, ranging from the hormonal signaling to the more recent concept where sugars are suggested to contribute significantly to the redox homeostasis mechanisms. These sugars include sucrose, RFOs, trehalose and fructans and are all directly involved in ROS scavenging. This involvement of sugars occurs in different cellular compartments (or different plant organs) where it contributes in an indirect manner compared to the classical anti-oxidative mechanisms. An integrated interaction is maintained between ROS and sugar-signaling pathways. Although we have only begun to understand the important role of sugars and its fate during environmental stress, the introduction of sugars might be of key interest in the future crop development. These sugars display characteristics of osmoprotectants (compatible solutes) and anti-oxidants (ROS scavenging). An interesting feature to investigate is the modulation of sugars and key metabolic enzymes during the development of environmentally tolerant crops. Furthermore, the use of direct genome-editing techniques such as CRISPR/Cas9 could hold great value in commercialization of new tolerant crop varieties. Interestingly, regulatory issues involving the use of CRISPR/Cas9 in crop plants could be of great advantage in crop commercialization strategies, from a genetically modified perspective. Recent examples confirmed that species modified using CRISPR/Cas9 will not be regulated as genetically modified organisms (GMOs), given that the methods developed for genome editing displayed no transgenic footprint of the modification tool in edited plants (Waltz 2016). This is a positive future outlook given that regulatory costs would be greatly reduced in such instances for crop development. Given that food security is such a global concern, technologies like CRISPR/Cas9 could be a feasible solution to crop improvement without the concern of GMO restrictions.

Lastly, a multitude of health beneficial phenolic and sugar compounds accumulate in plant vacuolar compartments as a part of the redox homeostasis mechanism. Many of these compounds are already present in commercial medicines, health supplements or used as food additives. Future studies could seek to implement environmental stresses in a positive light, not only to improve stress tolerance in crops, but also to enhance food quality and establish the concept of producing functional foods (foods that go beyond basic human nutrition).

References

Abebe T, Melmaiee K, Berg V, Wise RP (2010) Drought response in the spikes of barley: gene expression in the lemma, palea, awn, and seed. Funct Integr Genomics 10:191–205

Agati G, Azzarello E, Pollastri S, Tattini M (2012) Flavonoids as antioxidants in plants: location and functional significance. Plant Sci 196:67–76

Alscher RG, Erturk N, Heath LS (2002) Role of superoxide dismutases (SODs) in controlling oxidative stress in plants. J Exp Bot 53:1331–1341

Arfan M, Athar HR, Ashraf M (2007) Does exogenous application of salicylic acid through the rooting medium modulate growth and photosynthetic capacity in two differently adapted spring wheat cultivars under salt stress? J Plant Physiol 164:685–694

Asada K (2006) Production and scavenging of reactive oxygen species in chloroplasts and their functions. Plant Physiol 141:391

Ashraf M, Foolad MR (2007) Roles of glycine betaine and proline in improving plant abiotic stress resistance. Environ Exp Bot 59:206–216

Bachmann M, Keller F (1995) Metabolism of the raffinose family oligosaccharides in leaves of *Ajuga reptans* L (inter- and intracellular compartmentation). Plant Physiol 109:991

Bailly C, Audigier C, Ladonne F, Wagner MH, Coste F, Corbineau F, Côme D (2001) Changes in oligosaccharide content and antioxidant enzyme activities in developing bean seeds as related to acquisition of drying tolerance and seed quality. J Exp Bot 52:701–708

Barillari J, Cervellati R, Costa S, Guerra MC, Speroni E, Utan A, Iori R (2006) Antioxidant and choleretic properties of *Raphanus sativus* L sprout (Kaiware Daikon) extract. J Agric Food Chem 54:9773–9778

Barnes J, Zheng Y, Lyons T (2002) Plant resistance to ozone: the role of ascorbate. In: Omasa K, Saji H, Youssefian S, Kondo N (eds) Air Pollution and plant biotechnology: prospects for phytomonitoring and phytoremediation. Springer, Japan, Tokyo, pp 235–252

Bartels D, Sunkar R (2005) Drought and salt tolerance in plants. Crit Rev Plant Sci 24:23–58

Baud S, Boutin J-P, Miquel M, Lepiniec L, Rochat C (2002) An integrated overview of seed development in *Arabidopsis thaliana* ecotype WS. Plant Physiol Biochem 40:151–160

Blackman SA, Obendorf RL, Leopold AC (1995) Desiccation tolerance in developing soybean seeds: the role of stress proteins. Physiol Plant 93:630–638

Bryant JP (1992) Woody plant-mammal interactions. In: Herbivores: their interactions with secondary plant metabolites, vol II, Ecological and Evolutionary Processes. pp 344–371

Caffrey M, Fonseca V, Leopold AC (1988) Lipid-Sugar interactions. Plant Physiol 86:754

Cattivelli L, Rizza F, Badeck F-W, Mazzucotelli E, Mastrangelo AM, Francia E, Marè C, Tondelli A, Stanca AM (2008) Drought tolerance improvement in crop plants: an integrated view from breeding to genomics. Field Crops Res 105:1–14

Chaves MM, Oliveira MM (2004) Mechanisms underlying plant resilience to water deficits: prospects for water-saving agriculture. J Exp Bot 55:2365–2384

Chaves MM, Pereira JS, Maroco J, Rodrigues ML, Ricardo CPP, OsÓRio ML, Carvalho I, Faria T, Pinheiro C (2002) How plants cope with water stress in the field? photosynthesis and growth. Ann Bot 89:907–916

Clarke PJ, Davison EA (2004) Emergence and survival of herbaceous seedlings in temperate grassy woodlands: recruitment limitations and regeneration niche. Austral Ecol 29:320–331

Cominelli E, Galbiati M, Vavasseur A, Conti L, Sala T, Vuylsteke M, Leonhardt N, Dellaporta SL, Tonelli C (2005) A guard-cell-specific MYB transcription factor regulates stomatal movements and plant drought tolerance. Curr Biol 15:1196–1200

Couée I, Sulmon C, Gouesbet G, El Amrani A (2006) Involvement of soluble sugars in reactive oxygen species balance and responses to oxidative stress in plants. J Exp Bot 57:449–459
Cruz de Carvalho MH (2008) Drought stress and reactive oxygen species. Plant Signaling Behav 3:156–165
Das K, Roychoudhury A (2014) Reactive oxygen species (ROS) and response of antioxidants as ROS-scavengers during environmental stress in plants. Front Environ Sci 2(53):1–10
Dat J, Vandenabeele S, Vranová E, Van Montagu M, Inzé D, Van Breusegem F (2000) Dual action of the active oxygen species during plant stress responses. Cell Mol Life Sci 57:779–795
De Tullio MC, Arrigoni O (2003) The ascorbic acid system in seeds: to protect and to serve. Seed Sci Res 13:249–260
del Río LA, Sandalio LM, Corpas FJ, Palma JM, Barroso JB (2006) Reactive oxygen species and reactive nitrogen species in peroxisomes production, scavenging, and role in cell signaling. Plant Physiol 141:330
Delorge I, Janiak M, Carpentier S, Van Dijck P (2014) Fine tuning of trehalose biosynthesis and hydrolysis as novel tools for the generation of abiotic stress tolerant plants. Front Plant Sci 5:147
Desikan R, Hancock JT, Bright J, Harrison J, Weir I, Hooley R, Neill SJ (2005) A role for ETR1 in hydrogen peroxide signaling in stomatal guard cells. Plant Physiol 137:831
Diallinas G, Kanellis AK (1994) A phenylalanine ammonia-lyase gene from melon fruit: cDNA cloning, sequence and expression in response to development and wounding. Plant Mol Biol 26:473–479
Dixon RA, Paiva NL (1995) Stress-induced phenylpropanoid metabolism. Plant Cell 7:1085
Drennan PM, Smith MT, Goldsworthy D, van Staden J (1993) The occurrence of trehalose in the leaves of the desiccation-tolerant angiosperm *Myrothamnus flabellifolius* welw. J Plant Physiol 142:493–496
Dugas DV, Monaco MK, Olson A, Klein RR, Kumari S, Ware D, Klein PE (2011) Functional annotation of the transcriptome of *Sorghum bicolor* in response to osmotic stress and abscisic acid. BMC Genom 12:514
Eberhard S, Finazzi G, Wollman F-A (2008) The dynamics of photosynthesis. Annu Rev Genet 42:463–515
Ehleringer JR, Cooper TA (1992) On the role of orientation in reducing photoinhibitory damage in photosynthetic-twig desert shrubs. Plant, Cell Environ 15:301–306
Elbein AD (2010) Chapter 11—Cytoplasmic carbohydrate molecules: trehalose and glycogen A2—Holst, Otto In: Brennan PJ, Itzstein Mv, Moran AP (eds) Microbial glycobiology. Academic Press, San Diego, pp 185–201
Fait A, Angelovici R, Less H, Ohad I, Urbanczyk-Wochniak E, Fernie AR, Galili G (2006) Arabidopsis seed development and germination is associated with temporally distinct metabolic switches. Plant Physiol 142:839
Fan H-F, Ding L, Du C-X, Wu X (2014) Effect of short-term water deficit stress on antioxidative systems in cucumber seedling roots. Bot Stud 55:46
Farrant JM, Pammenter NW, Berjak P (1989) Germination-associated events and the desiccation sensitivity of recalcitrant seeds—a study on three unrelated species. Planta 178:189–198
Fayez KA, Bazaid SA (2014) Improving drought and salinity tolerance in barley by application of salicylic acid and potassium nitrate. J Saudi Soc Agric Sci 13:45–55
Ferreira THS, Tsunada MS, Bassi D, Araújo P, Mattiello L, Guidelli GV, Righetto GL, Gonçalves VR, Lakshmanan P, Menossi M (2017) Sugarcane water stress tolerance mechanisms and its implications on developing biotechnology solutions. Front Plant Sci 8:1077
Finch-Savage WE (1992) Embryo water status and survival in the recalcitrant species *Quercus robur* L: evidence for a critical moisture content. J Exp Bot 43:663–669
Fini A, Brunetti C, Di Ferdinando M, Ferrini F, Tattini M (2011) Stress-induced flavonoid biosynthesis and the antioxidant machinery of plants. Plant Signaling Behav 6:709–711
Fortes AM, Agudelo-Romero P, Silva MS, Ali K, Sousa L, Maltese F, Choi YH, Grimplet J, Martinez- Zapater JM, Verpoorte R, Pais MS (2011) Transcript and metabolite analysis in Trin-

cadeira cultivar reveals novel information regarding the dynamics of grape ripening. BMC Plant Biol 11:149

Foyer C, Noctor G (2003) Redox sensing and signalling associated with reactive oxygen in chloroplasts, peroxisomes and mitochondria. Physiol Plant 119:355–364

Foyer CH, DescourviÈRes P, Kunert KJ (1994) Protection against oxygen radicals: an important defence mechanism studied in transgenic plants. Plant, Cell Environ 17:507–523

Gapińska M, Skłodowska M, Gabara B (2008) Effect of short- and long-term salinity on the activities of antioxidative enzymes and lipid peroxidation in tomato roots. Acta Physiol Plant 30:11

Garnczarska M, Wojtyla Ł (2008) Differential response of antioxidative enzymes in embryonic axes and cotyledons of germinating lupine seeds. Acta Physiol Plant 30:427–432

Garrido I, Uriarte D, Hernández M, Llerena J, Valdés M, Espinosa F (2016) The evolution of total phenolic compounds and antioxidant activities during ripening of grapes (*Vitis vinifera* L, cv Tempranillo) grown in semiarid region: effects of cluster thinning and water deficit. Int J Mol Sci 17:1923

Gil-Humanes J, Wang Y, Liang Z, Shan Q, Ozuna CV, Sánchez-León S, Baltes NJ, Starker C, Barro F, Gao C, Voytas DF (2017) High-efficiency gene targeting in hexaploid wheat using DNA replicons and CRISPR/Cas9. Plant J 89:1251–1262

Gill RA, Jackson RB (2000) Global patterns of root turnover for terrestrial ecosystems. New Phytol 147:13–31

Gill CIR, Haldar S, Porter S, Matthews S, Sullivan S, Coulter J, McGlynn H, Rowland I (2004) The effect of cruciferous and leguminous sprouts on genotoxicity, *in vitro* and *in vivo*. Cancer Epidemiol Biomark Prev 13:1199

Grace SC, Logan BA (2000) Energy dissipation and radical scavenging by the plant phenylpropanoid pathway. Philos Trans R Soc Lond B Biol Sci 355:1499

Griggs D, Stafford-Smith M, Gaffney O, Rockström J, Öhman MC, Shyamsundar P, Steffen W, Glaser G, Kanie N, Noble I (2013) Sustainable development goals for people and planet. Nature 495:305

Hameed A, Bibi N, Akhter J, Iqbal N (2011) Differential changes in antioxidants, proteases, and lipid peroxidation in flag leaves of wheat genotypes under different levels of water deficit conditions. Plant Physiol Biochem 49:178–185

Hanson J, Smeekens S (2009) Sugar perception and signaling—an update. Curr Opin Plant Biol 12:562–567

Hincha DK, Zuther E, Heyer AG (2003) The preservation of liposomes by raffinose family oligosaccharides during drying is mediated by effects on fusion and lipid phase transitions. Biochimica et Biophysica Acta (BBA)—Biomembranes 1612:172–177

Hoekstra FA, Golovina EA, Buitink J (2001) Mechanisms of plant desiccation tolerance. Trends Plant Sci 6:431–438

Holländer-Czytko H, Grabowski J, Sandorf I, Weckermann K, Weiler EW (2005) Tocopherol content and activities of tyrosine aminotransferase and cystine lyase in Arabidopsis under stress conditions. J Plant Physiol 162:767–770

Horbowicz M, Obendorf RL (1994) Seed desiccation tolerance and storability: dependence on flatulence-producing oligosaccharides and cyclitols—review and survey. Seed Sci Res 4:385–405

Hu H, Dai M, Yao J, Xiao B, Li X, Zhang Q, Xiong L (2006) Overexpressing a NAM, ATAF, and CUC (NAC) transcription factor enhances drought resistance and salt tolerance in rice. Proc Natl Acad Sci 103:12987–12992

Huang X-Y, Chao D-Y, Gao J-P, Zhu M-Z, Shi M, Lin H-X (2009) A previously unknown zinc finger protein, DST, regulates drought and salt tolerance in rice via stomatal aperture control. Genes Dev 23:1805–1817

Hummel I, Pantin F, Sulpice R, Piques M, Rolland G, Dauzat M, Christophe A, Pervent M, Bouteillé M, Stitt M, Gibon Y, Muller B (2010) Arabidopsis plants acclimate to water deficit at low cost through changes of carbon usage: an integrated perspective using growth, metabolite, enzyme, and gene expression analysis. Plant Physiol 154:357

Igamberdiev AU, Seregélyes C, Manac'h N, Hill RD (2004) NADH-dependent metabolism of nitric oxide in alfalfa root cultures expressing barley hemoglobin. Planta 219:95–102
Jain M (2015) Function genomics of abiotic stress tolerance in plants: a CRISPR approach. Front Plant Sci 6:375
Jung JH, Altpeter F (2016) TALEN mediated targeted mutagenesis of the caffeic acid O-methyltransferase in highly polyploid sugarcane improves cell wall composition for production of bioethanol. Plant Mol Biol 92:131–142
Kameli A, Losel DM (1996) Growth and sugar accumulation in Durum wheat plants under water stress. New Phytol 132:57–62
Ke D, Sun G (2004) The effect of reactive oxygen species on ethylene production induced by osmotic stress in etiolated mungbean seedling. Plant Growth Regul 44:199–206
Keleş Y, Öncel I (2002) Response of antioxidative defence system to temperature and water stress combinations in wheat seedlings. Plant Sci 163:783–790
Khan W, Prithiviraj B, Smith DL (2003) Photosynthetic responses of corn and soybean to foliar application of salicylates. J Plant Physiol 160:485–492
Klapheck S (1988) Homoglutathione: isolation, quantification and occurrence in legumes. Physiol Plant 74:727–732
Koch KE, Wu Y, Xu J (1996) Sugar and metabolic regulation of genes for sucrose metabolism: potential influence of maize sucrose synthase and soluble invertase responses on carbon partitioning and sugar sensing. J Exp Bot 47:1179–1185
Koster KL, Leopold AC (1988) Sugars and desiccation tolerance in seeds. Plant Physiol 88:829
Kranner I, Roach T, Beckett R, Whitaker C, Minibayeva F (2010) Early production of reactive oxygen species during seed germination and early seedling growth in *Pisum sativum*. J Plant Physiol 167:805–811
Kwak JM, Mori IC, Pei ZM, Leonhardt N, Torres MA, Dangl JL, Bloom RE, Bodde S, Jones JDG, Schroeder JI (2003) NADPH oxidase AtrbohD and AtrbohF genes function in ROS-dependent ABA signaling in Arabidopsis. EMBO J 22:2623
Kwak JM, Nguyen V, Schroeder JI (2006) The role of reactive oxygen species in hormonal responses. Plant Physiol 141:323
Lakshmanan P, Robinson N (2014) Stress physiology: Abiotic stresses, in sugarcane: physiology, biochemistry, and functional biology. In: Botha FC (ed) Moore PH. Wiley, Chichester, pp 411–434
Lamb C, Dixon RA (1997) The oxidative burst in plant disease resistance. Annu Rev Plant Physiol Plant Mol Biol 48:251–275
Li W-X, Oono Y, Zhu J, He X-J, Wu J-M, Iida K, Lu X-Y, Cui X, Jin H, Zhu J-K (2008) The NFYA5 transcription factor is regulated transcriptionally and posttranscriptionally to promote drought resistance. Plant Cell 20:2238
Li J, Sun Y, Du J, Zhao Y, Xia L (2017) Generation of targeted point mutations in rice by a modified CRISPR/Cas9 system. Mol Plant 10:526–529
Lin T-P, Huang N-H (1994) The relationship between carbohydrate composition of some tree seeds and their longevity. J Exp Bot 45:1289–1294
Liu R, Xu S, Li J, Hu Y, Lin Z (2006) Expression profile of a PAL gene from *Astragalus membranaceus* var Mongholicus and its crucial role in flux into flavonoid biosynthesis. Plant Cell Rep 25:705–710
Ludlow MM, Ng TT (1974) Water stress suspends leaf ageing. Plant Sci Lett 3:235–240
Manach C, Scalbert A, Morand C, Rémésy C, Jiménez L (2004) Polyphenols: food sources and bioavailability. Am J Clin Nutr 79:727–747
Mittler R (2002) Oxidative stress, antioxidants and stress tolerance. Trends Plant Sci 7:405–410
Mohan C (2016) Genome editing in sugarcane: challenges ahead. Front Plant Sci 7:1542
Munné-Bosch S, Schwarz K, Alegre L (2001) Water deficit in combination with high solar radiation leads to midday depression of a-tocopherol in field-grown lavender (*Lavandula stoechas*) plants. Funct Plant Biol 28:315–321
Nakashima K, Ito Y, Yamaguchi-Shinozaki K (2009) Transcriptional regulatory networks in response to abiotic stresses in Arabidopsis and grasses. Plant Physiol 149:88

Nandi B, Roy S, Bhattacharya S, Sinha Babu SP (2004) Free radicals mediated membrane damage by the saponins acaciaside A and acaciaside B. Phytotherapy Res 18:191–194

Nishizawa A, Yabuta Y, Yoshida E, Maruta T, Yoshimura K, Shigeoka S (2006) Arabidopsis heat shock transcription factor A2 as a key regulator in response to several types of environmental stress. Plant J 48:535–547

Nishizawa A, Yabuta Y, Shigeoka S (2008) Galactinol and raffinose constitute a novel function to protect plants from oxidative damage. Plant Physiol 147:1251

Noctor G, Veljovic-Jovanovic S, Driscoll S, Novitskaya L, Foyer CH (2002) Drought and oxidative load in the leaves of C3 plants: a predominant role for photorespiration? Ann Bot 89:841–850

Oh M-M, Rajashekar CB (2009) Antioxidant content of edible sprouts: effects of environmental shocks. J Sci Food Agric 89:2221–2227

Oh M-M, Carey EE, Rajashekar CB (2010) Regulated water deficits improve phytochemical concentration in lettuce. J Am Soc Hortic Sci 135:223–229

Oh JE, Kwon Y, Kim JH, Noh H, Hong S-W, Lee H (2011) A dual role for MYB60 in stomatal regulation and root growth of *Arabidopsis thaliana* under drought stress. Plant Mol Biol 77:91–103

Pallanca JE, Smirnoff N (1999) Ascorbic acid metabolism in pea seedlings. A comparison of D-glucosone, L-sorbosone, and L-galactono-1,4-lactone as ascorbate precursors. Plant Physiol 120:453

Pego JV, Kortstee AJ, Huijser C, Smeekens SCM (2000) Photosynthesis, sugars and the regulation of gene expression. J Exp Bot 51:407–416

Peshev D, Vergauwen R, Moglia A, Hideg É, Van den Ende W (2013) Towards understanding vacuolar antioxidant mechanisms: a role for fructans? J Exp Bot 64:1025–1038

Peters S, Mundree SG, Thomson JA, Farrant JM, Keller F (2007) Protection mechanisms in the resurrection plant *Xerophyta viscosa* (Baker): both sucrose and raffinose family oligosaccharides (RFOs) accumulate in leaves in response to water deficit. J Exp Bot 58:1947–1956

Petridis A, Therios I, Samouris G, Tananaki C (2012) Salinity-induced changes in phenolic compounds in leaves and roots of four olive cultivars (*Olea europaea* L) and their relationship to antioxidant activity. Environ Exp Bot 79:37–43

Peukert M, Thiel J, Peshev D, Weschke W, Van den Ende W, Mock H-P, Matros A (2014) Spatio-temporal dynamics of fructan metabolism in developing barley grains. Plant Cell 26:3728

Pilon-Smits EAH, Ebskamp MJM, Paul MJ, Jeuken MJW, Weisbeek PJ, Smeekens SCM (1995) Improved performance of transgenic fructan-accumulating tobacco under drought stress. Plant Physiol 107:125

Pilon-Smits EAH, Terry N, Sears T, van Dun K (1999) Enhanced drought resistance in fructan-producing sugar beet. Plant Physiol Biochem 37:313–317

Pinheiro C, Chaves MM (2011) Photosynthesis and drought: can we make metabolic connections from available data? J Exp Bot 62:869–882

Pinheiro C, Chaves MM, Ricardo CP (2001) Alterations in carbon and nitrogen metabolism induced by water deficit in the stems and leaves of *Lupinus albus* L. J Exp Bot 52:1063–1070

Poorter H, Nagel O (2000) The role of biomass allocation in the growth response of plants to different levels of light, CO_2, nutrients and water: a quantitative review. Funct Plant Biol 27:1191

Quan L-J, Zhang B, Shi W-W, Li H-Y (2008) Hydrogen peroxide in plants: a versatile molecule of the reactive oxygen species network. J Integr Plant Biol 50:2–18

Ramel F, Sulmon C, Bogard M, Couée I, Gouesbet G (2009) Differential patterns of reactive oxygen species and antioxidative mechanisms during atrazine injury and sucrose-induced tolerance in *Arabidopsis thaliana* plantlets. BMC Plant Biol 9:28

Ren B, Yan F, Kuang Y, Li N, Zhang D, Lin H, Zhou H (2017) A CRISPR/Cas9 toolkit for efficient targeted base editing to induce genetic variations in rice. Sci China Life Sci 60:516–519

Reymond M, Muller B, Leonardi A, Charcosset A, Tardieu F (2003) CombiningqQuantitative trait loci analysis and an ecophysiological model to analyze the genetic variability of the responses of maize leaf growth to temperature and water deficit. Plant Physiol 131:664

Roberts EH (1973) Predicting the storage life of seeds. Seed Sci Technol (1):499–514

Rodrigues FA, Da Graça JP, De Laia ML, Nhani-Jr A, Galbiati JA, Ferro MIT, Ferro JA, Zingaretti SM (2011) Sugarcane genes differentially expressed during water deficit. Biol Plant 55:43–53
Rushton PJ, Somssich IE, Ringler P, Shen QJ (2010) WRKY transcription factors. Trends Plant Sci 15:247–258
Sah SK, Kaur G, Wani SH (2016) Metabolic engineering of compatible solute trehalose for abiotic stress tolerance in plants. In: Iqbal N, Nazar R, Khan NA (eds) Osmolytes and plants acclimation to changing environment: emerging omics technologies. Springer India, New Delhi, pp 83–96
Sairam RK, Tyagi A (2004) Physiology and molecular biology of salinity stress tolerance in plants. Curr Sci 86:407–421
Schramm F, Larkindale J, Kiehlmann E, Ganguli A, Englich G, Vierling E, Von Koskull-Döring P (2008) A cascade of transcription factor DREB2A and heat stress transcription factor HsfA3 regulates the heat stress response of Arabidopsis. Plant J 53:264–274
Schubert S, Serraj R, Plies-Balzer E, Mengel K (1995) Effect of drought stress on growth, sugar concentrations and amino acid acumulation in N2-fixing Alfalfa (*Medicago sativa*). J Plant Physiol 146:541–546
Shan Q, Wang Y, Li J, Zhang Y, Chen K, Liang Z, Zhang K, Liu J, Xi JJ, Qiu J-L, Gao C (2013) Targeted genome modification of crop plants using a CRISPR-Cas system. Nat Biotechnol 31:686
Shannon MC (1997) Adaptation of plants to salinity. In: Sparks DL (ed) Advances in agronomy. Academic Press, pp 75–120
Shao HB, Liang ZS, Shao MA, Sun Q (2005) Dynamic changes of anti-oxidative enzymes of 10 wheat genotypes at soil water deficits. Colloids Surf, B 42:187–195
Shi J, Gao H, Wang H, Lafitte HR, Archibald RL, Yang M, Hakimi SM, Mo H, Habben JE (2017) ARGOS8 variants generated by CRISPR-Cas9 improve maize grain yield under field drought stress conditions. Plant Biotechnol J 15:207–216
Sinhababu A, Kar RK (2003) Comparative responses of three fuel wood yielding plants to PEG-induced water stress at seedling stage. Acta Physiol Plant 25:403–409
Smeekens S (1998) Sugar regulation of gene expression in plants. Curr Opin Plant Biol 1:230–234
Sofo A, Tuzio AC, Dichio B, Xiloyannis C (2005) Influence of water deficit and rewatering on the components of the ascorbate–glutathione cycle in four interspecific *Prunus* hybrids. Plant Sci 169:403–412
Stockinger EJ, Gilmour SJ, Thomashow MF (1997) Arabidopsis thaliana CBF1 encodes an AP2 domain-containing transcriptional activator that binds to the C-repeat/DRE, a cis-acting DNA regulatory element that stimulates transcription in response to low temperature and water deficit. Proc Natl Acad Sci 94:1035–1040
Svitashev S, Schwartz C, Lenderts B, Young JK, Mark Cigan A (2016) Genome editing in maize directed by CRISPR–Cas9 ribonucleoprotein complexes. Nat Commun 7:13274
Tanaka Y, Sano T, Tamaoki M, Nakajima N, Kondo N, Hasezawa S (2005) Ethylene inhibits abscisic acid-induced stomatal closure in Arabidopsis. Plant Physiol 138:2337
Tommasi F, Paciolla C, Arrigoni O (1999) The ascorbate system in recalcitrant and orthodox seeds. Physiol Plant 105:193–198
Turner NC, Wright GC, Siddique KHM (2001) Adaptation of grain legumes (pulses) to water-limited environments. In: Advances in agronomy. Academic Press, pp 193–231
Usadel B, Bläsing OE, Gibon Y, Retzlaff K, Höhne M, Günther M, Stitt M (2008) Global transcript levels respond to small changes of the carbon status during progressive exhaustion of carbohydrates in Arabidopsis rosettes. Plant Physiol 146:1834
Valluru R, Van den Ende W (2008) Plant fructans in stress environments: emerging concepts and future prospects. J Exp Bot 59:2905–2916
Valluru R, Van den Ende W (2011) Myo-inositol and beyond—emerging networks under stress. Plant Sci 181:387–400
Valluru R, Lammens W, Claupein W, Van den Ende W (2008) Freezing tolerance by vesicle-mediated fructan transport. Trends Plant Sci 13:409–414
Van den Ende W, Valluru R (2009) Sucrose, sucrosyl oligosaccharides, and oxidative stress: scavenging and salvaging? J Exp Bot 60:9–18

Vereyken IJ, Chupin V, Demel RA, Smeekens SCM, De Kruijff B (2001) Fructans insert between the headgroups of phospholipids. Biochimica et Biophysica Acta (BBA)—Biomembranes 1510:307–320

Waltz E (2016) CRISPR-edited crops free to enter market, skip regulation. Nat Biotechnol 34:582

Wingler A (2002) The function of trehalose biosynthesis in plants. Phytochemistry 60:437–440

Yamaguchi T, Blumwald E (2005) Developing salt-tolerant crop plants: challenges and opportunities. Trends Plant Sci 10:615–620

Yamaguchi-Shinozaki K, Shinozaki K (2006) Transcriptional regulatory networks in cellular responses and tolerance to dehydration and cold stresses. Annu Rev Plant Biol 57:781–803

Zhang X, Zhang L, Dong F, Gao J, Galbraith DW, Song C-P (2001) Hydrogen peroxide is involved in abscisic acid-induced stomatal closure in *Vicia faba*. Plant Physiol 126:1438–1448

Zong Y, Wang Y, Li C, Zhang R, Chen K, Ran Y, Qiu J-L, Wang D, Gao C (2017) Precise base editing in rice, wheat and maize with a Cas9-cytidine deaminase fusion. Nat Biotechnol 35:438

Redox Homeostasis in Plants Under Arsenic Stress

Seema Mishra, Sanjay Dwivedi, Shekhar Mallick and Rudra Deo Tripathi

Abstract Cellular and intracellular redox balance is a key factor for smooth functioning of enzymes and protein involved in essential metabolic pathways. Metalloid arsenic (As) is toxic to almost all forms of life and is a serious environmental concern due to its wide natural and anthropogenic contamination around the world. Arsenic severely impacts growth of plants and yield of crops grown in As-contaminated soils. Additionally, it gets accumulated in cereal grains, leading to contamination of food, and also hampers the nutrient (essential amino acids and minerals) accumulation grains. Redox imbalance and oxidative stress has been shown to be major cause of As toxicity. Some studies have reported that As also initiates nitrosative stress by inducing reactive nitrogen species. Therefore, understanding the mechanism of redox homeostasis in plants under As stress is important to develop tolerant plants. In the current chapter, the response of known pathways and molecules involved in maintaining cellular redox balance in plant has been discussed in relation to As stress tolerance. Furthermore, the recent studies showing the role of various plant growth regulators (signaling molecules) and glutaredoxins in As stress tolerance has also been discussed.

1 Introduction

Geological processes, industrial application, past agricultural use of arsenicals, and contaminated groundwater have resulted in environmental arsenic (As) pollution

S. Mishra (✉) · S. Dwivedi · S. Mallick · R. D. Tripathi (✉)
Plant Ecology and Environmental Science Division, CSIR-National Botanical Research Institute
Rana Pratap Marg Lucknow, 226001 Lucknow, India
e-mail: seema_mishra2003@yahoo.co.in

R. D. Tripathi
e-mail: tripathi_rd@gmail.com

S. Mishra
The Biology Centre of the Czech Academy of Sciences, Institute of Plant Molecular Biology
České Budějovice, České Budějovice 37005, Czech Republic

DDU Gorakhpur University, Gorakhpur 273009, India

S. K. Panda and Y. Y. Yamamoto (eds.), *Redox Homeostasis in Plants*,
Signaling and Communication in Plants, https://doi.org/10.1007/978-3-319-95315-1_9

at a massive scale. Nearly 17.15 Mt As is released annually from the lithosphere through volcanic activities, and about 80 Mt As is deposited into arable land per year through anthropogenic activities (Matschullat 2000). Apart from these, naturally As-enriched groundwater is a serious concern in many countries of the world. Several Asian countries, such as Bangladesh, Bhutan, India, China, Mongolia, Nepal, Cambodia, Myanmar, Afghanistan, Korea, and Pakistan, some parts of USA, and several Latin American countries such as Argentina, Brazil, Chile, and Mexico are naturally contaminated with As (Berg et al. 2001; Chakraborti et al. 2002; Mandal and Suzuki 2002; Nordstrom 2002; Armienta and Segovia 2008; Schlebusch et al. 2015; Mishra et al. 2016a). In India, 18 states and 3 union territories are As-affected from different sources (Mishra et al. 2016a). Arsenic is a carcinogen to both humans and animals causing skin, lung, and lever cancer (Anawar et al. 2002; Meharg and Rahman 2003; Ng et al 2003; Das et al. 2004). The non-carcinogenic health hazards due to chronic As exposure include hyper- and hypopigmentation, keratosis, hypertension, cardiovascular diseases, and diabetes (Anawar et al. 2002; Meharg and Rahman 2003; Das et al. 2004). Groundwater is extensively used in irrigation, thus contaminating soil and subsequently the crops, vegetable, and fodder. Though past use of arsenicals as pesticide and some fertilizers have also been shown to result in significant uptake of As in edible parts of crops, the most prominent route of As accumulation in plants is the use of As-contaminated irrigation water (Zhao et al. 2010; Punshon et al. 2017). Therefore, high concentration of As in groundwater and its transfer to the food chain is one of the major challenges of the twenty-first century. Though cereal grains such as rice are a major source of arsenic to human, several medicinal plants are also shown to accumulate significant amount of arsenic (Mishra et al. 2013a; Kumari et al. 2018). Arsenic, upon accumulation, possesses high toxicity to plants as well. It reduces seed germination, seedling growth, pigments, and hamper photosynthesis (Mishra et al. 2014, 2016b) and also severely affects yield attributing characters like reduction in number of tiller, panicles, and seed weight (Dwivedi et al. 2010; Bag et al. 2014). Therefore, As-induced yield loss is another side of As contamination. Impairment of transporter for nutrient uptake, uptake competition, inhibition of vital enzymes and proteins, and increased production of reactive oxygen species (ROS) have been reported as the main mechanisms of As toxicity to plants (Requejo and Tena 2005; Hartley-Whitaker et al. 2001a; Mishra et al. 2008). Arsenic exposure has been reported to increase reactive nitrogen species (RNS) in plants as well (Corpas et al. 2007; Leterrier et al. 2012; Singh et al. 2016; Farnese et al. 2017). Removal of As from cytosol and/or conversion to less toxic species are the primary plant responses to protect vital cellular organelles from As toxicity, while removal of ROS is considered as secondary response of the plant under As stress. Since conversion of As to less toxic species, through complexation with thiolic ligands or transformation of inorganic As to methylated As, as well as neutralization of ROS consumes cellular reductants, balancing cellular redox state is crucial under As stress. The current review emphasizes the pathways and molecules involved in cellular redox homeostasis under As stress.

2 Uptake and Toxicity of Arsenic in Plants

In soil and water, As is normally present as inorganic arsenate (AsV) and arsenite (As III) depending upon the pH and redox potential of the environment (Cullen and Reimer 1989). A small to significant amount of methylated As species (dimethylarsinic acid, DMA; monomethylarsonic acid, MMA) may also occur in soil due to microbial Methylation of inorganic As or due to use of methylated arsenicals as pesticide (Bentley and Chasteen 2002; Zhao et al. 2010). Arsenate is taken up by the plants through phosphate transporters (Asher and Reay 1979). Most phosphate transporters (Phts) show slightly higher affinity for P compared to As, but newly identified PvPht1;3 from Pteris vittata demonstrated relatively higher affinity for As than other isoforms such as PtPht;5 (DiTusa et al. 2016). This appears hyperaccumulator-specific adaptation leading to enhanced arsenic accumulation in *P. vittata* (Lindsay and Maathuis 2017). Arsenite is taken up by nodulin 26-like intrinsic (NIPs) aquaporin channels (Ma et al. 2008). Till date several phosphate transporters involved in AsV transport have been identified in different plant species such as OsPHT1; 1 (Kamiya et al. 2013), OsPHT1; 8 (Wu et al. 2011) in rice and AtPHT1; 1, AtPHT1; 4, AtPHT1; 5, AtPHT1; 7, AtPHT1; 8, AtPHT1; 9 in A. thaliana (Shin et al. 2004; Catarecha et al. 2007; Remy et al. 2012; LeBlanc et al. 2013; Fontenot et al. 2015). Whereas AtNIP1; 1, AtNIP1; 2, AtNIP5; 1 (Kamiya and Fujiwara 2009), AtNIP3; 1 (Xu et al. 2015), AtNIP6; 1 (Bienert et al. 2008), AtNIP7; 1 (Isayenkov and Maathuis 2008) in *A. thaliana* and OsNIP1; 1, OsNIP2; 2 (OsLsi6), OsNIP3; 1 (Ma et al. 2008), OsNIP3; 2 (Bienert et al. 2008; Chen et al. 2017), OsNIP3; 3 (Katsuhara et al. 2014) in rice have been identified as AsIII transporters. Rice Lsi1 (OsNIP2; 1) mediates the uptake of AsIII (Ma et al. 2008) and organic arsenical (DMA, MMA), whereas OsLsi2 is involved in efflux of AsIII and xylem loading (Ma et al. 2008).

Arsenic and other nonessential and essential (at high doses) metals and metalloids represent one of the most damaging abiotic stresses in plants. They cause toxicity by targeting crucial molecules and vital processes in the plant cell. The chemical analogy of As species to essential/ beneficial elements not only facilitates their uptake through shared transporters, but it is also one of the major causes of toxicity. Arsenate, competing with P for uptake resulting P deficiency (Tripathi et al. 2007; Mishra et al. 2014), may replace P in phosphorylation reactions and interferes with the energy and P metabolism, such as by hampering oxidative phosphorylation and ATP synthesis, whereas AsIII, due to it high affinity to sulfhydryl group, binds to the vital enzymes and proteins and hampers their activity. The high reactivity of AsIII makes it systemically more poisonous than the As(V). Inside the cell, AsV is also mostly reduced to arsenite (Pickering et al. 2006; Raab et al. 2007; Mishra et al. 2013b, 2017); therefore, AsV-exposed plant also experiences a part of toxicity in the form of AsIII. A study on As localization at subcellular level, in aquatic plant *Ceratophyllum demersum*, revealed that at sublethally toxic (up to 1 μM) concentration, As was predominantly accumulated in the nucleus of plant cell. Only at lethal doses it filled vacuole and other cellular compartment. This indicates that As may interfere with nucleic acid synthesis by replacing P (Mishra et al. 2016b). In an earlier study, As inhibited the growth of *C. demersum* even at sub-micromolar concentration. The reduction in growth was correlated to the hampered pigment biosynthesis which occurred prior to any effect on photosynthetic parameters or to oxidative stress (Mishra at al. 2014; Mishra et al. 2016b). However at lethal concentration, plant suffered with severe oxidative stress due to impairment of photosynthesis. In several past and recent studies, oxidative stress and altered redox state have been the most prominent As stress response in plants (Singh et al. 2006; Talukdar 2013; Talukdar and Talukdar 2013; Zutshi et al. 2014).

3 Oxidative and Nitrosative Stress in Arsenic-Exposed Plants

Arsenic may cause the overproduction of ROS-inducing oxidative stress primarily by two ways: (1) Similar to Fe and Cu which are redox-active and generate ROS directly through redox reactions, As also exhibits two oxidation states in biological systems and thus may generate ROS during the interconversion of arsenic from one ionic form to the other (Mylona et al. 1998). (2) Arsenic may increase production of ROS indirectly by stimulating ROS-producing enzymes like NADPH oxidases, displacing essential cations from the binding sites of enzymes and inhibiting the activities of enzymes involved in electron transport chain in mitochondria and chloroplast, inactivation of antioxidative enzymes in different cell compartments, and inhibition of pathways involved in biosynthesis of molecules essential for redox homeostasis in cells. The major sites of ROS production are membranes, mitochondria, chloroplast, and peroxisomes. The ROS mainly comprise of free radicals like superoxide radicle ($O_2^{\bullet -}$), hydroxyl radicals ($OH^{\bullet}$), and non-radicals like hydrogen peroxide (H_2O_2) and singlet oxygen ($^1O_2$). Initially, the ROS was thought to be toxic by-products of aerobic metabolism. However, in recent years, the role of ROS as an important signaling molecule in plants has become apparent. Thus, at normal physiological level, ROS play essential roles, yet its enhanced generation has deleterious effects on functioning of cell (Cuypers et al. 2009). The increase in the level of ROS in plants causes oxidative damage to biomolecules such as lipids, proteins, and nucleic acids. Membrane damage is a common phenomenon under abiotic stress, including As, occurring as a result of oxidation of membrane lipids initiated by ROS or by the action of lipoxygenase (LOX) enzyme. Accumulation of malondialdehyde (MDA) is a good indicator of membrane damage, and an increase in MDA under As stress has been reported in several studies.

Arsenic stress has been shown to increase the level of nitric oxide (NO), a gaseous free radical, and belongs to a family of related molecules named reactive nitrogen species (RNS) such as S-nitrosoglutathione (GSNO), peroxynitrite ($ONOO^-$), dinitrogen trioxide (N_2O_3), and nitrogen dioxide (NO_2). NO and NO-derived products can have toxic physiological consequences, such as increased nitration (Chaki et al. 2009, 2011; Corpas et al. 2011). The phenomenon of increased nitration is known as

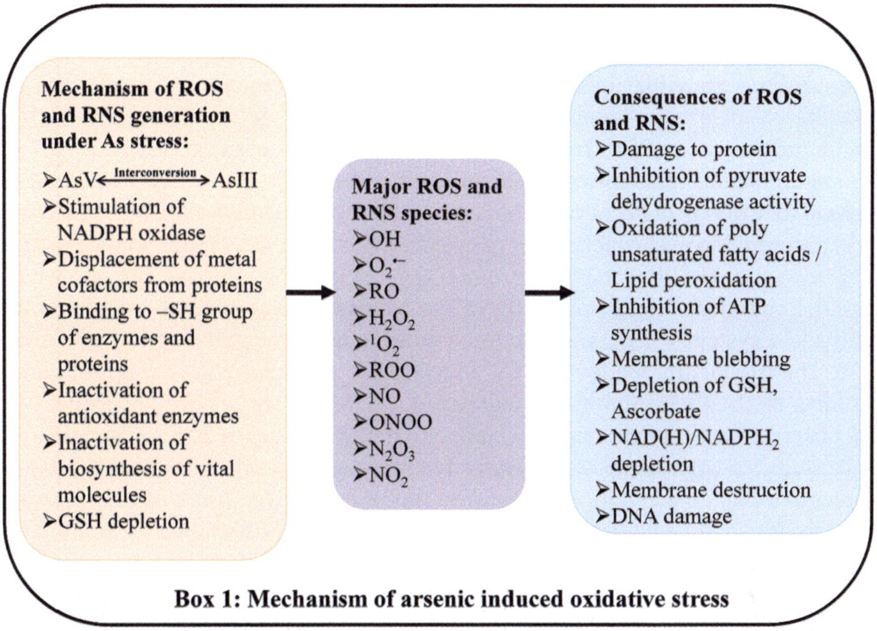

Box 1 Mechanism of arsenic induced oxidative stress

nitrosative stress which results in deactivation and damage to specific proteins, lipids, and nucleic acids (Valderrama et al. 2007; Corpas et al. 2009; Leterrier et al. 2012). The ROS and RNS induced under As stress and their deleterious effects in plants have been summarized in Box 1. In aquatic plant *Ceratophyllum demersum*, up to moderate As (50 μM) exposure concentration plant did not experience any significant toxicity measured as plant biomass and photosynthetic pigments. However, at higher concentrations (250 μM) and longer exposure duration (7 d), a significant increase in the level of H_2O_2, $O_2^{\bullet-}$, and MDA occurred along with increase in ion leakage showing membrane damage. The increased oxidative stress was concomitant with the reduction of plant biomass production (Mishra et al. 2008). Similar results were obtained in the medicinal plant *Bacopa monnieri* exposed to different concentrations of As. The level of ROS and MDA increased at higher and/or longer exposure to As. Roots experienced more oxidative stress than the leaves (Mishra et al. 2013a). In contrast, in *Hydrilla verticillata* exposed to As (either AsV or AsIII), a continuous concentration- and duration-dependent increase in the level of H_2O_2 and MDA was observed (Srivastava et al. 2007). Hartley-Whitaker et al. (2001a) also reported that increased oxidative stress, resulting in lipid peroxidation, was the main reason of toxicity in the As non-tolerant *Holcus lanatus* plant. Further, the proteomic analysis of maize roots revealed that oxidative stress was a main factor contributing to arsenic toxicity in plants (Requejo and Tena 2005). Arsenic exposure significantly decreased germination percentage, radicle length, dry mass, and activities of antioxidative enzymes in soybean (*Glycine max*), while it stimulated the levels of As and ROS (Chandrakar et al. 2016). The activity of LOX was increased several folds under As stress in soybean seedlings, showing higher lipolytic activity in cellular membranes and increased oxidation of membrane-bound fatty acids under As stress (Chandrakar et al. 2016). A high increase in LOX activity was also observed in As-stressed watercress (*Nasturtium officinale)* (Namdjoyan and Kermanian 2013). *Spirodela intermedia* plants exposed to different concentration of As showed a concentration-dependent increase in As accumulation. The increase in plant As level boosted the production of reactive oxygen species (ROS), such as H_2O_2 and $O_2^{\bullet -}$, within 24 h of As exposure which caused a significant increase in MDA. The increased level of ROS stimulated the antioxidant system which was able to control the overproduction of H_2O_2, but not $O_2^{\bullet -}$. Thus, the increased level of $O_2^{\bullet -}$ resulted in the damage to root tips of plants (da-Silva et al. 2017). Significant increase in MDA content upon As exposure was also observed in *A. thaliana* upon accumulation of As (Odjegba 2012). Though most of the studies are performed with relatively higher As concentrations, that is submillimeter to molar concentrations, and lower time duration. Exposure of *C. demersum* to environmentally relevant As concentration for longer duration in a continuous flow system to mimic natural conditions revealed that at sublethal As concentration (up to 1 μM) the main reason of growth retardation was inhibition of photosynthetic pigment which led to less efficient exciton transfer toward the PSIIRC. Only at lethal concentration ($\geq$5 μM), significant increase in ROS was observed (Mishra et al. 2014). The sunflower plants exposed to 10 μM As accumulated high concentration of As in root and leaves which resulted in increased lipid peroxidation, electrolyte leakage enhanced the generation of H_2O_2 and hampered biomass production (Saidi et al. 2017). Arsenic also caused oxidative stress in tomato fruit which was evident by an increase in H_2O_2 and MDA content (Marmiroli et al. 2017). Peanut plant exposed to 20 μM As exhibited significant oxidative stress depicted by increase in $O_2^{\bullet -}$ and H_2O_2 in root epidermis and lipid peroxidation including the increased activity of NADPH oxidase. A higher concentration of As resulted in oxidative burst particularly associated with NADPH oxidase activity (Bianucci et al. 2017). Similarly in medicinal plant, *Artemisia annua,* the level of H_2O_2, O^{2-}, and MDA was induced under As stress, while biomass production and chlorophyll content were severely decreased (Kumari et al. 2018).

4 Redox Homeostasis Under Arsenic Stress

As stated earlier, ROS play an important role in plants as a signaling molecules, controlling processes like growth, development, and biotic and abiotic stress response. Elevated production of ROS, however, has detrimental effects on plants. Therefore, for the normal cellular functioning, the maintenance of the delicate balance between ROS production and elimination is necessary. Plants have evolved a fine machinery to regulate redox homeostatic both under normal and stress conditions. The general defense response of plant against overdose of As includes complexation of As with thiols, sequestration of complexed/uncomplexed As into vacuole, uptake inhibition of As, effluxing As out of cell, and intensification of antioxidative mechanisms. The antioxidant machinery is primarily constituted by two components which work in a coordinated way during stressful conditions to maintain cellular redox homeostasis: first, the enzymatic components comprising of the superoxide dismutase (SOD), catalase (CAT), ascorbate peroxidase (APX), guaiacol-type peroxidases, glutathione reductase (GR), glutathione peroxidase (GPX), monodehydroascorbate reductase (MDHAR), and dehydroascorbate reductase (DHAR); second, the non-enzymatic antioxidants like ascorbic acid (AA), reduced glutathione (GSH), α-tocopherol, carotenoids, phenolics, flavonoids, and the osmolyte proline. Small oxidoreductases, like glutaredoxins and thioredoxins, and various plant hormones also play significant role in maintaining cellular redox state of the cell under different kind of stresses including As stress. Different pathways and molecules involved in redox homeostasis are summarized in Fig. 1.

5 Role of Antioxidant Enzymes in Redox Homeostasis Under Arsenic Stress

The enzymatic defense system of the plant to reactive oxygen species constitutes SOD, CAT, APX, GPX, GR, MDHAR, and DHAR working in various cell compartments. SOD (EC 1.15.1.1) constitutes the first line of defense converting superoxide radical to hydrogen peroxide (H_2O_2). SOD is localized at all possible sites of $O_2^{\bullet -}$ generation in the cell such as MnSOD in mitochondria, Fe and Cu-Zn SODs in chloroplasts, Cu-Zn and Mn SOD in peroxisomes, and Cu-Zn SOD in cytosol (Alscher et al. 2002). The H_2O_2 formed is then reduced to water and oxygen either by APX (EC 1.11.1.11) in ascorbate–glutathione cycle or by GPX (1.11.1.9), CAT (EC 1.11.1.6) and guaiacol-type peroxidases (EC 1.11.1.7). These enzymes are present in all subcellular compartments (Mittler 2002). GPX is characterized as plasma membrane-bound peroxidase (Mika and Lüthje 2003). The APX works in ascorbate–glutathione cycle in which APX is a key enzyme. APX utilizes ascorbate as specific electron donor to reduce H_2O_2 to water. The ascorbate–glutathione cycle is main hydrogen peroxide-detoxification system in chloroplasts, cytosol, mitochondria, and peroxisomes (Asada 1992, 1999; Noctor and Foyer 1998; Mittler et al. 2004). The monodehydroascorbate and dehydroascorbate produced in ascorbate–glutathione cycle are reduced back to ascorbate via monodehydroascorbate reductase (EC 1.6.5.4; MDHAR) and dehydroascorbate reductase (EC 1.8.5.1) using NADH and GSH (reduced glutathione) as reductant, respectively. GR is another important enzymatic constituent of this orchestrated antioxidant defense system that helps in maintaining a high GSH/GSSG ratio. GR reduces back the oxidized glutathione (GSSG), formed in ascorbate–glutathione cycle or glutaredoxin–glutathione cycle, to GSH. Glutathione peroxidase (GPX; EC 1.11.1.9) plays role in detoxification of H_2O_2 and organic hydroperoxides and lipid peroxides. Apart from these,

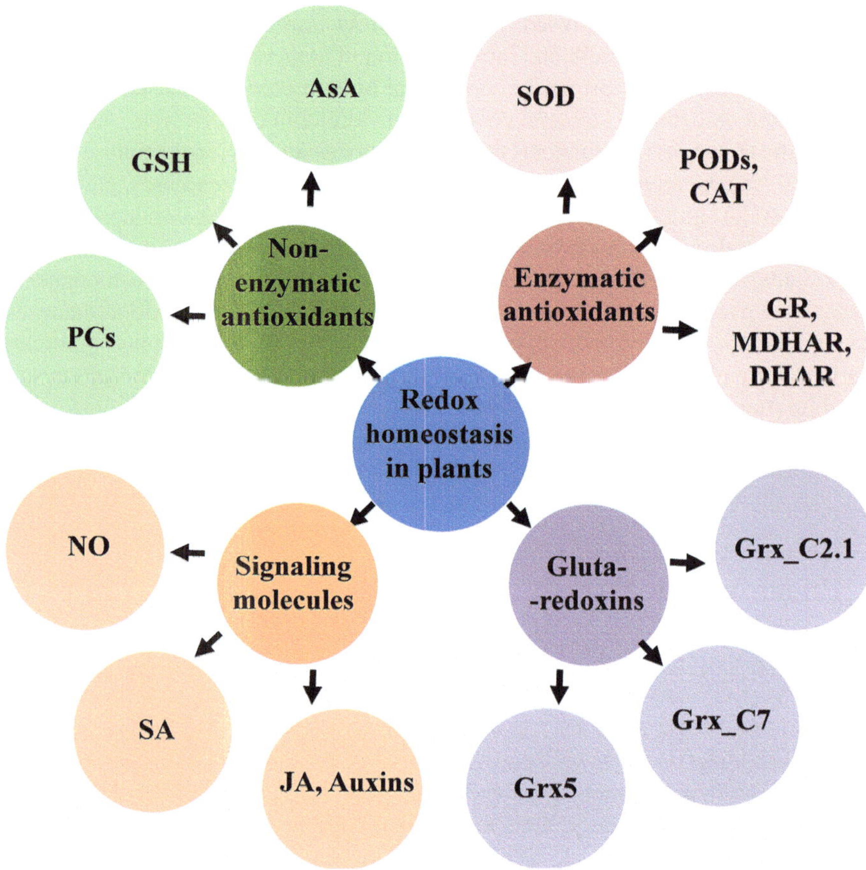

Fig. 1 Molecules and enzymes involved in redox homeostasis under arsenic stress. SOD: superoxide dismutase, PODs: peroxidases, CAT: catalase, GR: glutathione reductase, MDHAR: monodehydroascorbate, DHAR: dehydroascorbate, AsA: ascorbate, GSH: reduced glutathione, PC: phytochelatin, SA: salicylic acid, NO: nitric oxide, JA: jasmonate. *Note* Grx5 identified from *Pteris vittata*, Grx_C7 and Grx_C2.1 identified from *Oryza sativa*

the activity of glutathione-S-transferases (GSTs; EC 2.5.1.18), an enzyme which catalyzes the conjugation of a range of toxicants or their metabolites to GSH during their detoxification, has also been reported to increase under As stress. Under normal physiological conditions, the level of ROS in different cell organelles is tightly regulated by the well-coordinated activity of these antioxidant enzymes. It has been observed that under stress conditions the activity of these enzymes increases several folds; thus, stimulation of antioxidant enzyme activity is also considered as a marker of oxidative stress.

In *C. demersum* plant at moderate concentration of As (up to 50 μM), the activity of SOD, APX, GPX, CAT, GR, and GST along with the enzymes of thiol biosynthesis increased significantly, while no significant increase in lipid peroxidation and ion leakage was observed. At higher concentration, on the other hand the activity of SOD declined leading to toxicity to plant (Mishra et al. 2008). Similarly, the activity of antioxidant enzymes, i.e., SOD, CAT, APX, POX (peroxidase), GPX, and GR, and the concentration of a non-enzymatic antioxidant, i.e., anthocyanins, were triggered in response to ROS accumulation in *Spirodela intermedia* plants under As exposure (da-Silva et al. 2017). Such antioxidant system was able to control the overproduction of H_2O_2 even at 2.0 mg As L^{-1}; however, the level of $O_2^{\bullet -}$ could not be controlled resulting in oxidative damage in roots. *H. verticillata* exposed to either AsV or AsIII resulted in significant increase in antioxidant enzymes (Srivastava et al. 2007). Arsenate-treated plants induced higher level of antioxidant enzymes than AsIII-exposed plants which caused more stimulation of thiols. The higher tolerance of As hyperaccumulator fern *Pteris vittata* was found to be associated with higher levels of antioxidant enzymes and GSH in comparison with an arsenic-sensitive fern *Pteris ensiformis* (Srivastava et al. 2005; Singh et al. 2006). Peanut plants treated with realistic groundwater doses of As (20 μM) exhibited significant increase in the SOD, CAT, GR, GST, and thiols (glutathione and thioredoxins) in response to increased ROS (Bianucci et al. 2017). The level of SOD, CAT, and GST increased till 100 μM As, while glutathione reductase (GR) and glutathione peroxidase/peroxiredoxin (GPX/PRX) were significantly decreased resulting in reduced growth and altered root architecture. Study showed that GST activity increased even at the lowest studies concentration and can be used as suitable biomarker of As toxicity in peanut plants (Bianucci et al. 2017). A significant increase in antioxidant enzymes (SOD, CAT, APX, and GR) was observed in response to As-induced ROS in rice, *Arabidopsis thaliana,* and in medicinal plants, *Artemisia annua* and *Bacopa monnieri* (Dwivedi et al. 2010; Odjegba 2012; Mishra et al. 2013a; Kumari et al. 2018).

6 Role of Non-enzymatic Antioxidants

Among non-enzymatic antioxidants, ascorbate, carotenoids, GSH and other thiols are important components. Glutathione (GSH), the low-molecular-weight thiol, plays a central role in cellular redox homeostasis. It constitutes the major fraction of non-protein-reduced sulfur. In unstressed conditions, the intracellular milieu is maintained to a reducing environment due to high concentrations of GSH. The redox-active thiol group in GSH plays an important role in defense of plants against oxidative stress. Inside the cell, GSH is found to exist interchangeably with the oxidized form GSSG (glutathione disulfide). Various ROSs are able to oxidize GSH to GSSG. GSSG is also produced in other reactions involving GSH. GR reduces GSSG back to GSH using NADPH. The proportion of glutathione in the reduced form reflects the relative rates of reduction and oxidation and is always greater than 0.9 under non-stress conditions (Noctor et al. 1998). The high concentration of GSH in the chloroplast stroma (around 5 mM) is thought to act as an important redox buffer, preventing inactivation of enzyme by protecting potentially susceptible protein thiol groups. Under As stress, the ROS synthesis is accelerated leading to altered redox balance in various cell compartments. GSH plays multifarious role under As stress, which are directly or indirectly related to the upkeep of redox homeostasis of the cell. GSH can directly scavenge ROS being itself less susceptible to attack by oxidants (Douglas 1987). It plays a key role in protecting membranes to damage by free radicals by trapping them in aqueous phase (Noctor and Foyer 1998) and as a part of ascorbate–glutathione cycle and glutathione/glutaredoxin cycle. GSH is also a substrate for the GST, which catalyzes the conjugation of GSH with various toxicants their metabolites for further metabolism or for sequestration in vacuole (Moons 2003). The activity of GST has been shown to significantly increase upon As exposure (Mishra et al. 2008; Bianucci et al. 2017). Further, GSH is the precursor of phytochelatins (PCs), which bind As, thus minimizing the cellular free As concentration (Grill et al. 1985; Raab et al. 2004, 2005; Mishra et al. 2008, 2013a,b). Furthermore, GSH is also used as reductant in enzymatic and non-enzymatic reduction of AsV to AsIII, which is essential for the complexation with PCs, and it acts as ligand to directly bind AsIII (Mishra et al. 2013b, 2017). Therefore, the role of GSH is more crucial to protect cells under As stress. PCs are low-molecular-weight peptides derived from GSH with the general structure (-Glu-Cys)n-Gly where $n = 2–11$ (Grill et al. 1985). Complexation with phytochelatins (PCs) is a well-known mechanism of arsenic detoxification in plants. Complexation and vacuolar sequestration of As–PC complex facilitate to reduce the concentration of free arsenic in the cytoplasm, thus reducing its toxicity (Grill et al. 1985; Raab et al. 2004, 2005; Mishra et al. 2008, 2013a, b; Song et al. 2010). The overexpression of phytochelatin genes *AtPCS1*, or *CePCS* or *CdPCS1* in *Nicotiana tabacum* resulted in an increase of As tolerance and accumulation (Wojas et al. 2010; Shukla et al. 2012). Induction of PCs, GSH, and other upstream thiols under As stress has been reported in several studies (Hartley-Whitaker et al. 2001b; Zhao et al. 2003; Cai et al. 2004; Raab et al. 2004, 2005; Srivastava et al. 2007; Mishra et al. 2008, 2013a, b, 2017; Kumari et al. 2018). Apart from As complexation, PCs also play role in redox homeostasis in plants. In several plants, a large amount of PCs, both reduced and oxidized, are present uncomplexed in the cell, suggesting a role of PCs in redox balancing under As stress (Raab et al. 2005; Mishra et al. 2013b, 2017; Kumari et al. 2018).

While GSH is the most important redox buffer, ascorbate is considered the most important reducing substrate for H_2O_2 detoxification in plant cells (Mehlhorn et al. 1996). In plants, ascorbate can accumulate to millimolar concentrations with the ascorbate pool representing over 10% of the soluble carbohydrate in leaves. Ascorbate is a major primary antioxidant in plants, and it reacts directly with $OH^{\bullet}$, $O_2^{\bullet -}$, and singlet oxygen ($^1O_2$). Ascorbate plays important role in photoprotection and the regulation of photosynthesis and in preserving the activities of enzymes that contain prosthetic transition metal ions (Noctor and Foyer 1998). Ascorbate is also crucial in maintaining α-tocopherol in the reduced state (Hess 1994). An increase in the level of these low-molecular-weight molecules under As stress has been observed in several studies. The greater tolerance of *P. vittata* to As has also been associated with greater levels of reduced GSH along with antioxidant enzymes in comparison with *P. ensi-*

formis which is As-sensitive (Singh et al. 2006; Srivastava et al. 2005). An increased accumulation of nonprotein thiol and cysteine was also observed in As-exposed rice seedling and in the plants grown on As-contaminated soil (Shri et al. 2009; Dwivedi et al. 2010). Tolerance of aquatic plants *C. demersum* and *H. verticillata* and medicinal plants *B. monnieri* and *Artemisia annua* has been associated to increased redox state mediated by coordinated action of thiols and enzymatic antioxidants.

7 Role of Glutaredoxin

Glutaredoxins (Grxs) are ubiquitous small GSH-dependent oxidoreductases of the thioredoxin (Trx) family which are involved in many cellular processes, such as protein folding and regulation, reduction of dehydroascorbate, repair of oxidatively damaged proteins, and sulfur metabolism. Grxs catalyze dithiol–disulfide exchange reactions or reduce protein-mixed glutathione disulfides by using the reducing power of GSH to catalyze disulfide reactions in the presence of NADPH-dependent GR (Holmgren et al. 2005; Meyer and Hell 2005). Grxs and Trxs are structurally similar; however, their function is regulated differentially. The oxidized form of Trx is reduced directly by NADPH and thioredoxin reductase, whereas Grx is reduced by glutathione (GSH) using electrons donated by NADPH. Grx can reduce disulfides through a dithiol mechanism by utilizing two cysteine residues from the active site, or through monothiol mechanism by using a single cysteine at the active site of the Grx molecule. Their crucial function in response to oxidative stress has been known in several organisms such as Escherichia coli, yeast, and human.

Though plant genomes contain many Grxs, their roles are not well characterized. At least 31 isoforms of Grxs are present in *A. thaliana* (Rouhier et al. 2004) and 48 in rice (Verma et al. 2016a). However, so far, only a few plant Grxs have been characterized based on their localization, expression, or structural similarities (Rouhier et al. 2004, 2006; Xing et al. 2005; Cheng et al. 2006) and until now only three plant Grxs have been reported to be linked to arsenic metabolism (Sundaram et al. 2008; Verma et al. 2016a). Sundaram et al. (2008) characterized Grx5 from arsenic hyperaccumulator fern *Pteris vittata* involved in arsenic tolerance. The expression of PvGrx5 in *E. coli* mutants; strain AW3110 (in which arsenic resistance genes of the ars operon were deleted); and strain WC3110 (in which a gene for the ArsC arsenate reductase was deleted) increased arsenic tolerance of these mutant strains. While its expression in a strain in which the ars operon was deleted and the gene for the GlpF aquaglyceroporin was disrupted (strain OSBR1), the As tolerance did not increase. This suggested that PvGrx5 had a role in cellular arsenic resistance independent of the ars operon genes but dependent on GlpF. PvGrx5 expressing AW3110 cells accumulated significantly lower levels of arsenite compared to vector controls (Sundaram et al. 2008).

Recently, Grx_C7 and Grx_C2.1 from rice were shown to be involved in the regulation of intracellular arsenite (AsIII). Their expression in *E. coli* and *S. cerevisiae* increased As tolerance and accumulation. The *S. cerevisiae* cells expressing OsGrxs also contained higher GSH pool. Purified OsGrxs displays in vitro GSH-disulfide oxidoreductase, glutathione reductase, and arsenate reductase activities (Verma et al. 2016b). Further, their overexpression in *A. thaliana* conferred arsenic (As) tolerance and reduced As accumulation in seeds and shoot tissues compared to WT plants (Verma et al. 2016a).

8 Role of Plant Hormones and Nitric Oxide in Arsenic-Induced ROS

Plant hormones such as salicylic acid (SA), abscisic acid (ABA), auxin, brassinosteroid (Br), gibberellic acid (GA), hydrogen sulfide (H_2S), and jasmonate (JA) are signaling molecules and are involved in plant defense against biotic and abiotic stresses (Bari and Jones 2009; del Amor and Cuadra-Crespo 2011; Ali et al. 2014; Ryu and Cho 2015). Nitric oxide (NO) is also a bioactive gaseous molecule and a free radical and have a multifunction role in several physiological processes and under biotic and abiotic stresses including As stress (Singh et al. 2009, 2016, 2017). These molecules have been shown to mitigate As-induced ROS and thus As-induced toxicity when supplied exogenously.

Exogenous supplementation of NO as sodium nitroprusside (SNP, 50 μM) to rice significantly ameliorated the As-induced (25 or 50 μM) decrease in root and coleoptile length of rice and reduced the As-induced oxidative stress measured in terms of MDA, $O_2^{\bullet-}$, and H_2O_2 content. Interestingly, NO addition also decreased the activities of antioxidant enzymes in comparison with As-alone-treated plants, suggesting a ROS scavenging activity of NO. Study concluded that exogenous NO provides resistance to rice against As toxicity and has an ameliorating effect against As-induced stress (Singh et al. 2009). Similar observations have been reported in relation to tall fescue, where the application of NO (100 mM SNP) reduced arsenic-induced oxidative damage in leaves (Jin et al. 2010). However, in contrast to the results of Singh et al. (2009), NO application increased the activities of antioxidant enzymes (viz. SOD, CAT, and APX) in tall fescue leaves. Arsenic stress also resulted in pronounced increases in endogenous NO production. Therefore, it was suggested that arsenic stress elevated endogenous NO level and that NO might act as a signaling molecule to enhance antioxidant enzyme activities, further protecting against injuries caused by arsenic toxicity (Jin et al. 2010). The role of NO signaling was analyzed in *Pistia stratiotes* leaves upon exogenous supplementation of NO as SNP (Farnese et al. 2017). Exposure to As resulted in significant increase in As accumulation and concentration of ROS which triggered changes in the primary metabolism of the plants. Photosynthesis and photorespiration decreased sharply, and the respiration process increased. The disruption of the cells was also observed under As stress. These damages were attenuated by NO supplementation. The study concluded that NO plays a central role as a signaling agent able to change the whole plant physiology (Farnese et al. 2017). Similar ameliorating effect of exogenous NO against

As-induced oxidative stress and growth inhibition was observed in rice plants by Singh et al. (2016). NO supplementation reduced the accumulation of As in roots and shoots of the rice plant and reduced the expression of OsLsi1 and OsLsi2, the As influx and efflux transporters, respectively. Further NO supplementation maintained the GSH/GSSG ratio and reverted the As-induced Fe deficiency in shoot by modulating the expression of various Fe transporters (Singh et al. 2016).

Though exogenous NO seems beneficial to plant against As stress, the increase of indigenous NO due to As stress, however, caused an increase in RNS and nitrosative stress in Arabidopsis seedlings (Leterrier et al. 2012). Arsenate provoked a significant reduction in growth parameters and an increase in lipid oxidation along with a significant increase in NO content combined with an associated increase in protein nitration. Thus, indicating a process of nitrosative stress. Arsenic also increased the S-nitrosoglutathione reductase (GSNOR) activity as well as a concomitant reduction in glutathione and S-nitrosoglutathione (GSNO) content. Results indicated that As exposure at 500 mM caused nitro-oxidative stress in Arabidopsis by affecting the glutathione reductase and the GSNOR activities (Leterrier et al. 2012). GSNO is thought to function as a mobile reservoir of NO bioactivity; thus, it could have great physiological importance for plants.

Salicylic acid (SA; ortho-hydroxy benzoic acid), a naturally occurring plant hormone, has diverse regulatory roles in plant metabolism in the regulation of plant growth, development, interaction with other organisms and in the responses to various environmental stresses (Raskin 1992; Yalpani et al. 1994; Senaratna et al. 2000). SA has been shown to modulate plant responses to a wide range of oxidative stresses (Shirasu et al. 1997). The studies related to response of SA to heavy metal stress indicated that SA as an internal signal molecule interacts with reactive oxygen species (ROS) signal pathway. It may directly scavenge ROS and/or modulate redox balance through activation of antioxidant responses. SA application alleviated the negative effect of As on *Helianthus annuus* seedlings exposed to 10 μM As (Saidi et al. 2017). SA application increased the growth of seedlings in comparison with As-alone treatment and led to decrease in oxidative damages. Furthermore, SA application led to higher activity of CAT, APX, and GPX and concomitantly decreased SOD and guaiacol peroxidase activities (Saidi et al. 2017). Likewise, exogenous SA remarkably enhanced the germination percentage of soybean (*G. max)* seedlings and the activities of SOD, CAT, and APX. It also increased the accumulation of proline while decreased the level of As, ROS, and LOX in As-stressed seedlings. Thus, SA supplementation conferred As-stressed tolerance to soybean by reducing As content and ROS production and regulating the antioxidant enzymes and proline accumulation (Chandrakar et al. 2016). Exogenous application of SA also alleviated the stress induced by As in Arabidopsis and in *Artemisia annua.* SA application reduced the extent of H_2O_2 and O^{2-} generation and lipid peroxidation, while reverting biomass and chlorophyll content to overcome oxidative stress. (Odjegba 2012; Kumari et al. 2018). Jasmonic acid and auxins also play beneficial role under various stresses including metal-stress-induced antioxidant responses (Keramat et al. 2010; Poonam et al. 2013; Yan et al. 2013). However, only few studies have been done in response to As stress. Methyl Jasmonate (MJ) has been shown to enhanced the redox states of ascorbate and GSH, and the related enzymes involved in the AsA–GSH cycle in *Brassica napus* under As stress. MJ application also reverted the As-induced decrease in root growth and enhanced reactive oxygen species (H_2O_2 and O_2) and lipid peroxidation, while it increased the activity of antioxidant enzymes (Farooq et al. 2018).

In a study, it was found that the mutants of auxin transporters *aux1*, *pin1* and *pin2* were significantly more sensitive to AsIII than the wild-type (WT) plants. Auxin transport inhibitors significantly reduced plant tolerance to AsIII in the WT, while exogenous supply of indole-3-acetic acid improved AsIII tolerance of *aux1* and not that of WT. Arsenite increased the levels of H_2O_2 in WT but not in *aux1*, suggesting a positive role for auxin transport through AUX1 on plant tolerance to AsIII stress via reactive oxygen species (ROS)-mediated signaling. The *aux1* mutant was also more sensitive to high-temperature stress and salinity in comparison with wild-type plants. This suggests that auxin transport influences a common element shared by plant tolerance to AsIII, salinity, and high-temperature stress (Krishnamurthy and Rathinasabapathi 2013).

9 Conclusion and Future Prospects

Current review of literature showed that increased generation of ROS and RNS is major mechanism of As toxicity in plants, which leads to various kinds of damages to the cell including protein, membrane, and DNA damage. Formation of RNS is generally thought to be a phenomenon taking place during anaerobic metabolism or in oxygen-deficient conditions. However, As stress also triggers the generation of RNS and nitrosative stress. Plants have evolved a complex and well-coordinated machinery to keep the level of ROS and RNS under control during As stress. This includes the enzymatic and non-enzymatic system which is directly involved in ROS and RNS quenching, and small proteins, such as Grxs, and various plant hormone and signaling molecules which are involved in increasing cellular redox balance by stimulating antioxidative system and/or by modifying As uptake and transport. The role of Grxs in redox homeostasis under As stress in plants is not well explored. A few recent studies, however, represent their essential role in As stress response. A detail understanding of mechanism of Grx in redox hemostasis under As stress and of plant hormones would lead to development of more stress-tolerant as well as low As accumulating crop plants in future.

Acknowledgement Seema Mishra and Rudra Deo Tripathi are grateful to CSIR for the award of Scientist Pool and Emeritus Scientist scheme, respectively. SM is also grateful to Department of Science and Technology, New Delhi (DST-SERB) for financial support under Fast Track Scheme (SB/YS/LS-381/2013).

References

Ali B, Qian P, Jin R, Ali S, Khan M, Aziz R, Tian T, Zhou WJ (2014) Physiological and ultra-structural changes in *Brassica napus* seedlings induced by cadmium stress. Biol Plant 58:131–138

Alscher RG, Erturk N, Heath LS (2002) Role of superoxide dismutases (SODs) in controlling oxidative stress in plants. J Exp Bot 53(372):1331–1341

Anawar HM, Akai J, Mostofa KM, Safiullah S, Tareq SM (2002) Arsenic poisoning in ground water: health risk and geochemical sources in Bangladesh. Environ. Intl. 27:597–604

Armienta MA, Segovia N (2008) Arsenic and fluoride in the groundwater of Mexico. Environ Geochem Health 30:345–353

Asada K (1992) Ascorbate peroxidase: a hydrogen peroxide-scavenging enzyme in plants. Physiol Planta 85:235–241

Asada K (1999) The water cycle in chloroplasts: scavenging of active oxygens and dissipation of excess photons. Annu Rev Plant Physiol Plant Mol Biol 50:601–639

Asher CJ, Reay PF (1979) Arsenic uptake by barley seedlings. Aust J Plant Physiol 6:459–466

Bag MK, Adhikari B, Dwivedi S, Tripathi RD (2014). Consequences of arsenate exposure on important yield-associated traits of rice (*Oryza sativa L.*) J Plant Sci Res 1(2):109

Bari R, Jones JD (2009) Role of plant hormones in plant defence responses. Plant Mol Biol 69:473–488

Bentley R, Chasteen TG (2002) Microbial methylation of metalloids: arsenic, antimony, and bismuth. Microbiol Mol Biol Rev 66(2):250–271

Berg M, Tran HC, Nguyen TC, Pham HV, Schertenleib R, Giger W (2001) Arsenic contamination of ground water and drinking water in Vietnam: a human health threat. Environ Sci Technol 35:2621–2626

Bianucci E, Furlan A, Tordable MDC, Hernandez LE, Carpena-Ruiz RO, Castro S (2017) Antioxidant responses of peanut roots exposed to realistic groundwater doses of arsenate: identification of glutathione S-transferase as a suitable biomarker for metalloid toxicity. Chemosphere 181:551–561

Bienert GP, Thorsen M, Schüssler MD (2008) A subgroup of plant aquaporins facilitate the bi-directional diffusion of As $(OH)_3$ and $Sb(OH)_3$ across membranes. BMC Plant Biol 6:26

Cai Y, Su J, Ma LQ (2004) Low molecular weight thiols in arsenic hyperaccumulator*Pteris vittata* upon exposure to arsenic and other trace elements. Environ Pollut 129:69–78

Catarecha P, Segura MD, Franco-Zorrilla JM, Garcia-Ponce B, Lanza M, Solano R, Paz-Ares J, Leyva A (2007) A mutant of the *Arabidopsis* phosphate transporter PHT1; 1 displays enhanced arsenic accumulation. Plant Cell 19:1123–1133

Chaki M, Fernández-Ocaña AM, Valderrama R, Carreras A, Esteban FJ, Luque F, Gómez-Rodríguez MV, Begara-Morales JC, Corpas FJ, Barroso JB (2009) Involvement of reactive nitrogen and oxygen species (RNS and ROS) insunflower mildew interaction. Plant Cell Physiol 50:265–279

Chaki M, Valderrama R, Fernández-Ocaña AM, Carreras A, Gómez-Rodríguez MV, Pedradas JR, Begara-Morales JC, Sánchez-Calvo B, Luque F, Leterrier M, Corpas FJ, Barroso JB (2011) Mechanical wounding induces a nitrosative stress by downregulation of GSNO reductase and a rise of Snitrosothiolsin sunflower (*Helianthus annuus*) seedlings. J Exp Bot 62:1803–1813

Chakraborti D, Rahman MM, Paul K, Chowdhury UK, Sengupta MK, Lodh D, Chanda CR, Saha KC, Mukherjee SC (2002) Arsenic calamity in the Indian sub-continent-what lesions have been learned? Talanta 58:3–22

Chandrakar V, Dubey A, Keshavkant S (2016) Modulation of antioxidant enzymes by salicylic acid in arsenic exposed *Glycine max* L. J Soil Sci Plant Nutri 16(3):662–676

Chen Y, Sun SK, Tang Z, Liu G, Moore KL, Maathuis FJM, Miller AJ, McGrath SP, Zhao FJ (2017) The nodulin 26-like intrinsic membrane protein OsNIP3;1 is involved in arsenite uptake by lateral roots in rice. J Exp Bot 68(11):3007–3016

Cheng NH, Liu JZ, Brock A, Nelson RS, Hirschi KD (2006) AtGRXcp, an *Arabidopsis* chloroplastic glutaredoxin, is critical for protection against protein oxidative damage. J Biol Chem 281:26280–26288

Corpas FJ, delRío LA, Barroso JB (2007) Need of biomarkers of nitrosative stressin plants. Trends Plant Sci 12:436–438
Corpas FJ, Chaki M, Leterrier M, Barroso JB (2009) Protein tyrosine nitration: a new challenge in plants. Plant Signal Behav 4:920–923
Corpas FJ, Leterrier M, Valderrama R, Airaki M, Chaki M, Palma JM, Barroso JB (2011) Nitric oxide imbalance provokes a nitrosative response inplants under abiotic stress. Plant Sci 181:604–611
Cullen WR, Reimer KJ (1989) Arsenic speciation in the environment. Chem Rev 89(4):713–764
Cuypers A, Smeets K, Vangronsveld J (2009) Heavy metal stress in plants. In: Hirt H (ed) Plant stress biology: from genomics to systems biology. Wiley-VCH Verlag GmbH & Co. KGaA, Weinheim, pp 161–178
Das HK, Mitra AK, Sengupta PK, Hossain A, Islam F, Rabbani GH (2004) Arsenic concentrations in rice, vegetables, and fish in Bangladesh: a preliminary study. Environ Int 30:383–387
da-Silva CJ, Canatto RA, Cardoso AA, Ribeiro C, Oliveira JA (2017) Arsenic-hyperaccumulation and antioxidant system in the aquatic macrophyte *Spirodela intermedia* W. Koch (Lemnaceae). Theo Exp Plant Physiol 29:203–213
del Amor FM, Cuadra-Crespo P (2011) Alleviation of salinity stress in broccoli using foliar urea or methyl-jasmonate: analysis of growth, gas exchange, and isotope composition. Plant Growth Regul 63:55–62
DiTusa SF, Fontenot EB, Wallace RW, Silvers MA, Steele TN, Elnagar AH, Dearman KM, Smith AP (2016) A member of the phosphate transporter 1 (Pht1) family from the arsenic-hyperaccumulating fern *Pteris vittata* is a high-affinity arsenate transporter. New Phytol 209:762–772
Douglas KT (1987) Mechanism of action of glutathione-dependent enzymes. In: Meister A (ed) Advances in Enzymology. Wiley, New York, pp 103–167
Dwivedi S, Tripathi RD, Tripathi P, Kumar A, Dave R, Mishra S, Singh R, Sharma D, Rai UN, Chakrabartya D, Trivedi PK, Adhikari B, Bag MK, Dhankher OP, Tuli R (2010) Arsenate exposure affects amino acids, mineral nutrient status and antioxidants in rice (*Oryza sativa* L.) genotypes. Environ Sci Technol 44:9542–9549
Farnese FS, Oliveira JA, Paiva EAS, Menezes-Silva PE, da Silva AA, Campos FV, Ribeiro C (2017) The involvement of nitric oxide in integration of plant physiological and ultrastructural adjustments in response to arsenic. Front Plant Sci 8:516
Farooq MA, Islam F, Yang C, Nawaz A, Athar HR, Gill RA, Ali B, Song W, Zhou W (2018) Methyl jasmonate alleviates arsenic-induced oxidative damage and modulates the ascorbate-glutathione cycle in oilseed rape roots. Plant Growth Regul 84:135–148
Fontenot EB, Ditusa SF, Kato N, Olivier DM, Dale R, Lin WY, Chiou TJ, Macnaughtan MA, Smith AP (2015) Increased phosphate transport of *Arabidopsis thaliana* Pht1; 1 by site-directed mutagenesis of tyrosine 312 may be attributed to the disruption of homomeric interactions. Plant, Cell Environ 38:2012–2022
Grill E, Winnacker E-L, Zenk MH (1985) Phytochelatins: the principal heavy-metal complexing peptides of higher plants. Science 230:674–676
Hartley-Whitaker J, Ainsworth G, Meharg AA (2001a) Copper- and arsenate-induced oxidative stress in *Holcus lanatus* L. clones with differential sensitivity. Plant, Cell Environ 24:713–722
Hartley-Whitaker J, Ainsworth G, Vooijs R, Ten Bookum W, Schat H, Meharg AA (2001b) Phytochelatins are involved in differential arsenate tolerance in *Holcus lanatus*. Plant Physiol 126:299–306
Hess JL (1994) Free radical scavenging. In: Alscher RG, Wellburn AR (eds) Plant responses to the gaseous environment. Chapman and Hall, London, pp 165–180
Holmgren A, Johansson C, Berndt C, Lonn ME, Hudemann C, Lillig CH (2005). Thiol redox control via thioredoxin and glutaredoxin systems. Biochem Soc Trans 33:1375–1377
Isayenkov SV, Maathuis FJ (2008) The *Arabidopsis thaliana* aquaglyceroporin AtNIP7; 1 is a pathway for arsenite uptake. FEBS Lett 582:1625–1628
Jin J-W, Xu Y-F, Huang Y-F (2010) Protective effect of nitric oxide against arsenic-induced oxidative damage in tall fescue leaves. African J Biotechnol 9(11):1619–1627

Kamiya T, Fujiwara T (2009) *Arabidopsis* NIP1; 1 transports antimonite and determines antimonite sensitivity. Plant Cell Physiol 50:1977–1981

Kamiya T, Islam MR, Duan GL, Uraguchi S, Fujiwara T (2013) Phosphated efficiency signaling pathway is a target of arsenate and phosphate transporter OsPT1 is involved in as accumulation in shoots of rice. Soil Sci Plant Nutr 59:580–590

Katsuhara M, Sasano S, Horie T, Matsumoto T, Rhee J, Shibasaka M (2014) Functional and molecular characteristics of rice and barley NIP aquaporins transporting water, hydrogenperoxide and arsenite. Plant Biotechnol. 31:213–219

Keramat B, Kalantari KM, Arvin MJ (2010) Effects of methyl jasmonate treatment on alleviation of cadmium damages in soybean. J Plant Nutr 33:1016–1025

Krishnamurthy A, Rathinasabapathi B (2013) Auxin and its transport play a role in plant tolerance to arsenite-induced oxidative stress in *Arabidopsis thaliana*. Plant, Cell Environ 36:1838–1849

Kumari A, Pandey N, Pandey-Rai S (2018) Exogenous salicylic acid-mediated modulation of arsenic stress tolerance with enhanced accumulation of secondary metabolites and improved size of glandular trichomes in *Artemisia annua* L. Protoplasma 255:139–152

LeBlanc MS, McKinney EC, Meagher RB, Smith AP (2013) Hijacking membrane transporters for arsenic phytoextraction. J Biotechnol 163:1–9

Leterrier M, Airaki M, Palma JM, Chaki M, Barroso JB, Corpas FJ (2012) Arsenic triggers the nitric oxide (NO) and S-nitrosoglutathione (GSNO)metabolism in *Arabidopsis*. Environ Pollut 166:136–143

Lindsay ER, Maathuis FJM (2017) New molecular mechanisms to reduce arsenic in crops. Trends Plant Sci 22(12):1016–1026

Ma JF, Yamaji N, Mitani N, Xu X-Y, Su Y-H, McGrath SP, Zhao F-J (2008) Transporters of arsenite in rice and their role in arsenic accumulation in rice grain. Proc Natl Acad Sci USA 105:9931–9935

Mandal BK, Suzuki KT (2002) Arsenic round the world: a review. Talanta 58:201–235

Marmiroli M, Mussi F, Imperiale D, Lencioni G, Marmiroli N (2017) Abiotic stress response to As andAsCSi, composite reprogrammingof fruit metabolites in tomatocultivars. Front Plant Sci 8:2201

Matschullat J (2000) Arsenic in the geosphere: a review. Sci Total Environ 249(1–3):297–312

Meharg AA, Rahman MM (2003) Arsenic contamination of Bangladesh paddy fields oils: implications for rice contribution to arsenic consumption. Environ Sci Technol 37:229–234

Mehlhorn H, Lelandais M, Korth HG, Foyer CH (1996) Ascorbate is the naturalsubstrate for plant peroxidases. FEBSLett. 378:203–206

Meyer AJ, Hell R (2005) Glutathione homeostasis and redox-regulation by sulfhydryl groups. Photosynth Res 86:435–457

Mika A, Lüthje S (2003) Properties of guaiacol peroxidase activities isolated from corn root plasma membranes. Plant Physiol 132(3):1489–1498

Mishra S, Srivastava S, Tripathi RD, Trivedi PK (2008) Thiol metabolism and antioxidant system complement each other during arsenate detoxification in *Ceratophyllum demersum* L. Aquat Toxicol 86:205–215

Mishra S, Srivastava S, Dwivedi S, Tripathi RD (2013a) Investigation of biochemical responses of *Bacopa monnieri* L. upon exposure to arsenate. Environ Toxicol 28(8):419–430

Mishra S, Wellenreuther G, Mattusch J, Stärk H-J, Küpper H (2013b) Speciation and distribution of arsenic in thenon-hyperaccumulator macrophyte *Ceratophyllum demersum*. Plant Physiol 163:1396–1408

Mishra S, Stärkb HJ, Küpper H (2014). A different sequence of events than previously reported leads to arsenic-induced damage in *Ceratophyllum demersum* L. Metallomics 6: 444–454

Mishra S, Alfeld M, Sobotka R, Andresen E, Falkenberg G, Küpper H (2016a) Analysis of sub-lethal arsenic toxicity to *Ceratophyllum demersum*: sub-cellular distribution of arsenic and inhibition of chlorophyll biosynthesis. J Exp Bot 67(15):4639–4646

Mishra S, Dwivedi S, Kumar A, Chauhan R, Awasthi S, Mattusch J, Tripathi RD (2016b) Current status of ground water arsenic contamination in India and recent advancements in removal techniques from drinking water. Int J Plant Environ 2(1–2):1–15

Mishra S, Mattusch J, Wennrich R (2017) Accumulation and transformation of inorganic and organic arsenic in rice and role of thiol-complexation to restrict their translocation to shoot. Sci Rep 7:40522

Mittler R (2002) Oxidative stress, antioxidants and stress tolerance. Trends Plant Sci 7:405–410

Mittler R, Vanderauwera S, Gollery M, Van Breusegem F (2004) Reactive oxygen gene network of plants. Trends Plant Sci 9:490–498

Moons A (2003) Osgstu3 and Osgstu4, encoding tau class glutathione S-transferases, are heavy metal- and hypoxic stress-induced and differentially salt stress-responsive in rice roots. FEBS Lett 553:427–432

Mylona PV, Polidoros AN, Scandalios JG (1998) Modulation of antioxidant responses by arsenic in maize. Free Rad Biol Med 25(4/5):576–585

Namdjoyan S, Kermanian H (2013) Exogenous nitric oxide (as sodium nitroprusside) ameliorates arsenic-induced oxidative stress in watercress (*Nasturtium officinale* R. Br.) plants. Sci Hortic 161:350–356

Ng JC, Wang J, Shraim A (2003) A global health problem caused by arsenic from natural sources. Chemosphere 52:1353–1359

Noctor G, Foyer CH (1998) Ascorbate glutathione: keeping active oxygen under control. Annu Rev Plant Physiol Plant Mol Biol 49:249–279

Noctor G, Arisi A-CM, Jouanin L, Kunert KJ, Rennenberg H, Foyer CH (1998) Glutathione: biosynthesis, metabolism and relationship to stress tolerance explored in transformed plants. J Exp Bot 49(321):623–647

Nordstrom DK (2002) Worldwide occurrences of arsenic in ground water. Science 29(6):2143–2144

Odjegba VJ (2012) Exogenous salicylic acid alleviates arsenic toxicity in *Arabidopsis thaliana*. Indian J Innov Dev 1(7):515–522

Pickering IJ, Gumaelius L, Harris HH, Prince RC, Hirsch G, Banks JA, Salt DE, George GN (2006) Localizing the biochemical transformations of arsenate in a hyper accumulating fern. Environ Sci Technol 40:5010–5014

Poonam S, Kaur H, Geetika S (2013) Effect of jasmonic acid on photosynthetic pigments and stress markers in *Cajanus cajan* (L.) Mill sp. seedlings under copper stress. Am J Plant Sci 4:817–823

Punshon T, Jackson BP, Meharg AA, Warczack T, Scheckel K, Guerinot ML (2017) Understanding arsenic dynamics in agronomic systems to predict and prevent uptake by crop plants. Sci Total Environ 58:209–220

Raab A, Feldmann J, Meharg AA (2004) The nature of arsenic-phytochelatincomplexes in *Holcus lanatus* and *Pteris cretica*. Plant Physiol 134:1113–1122

Raab A, Schat H, Meharg AA, Feldmann J (2005) Uptake, translocationand transformation of arsenate and arsenite in sunflower (*Helianthus annuus*): formation of arsenic-phytochelatin complexes during exposure to high arsenic concentrations. New Phytol 168:551–558

Raab A, Ferreira K, Meharg AA, Feldmann J (2007) Can arsenic-phytochelatin complex formation be used as an indicator for toxicity in *Helianthus annuus*? J Exp Bot 58:1333–1338

Raskin I (1992) Salicylate, a new plant hormone. Plant Physiol 99:799–803

Remy E, Cabrito TR, Batista RA, Teixeira MC, Sa-Correia I, Duque P (2012) The Pht1; 9 and Pht1; 8 transporters mediate inorganic phosphate acquisition by the *Arabidopsis thaliana* root during phosphorus starvation. New Phytol 195:356–371

Requejo R, Tena M (2005) Proteome analysis of maize roots reveals that oxidative stress is a main contributing factor to plant arsenic toxicity. Phytochemistry 66:1519–1528

Rouhier N, Gelhaye E, Jacquot JP (2004) CMLS cell Mol. Life Sci 61:1266–1277

Rouhier N, Couturier J, Jacquot JP (2006) Genome-wide analysis of plant glutaredoxin systems. J Exp Bot 57:1685–1696

Ryu H, Cho YG (2015) Plant hormones in salt stress tolerance. J Plant Biol 58:147–155

Saidi I, Yousfi N, Borgi MA (2017) Salicylic acid improves the antioxidant ability against arsenic-induced oxidative stress in sunflower (*Helianthus annuus*) seedling. J Plant Nutri 40(16):2326–2335

Schlebusch CM, Gattepaille LM, Engstrom K, Vahter M, Jakobsson M, Broberg K (2015) Human adaptation to arsenic-rich environments. Mol Biol Evol 32(6):1544–1555

Senaratna T, Touchell D, Bunn E, Dixon K (2000) Acetyl salicylic acid (aspirin) and salicylic acid induce multiple stress tolerance in bean and tomato plants. Plant Growth Regul 30:157–161

Shin H, Shin HS, Gary R, Harrison MJ (2004) Phosphate transportin *Arabidopsis*: Pht1;1 and Pht1;4 play a major role in phosphate acquisition from both low-and high-phosphate environments. Plant J 39:629–642

Shirasu K, Nakajima A, Rajshekar K, Dixon RA, Lamb C (1997) Salicylic acidpotentiates an agonist-dependent gain control that amplifies pathogen signalin the activation of defence mechanism. Plant Cell 9:261–270

Shri M, Kumar S, Chakrabarty D, Trivedi PK, Mallick S, Misra P, Shukla D, Mishra S, Srivastava S, Tripathi RD, Tuli R (2009) Effect of arsenic on growth, oxidative stress, and antioxidant system in rice seedlings. Ecotoxicol Environ Saf 72:1102–1110

Shukla D, Kesari R, Mishra S, Dwivedi S, Tripathi RD, Nath P, Trivedi PK (2012) Expression of phytochelatin synthase from aquatic macrophyte *Ceratophyllum demersum* L. enhances cadmium and arsenic accumulation in tobacco. Plant Cell Rep 31:1687–1699

Singh N, Ma LQ, Srivastava M, Rathinasabapathi B (2006) Metabolic adaptations to arsenic-induced oxidative stress in *Pteris vittata* L. and *Pteris ensiformis* L. Plant Sci 170:274–282

Singh HP, Kaur S, Batish DR, Sharma VP, Sharma N, Kohli RK (2009) Nitric oxide alleviates arsenic toxicity by reducing oxidative damage in the roots of*Oryza sativa* (rice). Nitric Oxide 20:289–297

Singh AP, Dixit D, Kumar A, Mishra S, Singh PK, Dwivedi S, Trivedi PK, Chakrabarty D, Mallick S, Pandey V, Dhankher OP, Tripathi RD (2016) Nitric oxide alleviated arsenic toxicity by modulation of antioxidants and thiol metabolism in rice (*Oryza sativa* L.). Front Plant Sci 6:1272

Singh AP, Dixit G, Kumar A, Mishra S, Kumar N, Dixit S, Singh PK, Dwivedi S, Trivedi PK, Pandey V, Dhankher OP, Norton GJ, Chakrabarty D, Tripathi RD (2017) A protective role for nitric oxide and silicic acid for arsenite phytotoxicity in rice (*O. sativa* L.). Plant Physiol Biochem 115:163–173

Song WY, Park J, Mendoza-Cózatl DG, Suter-Grotemeyer M, Shim D, Hörtensteiner S, Geisler M, Weder B, Rea PA, Rentsch D, Schroeder JI, Lee Y, Martinoia E (2010) Arsenic tolerance in *Arabidopsis* is mediated by two ABCC-type phytochelatin transporters. Proc Natl Acad Sci USA 107:21187–21192

Srivastava M, Ma LQ, Singh N, Singh S (2005) Antioxidant responses of hyper accumulator and sensitive fern species to arsenic. J Exp Bot 56:1335–1342

Srivastava S, Mishra S, Tripathi RD, Dwivedi S, Trivedi PK, Tandon PK (2007) Phytochelatins and antioxidant systems respond differentially during arsenite and arsenate stress in *Hydrilla verticillata* (L.f.) Royle. Environ Sci Technol 41:2930–2936

Sundaram S, Rathinasabapathi B, Ma LQ., Rosen BP (2008) An arsenate-activated glutaredoxin from the arsenic hyper accumulator fern *Pteris vittata* L. regulates intracellular arsenite. J Biol Chem 283(10):6095–6101

Talukdar D (2013) Arsenic-induced oxidative stress in the common bean legume, *Phaseolus vulgaris* L. seedlings and its amelioration by exogenous nitric oxide. Physiol Mol Biol Plants 19(1):69–79

Talukdar T, Talukdar D (2013) Response of antioxidative enzymes to arsenic-induced phytotoxicity in leaves of a medicinal daisy, *Wedelia chinensis* Merrill. J Nat Sci Biol Med 4(2):383–388

Tripathi RD, Srivastava S, Mishra S, Singh N, Tuli R, Gupta DK, Maathuis FJM (2007) Arsenic hazards: strategies for tolerance and remediation by plants. Trends Biotechnol 25:158–165

Valderrama R, Corpas FJ, Carreras A, Fernández-Ocaña A, Chaki M, Luque F, Gómez-Rodríguez MV, Colmenero-Varea P, del Río LA, Barroso JB (2007) Nitrosative stress in plants. FEBS Lett 581:453–461

Verma PK, Verma S, Meher AK, Pande V, Mallick S, Bansiwal AK, Tripathi RD, Dhankher OP, Chakrabarty D (2016a) Overexpression of rice glutaredoxins (OsGrxs) significantly reducesarsenite accumulation by maintaining glutathione pool andmodulating aquaporins in yeast. Plant Physiol Biochem 106:208–217

Verma PK, Verma S, Pande V, Mallick S, Tripathi RD, Dhankher OP, Chakrabarty D (2016b) Overexpression of rice glutaredoxinOsGrx_C7 and OsGrx_C2.1 reduces intracellular arsenic accumulation and increases tolerance in *Arabidopsis thaliana*. Front Plant Sci 7:740

Wojas S, Clemens S, Skłodowska A, Antosiewicz DA (2010) Arsenic responseof AtPCS1- and CePCS-expressing plants—effects of external As(V) concentration on As-accumulation pattern and NPT metabolism. J Plant Physiol 167:169–175

Wu ZC, Ren HY, McGrath SP, Wu P, Zhao FJ (2011) Investigating the contribution of the phosphate transport pathway to arsenic accumulation in rice. Plant Physiol 157:498–508

Xing S, Rosso MG, Zachgo S (2005) Development (Camb.) 132:1555–1565

Xu W, Dai1 W, Yan H, Li S, Shen H, Chen Y, Xu H, Sun Y, He Z, Ma M (2015) *Arabidopsis* NIP3;1 plays an important role inarsenic uptake and root-to-shoot translocation under arsenite stress conditions. Mole Plant 8:722–733

Yalpani N, Enyedi AJ, Leon J, Raskin I (1994) Ultravioletlight and ozone stimulate accumulation of salicylic acidand pathogenesis-related proteins and virus resistance intobacco. Planta 193:373–376

Yan Z, Chen J, Li X (2013) Methyl jasmonates modulator of Cd toxicity in *Capsicum frutescens* var. *fasciculatum* seedlings. Ecotoxicol Environ Saf 98:203–209

Zhao FJ, Wang JR, Barker JHA, Schat H, Bleeker PM, McGrath SP (2003) The role of phytochelatins in arsenic tolerance in the hyperaccumulator*Pteris vittata*. New Phytol 159:403–410

Zhao FJ, McGrath SP, Meharg AA (2010) Arsenic as a food chain contaminant: Mechanisms of plant uptake and metabolism and mitigation strategies. Annu Rev Plant Biol 61:535–559

Zutshi S, Bano F, Ningthoujam M, Habib K, Fatma T (2014) Metabolic adaptations to arsenic-induced oxidative stress in *Hapalosiphon fontinalis-339*. Int J Innov Res Sci Eng Technol 3(2):9386–9394

Printed by Printforce, the Netherlands